Nuclear Power from Underseas to Outer Space

John W. Simpson

Nuclear Power from Underseas to Outer Space

John W. Simpson

American Nuclear Society
La Grange Park, Illinois USA

Library of Congress Cataloging-in-Publication Data

Simpson, John W. (John Wistar), 1914–
 Nuclear power from underseas to outer space / John W. Simpson.
     p.   cm.
 Includes index.
 ISBN 0-89448-559-8
 1. Nuclear engineering—United States—History. 2. Westinghouse Electric Corporation—History. I. Title.
 TK9023.S49 1994
 621.48'0973—dc20   94-38815
                                                                                    CIP

ISBN: 0-89448-559-8
Library of Congress Catalog Card Number: 94-38815
ANS Order Number: 690042

Copyright © 1995 American Nuclear Society
555 North Kensington Avenue
La Grange Park, Illinois 60525 USA

All rights reserved. No part of this book
may be reproduced in any form without the written
permission of the publisher.

Printed in the United States of America

CONTENTS

Foreword *vii*

Preface *xi*

ONE   The Oak Ridge Years 1
TWO   Westinghouse Enters the Field 15
THREE   Building and Operation of the Mark I in Idaho 43
FOUR   Nuclear Propulsion Moves Forward 66
FIVE   Shippingport 97
SIX   Astronuclear Years 114
SEVEN   Early Commercial Nuclear Power 156
EIGHT   Commercial Nuclear Power 1963–1993 185
NINE   Offshore Plants, Fast Breeders, and Fusion 213
TEN   The GOCOs 237
ELEVEN   Other Nuclear Activities 254
TWELVE   The Naval Reactors Technical Story 283
THIRTEEN The Shippingport Technical Story 363
FOURTEEN The Astronuclear Technical Story 388
FIFTEEN   The Commercial Nuclear Technical Story 431
SIXTEEN   Conclusions and Outlook 454

Index 459

FOREWORD

For the countless thousands who have devoted part or all of their lives to the development of nuclear technology for propulsion or power, John Simpson's somewhat personalized history of nuclear power provides rewarding reading. We are fortunate indeed to have captured, while memory still serves, this material on the early development of nuclear technologies. The simple fact is that the pioneers are beginning to age—and will, regrettably, not always be available to us. So the history must now be captured in a permanent form. This volume does just that. For the many nuclear buffs, this will be a fun book.

Few men could have provided such a history as well as John Simpson, a great figure in the Westinghouse atomic program. Partly through good fortune and substantially through drive and skill, Westinghouse became the leading firm in the exploitation of nuclear energy. John Simpson, who has never been charged with excessive shyness, is happy to tell the story of nuclear power from his own perspective and the perspective of his company. This is not a detached or objective recounting of the history. It is the story of Westinghouse, First in Atomic Power, as the company advertisement and John Simpson both proudly state. General Electric and others will have to look elsewhere to receive their appropriate plaudits. As he himself recounts, in his annual briefings to utility executives a GE man was heard to say: "There's that John Simpson again telling them what they ought to do and the SOBs are believing him."

For nuclear buffs, this will be a walk through nostalgia land. Simpson recaptures the thrill of creating something new—a technology that will effectively serve the purposes of both security and economy. Perhaps it is the mark of our own age that there is a deep-seated public skepticism—a skepticism that has affected technology along with everything else in business or in government. As a consequence, all too many of our citizens have missed the sheer joy of these great achievements in technology—achievements that undergird the standard of living of ourselves and others in advanced industrial countries. Simpson manages to recreate that atmosphere of sheer joy—not through deliberate calculation, but through sheer enthusiasm.

Moreover, this volume should captivate more than the nuclear buffs. It should appeal to students of naval warfare, to midshipmen at Annapolis, to senior naval officers. It chronicles how one of the great changes in naval technology—as significant perhaps as the development of the ironclad or of steam power itself—came about. For it was submarine nuclear propulsion that provided unequivocal assurance of retaliatory capability—and thus reduced uncertainties surrounding the nuclear confrontation. Finally, this book should interest both government officials and corporate leaders as an example of successful teamwork in innovation between government and industry.

Simpson has provided a long and entertaining history. He recounts not only the logic and the technology, but also those adventitious elements that played so critical a role in determining who was drawn into this great enterprise. For example, it was sheer chance that Westinghouse happened to be experimenting with uranium for light bulb filaments, which led the government to turn to Westinghouse to produce the many tons of pure uranium required during the Manhattan Project. Courage and vision were required as well as the willingness to deal with the element of chance involved, and Westinghouse officials such as Gwilym Price and Charlie Weaver became convinced of the future of nuclear power and the need for Westinghouse to be an aggressive part of its development. As Westinghouse was feeling its way into the nuclear business, it was pure chance that Bettis Field was abandoned and became available as the site for the laboratory from which so much history subsequently flowed.

As part of his great adventure, Simpson pays appropriate tribute to the many colorful pioneers in the early days of commercial nuclear power—from the utility industry, the Bill Websters and McGregor Smiths—and above all to the personality of Hyman Rickover. Never easy, regularly colorful, Rickover runs like a thread through these memoirs. Rickover was, of course, both dedicated and unscrupulous. Simpson's own blend of admiration and ambivalence comes through. Though for the most part he is inclined to disguise his ambivalence and to pull his punches, from time to time he reveals such sentiments as "his staff could cheerfully have murdered him."

Rickover's *modus operandi* was to make his subordinates and dependents uncomfortable—in order to see what would issue forth. His "surprise" visits and, frequently, "surprise" agendas were designed to avoid a prepared and therefore contrived response. His statements such as "the trouble with atomic power is that too many physicists are making decisions; in fact, there are just too many physicists working on atomic power" were designed as much to annoy the physicists as to give comfort to the engineers. Rickover, as Simpson proudly states, demanded three things: excellence, speed, total dedication. But he also demanded an astonishing degree of self-abasement, which made many long to escape from his control.

For some 35 years, Rickover was the embodiment of nuclear propulsion. From the recommendation to the Navy in 1944 that it give highest priority to nuclear propulsion, it was Rickover who remained, for better or worse, the embodiment of that priority. Over time, he acquired such influence in Congress that he became virtually untouchable. Simpson became one of his chosen instruments. From the construction of the Mark I prototype in Idaho to the launching of the *Nautilus* (which embodied Westinghouse's lead in nuclear power), with the First Lady Mamie Eisenhower herself there to swing the bottle of champagne, nuclear propulsion became an essential ingredient in the nation's deterrent. As the vulnerability of SAC bases became a source of growing national concern, President Eisenhower pointed to this remarkable achievement when he paid special tribute to "the submarine-based Polaris missile system which can be so effectively hidden in the depths of the sea."

The story of nuclear power, which to a large extent was to evolve from the design of nuclear propulsion plants, is perhaps more familiar but equally well told. President Eisenhower saw fit to break ground for Shippingport, which to a great public fanfare, came on line in 1957. The cost of electric power at Shippingport was ten times the cost for power at conventional plants. During the next 15 years, however, those costs were sharply reduced as the industry moved rapidly down the learning curve. The early turnkey and subsequent commercial plants brought the kilowatt-hour cost down below that of fossil fuel plants. But that was before the industry was engulfed by opposition from environmentalists (whose campaign tactics came to resemble those of guerrilla warfare), growing public concern over safety, and an exploding regulatory burden, which became steadily more onerous.

Nonetheless, the Shippingport power costs might have provided an early warning to the proponents of nuclear power that its future need not necessarily be strewn with roses or crowned with inevitable success. Still in the midst of all the enthusiasm of that era, the eyes of true faith saw little reason to harbor doubts about the inevitable triumph of nuclear power. A subsequent Westinghouse plant, Yankee-Rowe, was described as "the nuclear success story of the year." Thus, the proponents of nuclear power were unprepared for the setbacks and the degree of public hostility that developed in the course of the 1970s and 1980s. Their skills at public relations did not come close to matching their technical skills.

While it goes beyond the substance of this history, it needs to be pointed out that in terms of both the legal structure and the commercial organization of nuclear power, this nation made just about every mistake in the book. It helps explain why nuclear power lags behind in the United States, while it has forged ahead in such countries as France and Japan—with different legal structures and commercial organizations, yet using designs licensed by U.S. companies.

By contrast, here in the United States, nuclear power has encountered manifold troubles, which many will regard as tragic. Simpson attributes some of the developing opposition to the end of the Vietnam War and the new role being sought by anti-Vietnam activists. I can personally testify to the accuracy of his observations. When questioned in congressional testimony about the vehement opponents of nuclear power, I responded as follows: "There are organizations that developed a capability for protest during the Vietnam War. Now that the war is over, there's the capability that must be exercised, if it is to remain effective. For the most part, these are the same people who followed Ho Chi Minh all through the Vietnam years." In short order, I received a call from the vice president's office urging me—i.e., warning me—to say no more on that particular subject.

Finally, a word should be said about the nuclear rocket. In war games during the 1960s, I had been intrigued by the SLAM (supersonic low altitude missile). However, the slow deployment of Soviet missile defenses and the antiballistic missile treaty made any such development for military purposes unnecessary. The development of NERVA (nuclear engine for rocket vehicle applications) was an outstanding technical success. Its utility, however, depended on the trip to Mars. When the Nixon administration curtailed the space program, the role for NERVA evaporated. Engineers tend to believe that what is a technical success should result in deployment. That is something of an illusion, as the nuclear rocket program, despite its technical success, so well demonstrates.

The overall history of nuclear power, from the perspective of the half century since the first successful chain reaction in the pile under the West Stands at Stagg Field in Chicago, provides a picture that is mixed: many exciting, indeed astonishing, technical successes and far fewer policy and commercial successes. This volume understandably focuses more on the technical and other successes. Beyond the outlines of the main story, it provides many little asides and insights. Suggestive of both future troubles and future hope is the vignette presented in the book about even so great a figure as Thomas Edison spreading alarms about the supposed dangers of alternating current. Such vignettes add spice to an already exciting tale. But, read on! As I said at the outset, for nuclear buffs this will be a fun book.

James Schlesinger
August 1994

PREFACE

This is a personal account of how nuclear power grew from a theoretical concept to become a part of history as seen by the author from his vantage point in an organization at the heart of the work. It is history. It is drama. It is the story of the people in his organization and their achievements as they created a new technology and harnessed it for man's practical use.

My role in this drama starts with the Daniels Pile in 1946 at Oak Ridge, Tennessee, where our goal was to develop a nuclear reactor to produce electricity. It was my introduction to the atom, and then the emphasis shifted to the development of nuclear propulsion for the Navy.

In a few fast-moving years, the story follows the development, design, construction, and operation of the prototype for the submarine *Nautilus*, the building of the *Nautilus* herself, and the emergence of the entire naval reactors program. During this time, I moved from assistant engineering manager to technical director and then director of the Bettis Atomic Power Laboratory, operated by Westinghouse for the U.S. Atomic Energy Commission.

Having proved that nuclear power could be used for naval propulsion, the government wanted to utilize it for peaceful uses. Rickover succeeded in getting the Shippingport project as an assignment for Bettis. Shippingport was the first reactor in the United States to generate electricity from nuclear power for commercial use. My role was project manager before becoming Bettis Laboratory director. After Shippingport, the Astronuclear Laboratory was created to get Westinghouse involved in the space program. We succeeded in getting the contract for the NERVA nuclear rocket and other space programs. At that time I was vice president responsible for both Astronuclear and the Westinghouse commercial nuclear program.

The latter chapters of the book cover commercial nuclear power, including PWRs and breeders and other miscellaneous activities. During this time I was electric utility group vice president, responsible for all Westinghouse electric utility business worldwide.

The first eleven chapters are mostly narrative and anecdotal focusing on *who* did *what* and why in this drama. The next four chapters are rather technical—they tell *how*—and then the book ends with a chapter listing conclusions and the outlook for the future.

There are times when the story is principally personal memoirs and sometimes it may be hard to distinguish between what I did and what the organization did. Because I retired in 1977 and was not directly involved in the Westinghouse activities after that, later technical progress receives less attention than it probably deserves. Also getting less than appropriate credit here is the work of the Argonne National Laboratory, the Naval Reactors Branch, the Atomic Energy Commission, the Department of Energy, and numerous contractors. Other nuclear reactor manufacturers also accomplished much during the same period, but the book covers mainly what Westinghouse did because that is what I saw and where I was directly involved. So be it.

The greatest joy of those years was my association with a most remarkable group of talented people, and I called on many of them to contribute to the research for this book. In 1975, I asked a number of people who had worked for me at Bettis to record what they felt were the most important technical accomplishments in which they were involved. There were many replies, but they lay dormant until early 1991. At that time I again contacted the original group plus several others as I began assembling data. I received much new material and in many cases the same general subject was covered by several people. My job was to determine the general form of the book and then edit all the material from my files and that which I received from the numerous contributors into a hopefully meaningful whole.

Major contributions were made by H.W. Arnold,* R.J. Creagan,* W.H. Esselman, R.N. Fillnow, H.B. Finger, A.F. Henry,* B. Lustman,* and D.E. Thomas. Significant contributions were also made by R.J. Coe, A.R. Collier, J.T. Conway, F.P. Cotter, R.C. Cunningham, J.E. Gray, G.W. Hardigg, W.M. Jacobi, W.E. Johnson,* P. Murray,* N.J. Palladino,* J.C. Rengel, L.H. Roddis,* P.N. Ross,* A. Squire,* T. Stern,* J.J. Taylor,* J.J. Theilacker, C.H. Weaver, and A.P. Zechella. Many others contributed and for their assistance I am grateful. I hope they understand that it's impossible to mention everybody.

Without the valuable assistance of these people, who devoted countless hours to the project, this book would not have been possible. They were also very important contributors to the development programs covered.

The manuscript was given a preliminary edit by J.W. Puckett, Jr. Then I asked T.K. Phares, former Westinghouse public relations executive and now a freelance writer, to act as rewrite man and editor to make the story as readable as possible for a broad audience. The manuscript then went for a technical review by the ANS Book Publishing Committee, chaired by Roger Tilbrook of ANL.

I would also like to thank my wife, Esther, for her help and patience throughout this project.

*John W. Simpson**
August 1994

---

*Members of the prestigious National Academy of Engineering.

CHAPTER 1

The Oak Ridge Years

Atomic Bomb Explodes over Hiroshima!

This may have been the most startling headline ever to scream across the front page of a newspaper. After this day in August 1945, the world would never be the same again. Nuclear weapons would dominate the strategic thinking of nations. International affairs would turn on the awesome potential of the atom.

That potential included not only devastating weapons, but the promise of limitless energy—if the atom could be harnessed. That was a big if, but the possibility was there. The world had entered the Atomic Age. To understand how this came to be, we must look back into history.

In 1905, the great physicist Albert Einstein propounded his theory $E = MC^2$: Energy equals mass times the speed of light squared. The fact that mass could be converted to energy was a startling scientific idea. At that time, however, it was only a fascinating theory with no practical application. There was no known practical method of converting mass to energy.

This changed dramatically in 1939, when Otto Hahn and Fritz Strassmann of the Kaiser Wilhelm Institute in Berlin, working from earlier experiments by Enrico Fermi of Italy, discovered what appeared to be fissioning (splitting) of the uranium atom, when it was bombarded by neutrons.

Early in 1939, Niels Bohr in Copenhagen was informed of these experiments and hastened to report this theory of uranium fissioning to American scientists during a trip to the United States. In a short time, the bare facts of the German discovery were verified at several American laboratories including Columbia University, Johns Hopkins University, the University of California, and the Carnegie Institution. Scientists found that when an atom of the 235 isotope of uranium is fissioned into two roughly equal parts, those parts have less mass than the whole atom did originally. The mass that is lost is converted into energy, a staggering amount of energy.

However, the scientists still lacked one important piece of the puzzle. Although theory predicted that there would be more than one neutron per fission, would there be a sufficient number for a self-sustaining chain reaction? That is, would each fission cause one or more additional fissions or would each have to be separately initiated? Soon H.L. Anderson, Enrico Fermi, H.B. Hanstein, Leo Szilard, and W.H. Zinn at Columbia University discovered that more than one neutron was emitted with each atom split. That extra neutron, they reasoned, was available to split another atom. So a chain reaction was possible with almost unlimited energy forthcoming in the process.

To explore this was going to require a major program, an expensive program. What to do?

The World War II storm was gathering. Some of the scientists, who were refugees from Nazi Germany and threatened areas, were concerned that this mighty new source of energy, with bomb potential, could fall into Hitler's hands. They decided that a letter from the world's most esteemed scientist—Einstein—might get the attention of President Franklin D. Roosevelt. Thus, Hungarian physicists Leo Szilard and Eugene Wigner, in consultation with Einstein, drafted a letter in July 1939 for Albert Einstein's signature informing the president that a powerful bomb could be developed using nuclear fission.

The letter caught FDR's attention. He appointed a Uranium Committee to determine the feasibility of an atomic bomb, but then bureaucratic malaise took over. For nearly a year there were meetings and talk, but little action. Finally, disquieting news from Europe forced the action. British Intelligence reported that leading German researchers were disappearing one by one into the laboratories of the Kaiser Wilhelm Institute in Berlin. Also, quantities of uranium from the mines of Czechoslovakia were funneling into the same place.

The Manhattan District

Roosevelt formed the Manhattan District of the Army Corps of Engineers to develop and build an atomic bomb. Major General Leslie Groves was put in charge of this most highly secret major project in American history.

It was an intriguing concept, but under what conditions could a chain reaction be possible, and if possible could it be controlled and could the reaction be fast enough for a useful bomb?

The first problem was whether uranium could be assembled in such a manner that at least one of the approximately two-and-a-half neutrons emitted from each fission could be made to cause another fission and thus produce a self-sustaining chain reaction. The second problem was where to get the necessary pure uranium metal. Westinghouse was about to enter the scene.

Westinghouse's Role

At that time one of the only sources of pure uranium metal in the world was the Westinghouse lamp division in Bloomfield, New Jersey. Researchers there had been experimenting with it as a possible material for making lamp bulb filaments.

One day Dr. Harvey C. Rentschler, director of lamp research at Bloomfield, got a call from Dr. Arthur H. Compton of the University of Chicago, where the Manhattan District was planning to build an atomic "pile," as the early reactors were called.

"How pure is your uranium?" Compton asked.

Rentschler replied that he thought it was pretty pure.

Without a flicker of excitement in his voice, Dr. Compton said, "How soon can you make us three tons?"

Rentschler was stunned. He had been working with this rare metal by the ounce. Now he was being asked for it by the ton. He stammered that it would "take a little time." Compton said he would come to Bloomfield.

Westinghouse top management, including A.W. Robertson, chairman, and George Bucher, president, conferred with Compton and Rentschler and they agreed to tackle the project. "But," demanded Dave Youngholm, lamp division vice president, "where in God's name can you make the stuff in this plant?"

They had to use the basement storage areas and the Bloomfield plant roof for emergency uranium production, but they got the job done. In the course of 18 months, the lamp division was producing uranium by the ton.

Compton later credited that achievement with advancing the possible date for dropping the bomb by at least a year. That turned out to be a critical year. Rentschler and his crew also brought the cost of pure uranium down from $750 a pound to $7.34.

The Manhattan District concentrated on developing and building an atomic pile, and on December 2, 1942—under the West Stands at Stagg Field of the University of Chicago—a chain reaction was achieved.

So far so good, but there were still a host of unanswered theoretical and engineering questions. Could enough of the fissionable isotope be obtained for a bomb? This turned out to be a Herculean task. The separation was done at Oak Ridge using both the gaseous diffusion and the electromagnetic processes. Also plutonium was produced at the Hanford Site in Washington for the second bomb.

Still the question remained: Could it be so configured that the chain reaction could proceed rapidly enough to cause a high-energy explosion? Could a device be assembled for producing electricity or other useful energy? The key question in the production of electricity was not whether the device could be made, but whether it would be economically competitive. We were at war—bombs were what was urgently needed. Peaceful nuclear energy would have to wait. But my story is about peaceful nuclear energy, so we will skip the story of creating the bomb. That story is well told in the Smythe report.[1]

It was at Oak Ridge that I was first indoctrinated into the mysteries of nuclear power, but my relationship with Hyman Rickover began seven years earlier, in 1939. He was then a lieutenant commander. I was part of a seven-man group in the Westinghouse switchgear division that produced over half of all the switchboards used by the Navy and Coast Guard during the war. By 1944, I had become manager of this group. My only Navy customer was Rickover, head of the Electrical Section of the Bureau of Ships and by the end of the war a captain.

Rickover had received reports from the British before we entered the war regarding problems with electrical equipment that was not shockproof or fireproof. Rickover's objective was to correct this deficiency. The true magnitude of his accomplishment was driven home to me after the war, when I was in Japan as a member of the Navy's Technical Mission. Not having such shockproof and fireproof equipment caused the Japanese Navy many problems.

We mounted a major effort to redesign almost all of the electrical equipment we were supplying to the Navy. Harold Edgerton was retained to use his stroboscopic photography to study equipment undergoing tests on shock stands. Not only must the equipment not break on test, it must continue to function properly. Circuit breakers must remain closed, relay and other instruments must not lose their calibration, etc. Also the equipment was redesigned to use fireproof materials.

How did we know that our equipment was shockproof and fireproof? Rickover had us assemble a complete switchboard and mount it on an eight-inch-thick steel armor plate slab at the Naval Proving Ground in Dahlgren, Virginia. An eight-inch shell was fired at the other side of the armor plate. After the first tests the equipment was in shambles.

---

[1]Henry D. Smythe, *Atomic Energy for Military Purposes 1940–45*, Princeton University Press, Princeton, NJ, 1945.

However, after a little over a year of development, the tests were repeated. This time the switchboard was energized at 450 volts and it worked fine. Later a switchboard was set up at the Westinghouse East Pittsburgh high-test lab. Copper wires were strung between the bus bars to cause an electrical short circuit and violent arcing. Despite the massive flames, the switchboard survived.

Later on as a final test, Rickover arranged for tests on board a destroyer. We dropped depth charges at slower and slower speeds. Things were working so well that I asked the skipper to back down over the depth charges. He was reluctant, but finally agreed. The equipment survived this test unscathed.

Working as we were on Navy wartime electrical apparatus, we were on the fringe of work being done for the secret Manhattan District project. None of us had any direct knowledge of its purpose, but it seemed to me that it had to involve an explosive or a fuel. The equipment being built at Westinghouse implied a chemical-type process, and certainly explosives and fuels were needed.

Equipment for the Manhattan District had a higher priority than my Navy equipment, so I spent a lot of time in the machine shop at night persuading cooperative machinists to do some work for me. I told the Manhattan people that they had better hurry, or we would have the war won before they accomplished their goal, whatever it was.

With the explosion of atomic bombs on Hiroshima and Nagasaki in 1945—which quickly ended the terrible war—the whole world knew the secret. When I went to Japan, immediately after the Japanese surrender to help evaluate the performance of Japan's naval electrical equipment, I visited Hiroshima and saw the destruction. It was a grim start to my nuclear career, which began only a few months later.

Nuclear Propulsion

After the discovery of fission in 1939, the Navy quickly realized that the atom offered a potential method for propulsion of naval vessels. But the Navy's nuclear thinking until the end of 1944 was focused on helping to achieve the immediate goal of the atomic bomb.

The Manhattan District had formed a committee chaired by Dr. Richard C. Tolman to recommend the direction in which the postwar program should move. The Tolman Committee recommended on December 28, 1944, that a high priority be given to the development of nuclear propulsion plants for naval vessels. This report read in part:

> The government should initiate and push, as an urgent project, research and development studies to provide power from nuclear sources for the propul-

sion of naval vessels. It might be advisable to authorize the initiation of these studies at once, without waiting for the postwar period, in order to utilize personnel already familiar with the pile theory and operation.

It would have been easier, of course, to develop nuclear propulsion for surface vessels than for submarines, because of the space problem on subs and the fact that they submerge to great depths. Surface nuclear propulsion could give added range and solve a few problems, such as eliminating exhaust fumes that interfere with flight operations on a carrier. But the advantages that nuclear power would bring to submarines were unique—the magic touchstone.

The prospect of a true submersible with unlimited range was irresistible to any submariner. The propulsion unit could be completely self-contained and not dependent on access to the atmosphere. The World War II subs had batteries, which permitted submerged operation, but only for 30 to 40 miles at high speed. Then they had to stop or surface. Nuclear power would make it possible to cruise submerged indefinitely. What a revolution in naval warfare a nuclear submarine would bring!

One daunting problem, however, was how to design a nuclear propulsion plant small enough to fit into even the largest submarine.

Enter Rickover

Although Rickover did an outstanding job for the Navy during the war, he had made many enemies. He antagonized the top executives of many of the largest U.S. companies by his unrelenting pressure and, in their opinion, his unreasonable demands. He had also made many enemies in the Navy because of his unorthodox approach to solving problems. He often stepped on the toes of high-ranking officers.

At his request, Rickover was given command in July 1945 of a large fleet-repair facility on Okinawa, but the war ended in August and he never got the opportunity to do the service and repair that would have been required had an invasion of Japan taken place. In October, a major typhoon struck the island and devastated the facility. The Navy decided not to rebuild it and ordered Rickover to the West Coast as inspector general of ship mothballing.

Despite the tremendous advantages the atom offered, most Navy officers thought that nuclear power was for the distant future, if ever. The difficulties were just too great. There was one exception—Rickover. So when the Navy decided to send a team to Oak Ridge, he was the only captain to respond. Captain Harry Burris had been recommended for the Oak Ridge assignment, and the Bureau of Ships was intending to send Rickover to the Massachusetts Institute of Technology (MIT) to study nuclear physics for three years. However, the Bureau chief, Vice Admiral Earle Mills, changed the plan. He decided Rickover should be the one to go to Oak Ridge. A fateful decision.

Rickover was not suited for a routine assignment in the Bureau of Ships. He had succeeded during the war because what he was doing was acknowledged as extremely important. I doubt if he could have done as well in peacetime. Mills did the United States a great favor by his decision.

Joining Rickover from the Bureau were three civilians and four junior officers, Lieutenant Commanders Dunford and Roddis and Lieutenants Dick and Libbey. Their assignment was to study reactor design. Rickover arranged, in addition, for them to accompany him to almost every nuclear installation so that all involved could get a broad understanding of the field.

Daniels Power Pile

Shortly after the war, Westinghouse CEO Gwilym Price hosted a two-day seminar in Pittsburgh on what to expect in the postwar world. Authorities from all over the country were invited to speak. One of the attendees was Captain Rickover, who had worked closely with Westinghouse during the war. He told Price that there was a great future in atomic energy and that Westinghouse should become more involved.

Westinghouse had been involved in atomic energy with industry's first atom smasher—the Van de Graaff generator at the company's research laboratory in Pittsburgh. Shortly after atomic fission with neutrons had been confirmed in 1939, William E. Shoupp and other Westinghouse scientists discovered the splitting of atoms by gamma rays using the Van de Graaff machine. This background proved valuable later, when some of these scientists were assigned to work at the company's Bettis Atomic Power Laboratory on the nuclear submarine.

Also during World War II, Westinghouse had produced about $75 million worth of special equipment for the electromagnetic separations project at Y-12 in Oak Ridge, Tennessee. This involvement by top management with the Manhattan District made them more receptive to the request to send engineers to Oak Ridge.

Price agreed with both of Rickover's points and decided that Westinghouse would participate in a project to build a peacetime reactor, then being planned by the Manhattan District, called the Daniels Power Pile.

Dr. Farrington Daniels was head of the Chemistry Department at the University of Wisconsin and had been director of the Metallurgical Laboratory, later to become Argonne National Laboratory. He had been working on a nitrogen fixation project and was familiar with both nuclear energy and high-temperature gases. He and some of his staff had developed the conceptual design of a high-temperature, gas-cooled, beryllium-moderated power reactor. At the urging of Charles Thomas, the CEO of Monsanto Chemical Company, which operated the Clinton Laboratory[2] at Oak Ridge, the Manhattan District initiated the Daniels Power Pile project. Reactors were called "piles" at first because at that time they consisted of a graphite moderator literally stacked in piles.

Assignment Oak Ridge

In March 1946, Dr. C. Rogers McCullough of Monsanto, director of the Daniels Power Pile project, visited Pittsburgh to get Westinghouse to send some engineers to Oak Ridge to do the engineering. I was selected, along with Bruce Ashcraft, Ernie Miller, N.J. Palladino, Sid Siegel, and A.H. Toepfer, partly because my switchgear group had been disbanded after the war ended, and I had no specific assignment at that time. I had also taken a course in atomic physics and had some advanced math courses in graduate school and was one of the youngest engineering managers in the company. Later P.N. Ross, who had worked for Rickover on loan from Westinghouse during the war, was added to the group.

While still in Pittsburgh, scientists from the bomb project gave us an unclassified briefing and said they had done the basic design for the Daniels Pile. We were to perform engineering design and procure the equipment. They thought that we would be finished in about a year, but might have to come back occasionally during construction. On hindsight, that was absolutely ridiculous, and it's hard to see how we could have been taken in. We probably thought that people who had accomplished the scientific marvel of the atomic bomb could do anything.

My temporary quarters were in the Guest House, until my dormitory room was assigned. The first afternoon, when I walked onto the porch, to my dismay whom should I meet but Rickover, my Navy customer from switchboard days.

My reactions were mixed at best. I had gotten along with Rickover reasonably well during the war. He had recommended me for a Navy Certificate of Commendation, and he was responsible for my inclusion in the Navy's team for technical evaluation of the Japanese naval electrical equipment. But I knew that wherever Rickover was, things would be hectic.

While I had high regard for Rickover's ability and believed the Navy had made a wise choice for the Oak Ridge assignment, I did not relish the idea of associating with him once more. Even though he was no longer my customer, I knew he would be difficult. It's a good thing I didn't then

---

[2]Clinton Laboratory later became Oak Ridge National Laboratory.

realize how long and close our association would be. It might have altered my plans, and that would have been a mistake.

Rickover didn't want to live in a dormitory room, so he arranged for a group of us to be assigned to a four-bedroom house. Joining him were Farrington Daniels, Sid Simon from the National Advisory Committee on Aeronautics, Harry Stevens from GE, and me.

Rickover probably selected Stevens and me because he intended to get GE and Westinghouse involved in the Navy program he was planning. As head of the Bureau of Ships electrical section, he had worked closely with both companies and believed they had the necessary size and depth of technical talent to handle a nuclear-propulsion development project.

The way I addressed Rickover kept changing, of course. I was never on a first name basis. During the war I had always addressed him as commander. When he was promoted from lieutenant commander to full commander, no change was needed. I soon adjusted to captain, and then admiral in 1953. At Bettis, he was always referred to as "the admiral." When you heard that term, nobody ever thought of anyone but Rickover, as though the Navy only had one admiral!

Life at Oak Ridge was comfortable and much less frantic and hectic than it was during the war years. Then, decisions had to be made and actions had to be taken immediately. Here, we were feeling our way into a new technology for which we were really unprepared. We drove ten miles to the laboratory in a car pool and since we seldom worked overtime, we also drove home together on most days. Living that way with Rickover, however, had its moments. It was never serene for long.

The county in which the reservation was located was dry and, of course, we regularly violated that prohibition and had devised a way to beat the system. We would drive to the next county to buy liquor and frequently there were spotters at the liquor store to take down license numbers and notify the police. When a car carrying liquor purchasers crossed into our dry county, the cops would pick them up. So we always took two cars and transferred the liquor before crossing the county line. To be on the safe side at our house, Rickover procured a large security safe in which we kept our liquor. He seldom drank any.

Rickover's complex personality surfaced often. One evening Sid Simon came back from a trip and announced that he was getting married. After dinner, Rickover spotted on the shelf a copy of the Emily Post book on etiquette. He picked it up and pretended to read the duties of the groom, including the fact that the groom must bear most of the expenses. It took us a little time to catch on that he was making it up as he went along.

At the dinner table, we had lively and interesting conversations covering a wide range of topics. Harry Stevens, the youngest of the group, liked to bait Rickover. He would say something like: "Do you think the day of the Navy has passed?" Or, "Why is the government so inefficient?" Rickover would counter, and Harry would make one or two more remarks. Then he would back off just before Rickover exploded.

Rickover also enjoyed needling people, particularly in the scientific ranks. At one meeting, with most of the top scientists from the bomb days, the subject was the future of atomic energy and its problems. Just as the meeting was about to conclude, the chairman asked Rickover for his comments.

"The trouble with atomic power," he said, "is that too many physicists are making decisions. In fact, there are just too many physicists working on atomic power." The physicists were angry, but the chemists all agreed. Such statements were always good for publicity. Rickover loved it.

He also made his presence felt with his aggressive manner at meetings or other functions. When talking to one of the top people, he would almost physically back him into a corner and keep talking. He wasn't really well known at that time of his life, but it didn't take long until everyone learned who he was.

My teacher-pupil relationship with Rickover was interesting while it lasted. I tried to teach him reactor physics in the evenings. While I certainly wasn't the best teacher, he felt comfortable with me. The problem was that he just didn't have the necessary technical background in math.

One evening he said to me, "John, I'm quitting your course."

I wanted to know why. "Surely," I said, "my course isn't too tough for you."

He ignored my attempt at humor. "No," he replied, "but I already know more than any of the company presidents or government officials I'll have to deal with. Any more detailed technical knowledge is a waste of time. I'd do better to spend my efforts on the big picture, not a lot of technical detail."

I thought his reasoning was correct. He was not a technical detail person. He was a master politician and an expert at getting things done; at deciding which technical option to back. He was also a good intuitive engineer. He believed in keeping things simple and never cutting corners.

Development on a Shoe String

As time went on, it became obvious to us that our group lacked the tools to make the Daniels Pile a reality. No one in the group had had reactor design experience and none had been involved with the bomb project.

The Daniels Pile was to be a 2100°F gas-cooled, beryllium-moderated reactor. Such a reactor would be a major challenge even today.

We were 30 technical people with wide and varied experience in many other fields. There was one theoretical physicist, one experimental physicist, a few aircraft designers, a couple of electrical engineers, several mechanical engineers and a few heat transfer men. But we had no laboratory and little money for consultants or for getting development done in other labs. Oak Ridge scientists gave us some help, but it was not a part of their responsibility. Also not having been involved in the bomb project we were outsiders.

We were attempting a revolutionary development on a shoe string. Harold Etherington, who later played a key role at the Argonne National Laboratory (ANL), was an important member of the group. As an example of how unprepared we were, they put me in charge of the instrument and control design, a field I had to learn from scratch.

We were taking courses in reactor physics and design. Except for our group, the prerequisite was a Ph.D. degree. The instructors were the best in the country and included Fred Seitz, later to become head of the National Science Foundation; Al Weinberg, later director of Oak Ridge National Laboratory (ORNL); and Eugene Wigner, a Nobel laureate.

I was technically on the Monsanto payroll, but I was still considered a Westinghouse employee and operated on a Westinghouse expense account. Westinghouse reimbursed Monsanto for my salary. Therefore I frequently went to Pittsburgh to report. Rickover often had been telling Price things that he should do, but Price had not worked with Rickover during the war and wasn't always ready to take Rickover's suggestions seriously. When I got back Rickover wanted to know why I had not persuaded Price to agree to do what had been suggested.

It was during this period that I received my first patent as coinventor, along with Elmer Wade of GE, of an analog computer to solve the multigroup theory equations for reactor design. These equations were just about the same as those for transmission line design. All the scientific training at Oak Ridge and in graduate school came in handy later. Even when I was not directly involved in actual design, I knew enough that the scientists were afraid to try to snow me for fear I might catch on.

Trying to understand enough about reactor physics and engineering was hard work, but finally we did come up with a fairly good conceptual design for the Daniels Pile. Fortunately we didn't have to build the reactor. After about a year, it became obvious that not only could the project not be completed on the timetable proposed, but without more basic research, it could not be done at all. I'm sure that with proper funding it could have been built. It probably would not have been done as was then proposed, however. The temperature undoubtedly would have been lowered, and the moderator might have been different.

From the Daniels Pile to *Nautilus*

The Manhattan District was dissolved on January 1, 1948, the Atomic Energy Commission (AEC) was formed, and Union Carbide took over the Oak Ridge National Laboratory contract. The AEC decided that further work on the gas-cooled reactor should be halted, pending an extensive survey of alternative reactor types. All future reactor development would be done at ANL in Chicago.

If the project had gone to completion, gas-cooled reactors would have had a head start for central station power, something that might well have changed the whole look of nuclear power in the world. But that was not to be.

In September 1947, a preliminary study report was submitted and the Daniels project was terminated shortly thereafter. The silver lining of this cloud was that we were now free to help Rickover proceed on nuclear submarine propulsion.

Even as we were officially winding up the Daniels Pile project, Rickover had us doing a conceptual design for a submarine propulsion plant using pressurized water. This work elaborated on an idea first proposed by Al Weinberg of Oak Ridge in April 1946 for a power production reactor. Some people credit an early report by Dr. Phil Abelson, on loan to the Naval Research Laboratory, with getting Rickover interested in nuclear propulsion. The formal fairly definitive, preliminary design report of a pressurized water reactor for submarine propulsion by the Daniels Pile group was issued in January or February of 1948. This became the point of departure for both the Argonne National Laboratory and Bettis in the design of the prototype and *Nautilus* reactors.

A small but steadily growing snowball now began to roll in the direction of nuclear submarine propulsion.

In 1947, Herbert Pomerantz of Oak Ridge and Albert Kaufman of MIT found that the 2% of hafnium that occurs in nature with zirconium could be removed, and that pure zirconium was a low absorber of neutrons. Those were two important breakthroughs. I was in Rickover's office when Sam Untermyer came in carrying a shiny piece of pure crystal bar zirconium. He said that they had removed the hafnium impurity, which would make excellent control rods, and the pure zirconium was ideal for fuel cladding.

In December of that year, the Navy Department formally stated the need for development of a nuclear-powered submarine and requested that the AEC take action for the early development, design, and construction of a suitable reactor.

On April 2, 1948, Vice Admiral E.W. Mills addressed the Underseas Warfare Conference in Washington and summarized the Navy's efforts to obtain action on the nuclear submarine. This speech marked the real beginning of the Nautilus power plant.

The following month the Military Liaison Committee went on record that a nuclear-powered submarine was of great strategic importance to the national defense. It recommended that formal status be authorized and that the work be done by industry.

The rapid-fire sequence of all of the events is somewhat confusing. Numerous important individuals and organizations were calling for this development in many forms, but, in most cases, they only amounted to expressions of opinion. There also were some expressions against proceeding with a nuclear-powered submarine. For example, the prestigious General Advisory Committee of the AEC, consisting of many of the country's most eminent scientists, advised against proceeding, calling the project technically premature.

In the spring of 1948, it was still in doubt whether the government would authorize the building of a nuclear sub. Meanwhile the AEC was transferring to ANL any who wished to pursue the development of nuclear propulsion; that included five of us from Westinghouse—N.J. Palladino, P.N. Ross, A.H. Toepfer, Ernie Miller, and me.

I decided not to go. Why? First, it seemed unlikely that ANL could successfully develop a submarine propulsion plant with the limited funds available and the lukewarm Navy interest. The inept handling of the Daniels Pile project by the government certainly did not inspire confidence in future programs. It should have been clear that, considering the thousands of the best scientific and technical people it took to develop and build the low-temperature Hanford production reactors, 30 or so engineers with no nuclear experience and limited funds could not develop the high-temperature gas-cooled reactor.

Westinghouse did not appear to have any near-term intentions with regard to nuclear power. Also, I had met and married Esther Slattery and we now were expecting our first child. My best course, I believed, was to return to the switchgear division at East Pittsburgh, one of the corporation's top operations and a place where I had a more predictable future. The nuclear field would still remain an option for later.

Phil Ross transferred to the Westinghouse heat transfer division where he worked briefly on the Navy's Project Wizard, the development of nonnuclear components and systems for a pressurized water reactor. This contract was terminated and the work folded into the Bettis contract. Joe

Palladino and Adolph Toepfer moved to ANL and Ernie Miller went back to the steam division.

Of course, that wasn't the end of the story—in fact, it was just the beginning. Despite the gloomy outlook, however, Rickover persisted and finally won out in his fight to build the world's first nuclear submarine. I was to come back to the nuclear world before very long.

CHAPTER 2
=========

Westinghouse Enters the Field

It was a cold, blustery day in January as Gwilym Price, Westinghouse president, stepped to the microphones on the speaker's platform in front of the flag-draped bow of the towering, gray submarine. His words expressed the hope of high-ranking spectators and a national radio audience.

"The *Nautilus*," he said, "is a testimonial to the ability and determination of free men to act in the defense of human rights and dignity. . . . It gives hope that this launching today will become the symbol of a future day when mass power for people everywhere will overcome and destroy those roots of war—poverty, insecurity and fear. May God so will it."

With that, Mamie Eisenhower, the First Lady, grasped a champagne bottle with both hands and swung it hard against the submarine hull. The giant boat slid slowly, then faster as it glided down the ways to a cacophony of cheers, factory sirens, ships' whistles, a Coast Guard band, and the loud horn of the *Nautilus* herself.

The world's first nuclear-powered vessel—once thought an impossible dream—was a reality. The atomic age has entered its second stage on this day in 1954, just 14 years after the first chain reaction at Stagg Field.

The tale of how, after the early work at Argonne National Laboratory (ANL), the remarkable persistence and vision of a little-known Navy captain and the equally remarkable dedicated work of a team of scientists and engineers from industry pulled off this tremendous achievement is one of the great stories of American technical leadership.

From this achievement, President Eisenhower told the United Nations, "the United States knows that peaceful power from atomic energy is no

Figure 2.1 Launching of the *Nautilus*.

dream, of the future. That capability, already proved, is here—now—today."

The entrance of Westinghouse into the nuclear field, which it was destined to dominate for years to come, resulted from the exercise of vision by its chief executive and the persuasion of the Navy officer who was to win the title of "father of the nuclear Navy." At the time, it attracted little attention. Few were looking and probably fewer cared, because it was just a small contract for heat transfer and auxiliary non-nuclear system development.

Captain Hyman Rickover had been trying for some time to bring Westinghouse into the nuclear development effort and Price was convinced that the company's future would be well served by becoming involved. After all, generating and harnessing power was the principal business of Westinghouse in the peacetime world. The only thing lacking to make the nuclear energy field attractive was money.

This changed on June 12, 1948, when funding became available. Westinghouse promptly met with Pentagon officials and agreed to perform engineering development for the auxiliary systems for a nuclear submarine propulsion plant. "Project Wizard," as it was called, amounted to only $830,000, but it clearly was an important beginning.

For Price it also was a roll of the dice—a gamble on the future. Rickover, then a special assistant for nuclear matters to Vice Admiral Mills, chief of the Bureau of Ships, had no assurance that he would continue to control the Navy's nuclear future or that he would succeed, if given a chance. Price bet that he would. It was a good bet.

ANL Becomes Reactor Development Center

Behind the scenes, early nuclear development involved several competing organizations. The Atomic Energy Commission (AEC) made a decision in 1948 to establish the Argonne National Laboratory outside Chicago as the U.S. center for development of all nuclear reactors. This decision caused a competitive uproar in other laboratories that had been designing reactors. Protest meetings were held, including one in the home of Alvin Weinberg, director of the Oak Ridge National Laboratory (ORNL). Even Sid Siegel, later the head of physics at the Westinghouse-operated Bettis Atomic Power Laboratory near Pittsburgh, Pennsylvania, attended this meeting. But, despite the objections, ANL inherited responsibility for nuclear reactor design.

In the overall nuclear program as foreseen by the AEC, there were to be four reactor development projects: the submarine thermal reactor (STR), an intermediate-energy spectrum breeder reactor, a materials testing reactor, and an experimental fast breeder reactor. The first of these, the STR, had top priority.

Shortly after the Naval Reactors Branch (NRB) was formed by the AEC, it established the STR as a formal project. The Argonne lab was assigned research and conceptual design and it then set up a Naval Reactors Division on December 3, 1948. This organization, directed by Harold Etherington, formerly with the Daniels Pile group, included many outstanding engineers from Westinghouse, General Electric, and Allis-Chalmers, plus Navy officers and civilians. Westinghouse engineers included Robert Creagan, Lloyd Kramer, Joe Palladino, Adolph Toepfer, and others.

Rickover Gets His Two Hats

The establishment by the Bureau of Ships of its Nuclear Power Division headed by Rickover on August 4, 1948, was the culmination of a long and bitter struggle—a story well told elsewhere.[1] The following month the AEC

---

[1] R.G. Hewlett and F. Duncan, *Nuclear Navy*, Chap. 3, University of Chicago Press, Chicago, 1974.

announced formation of its division of reactor development and Dr. Lawrence Hafstad, executive secretary of the Department of Defense Research and Development Board, was named director.

As head of the Navy's nuclear power division and the AEC's naval reactors branch of the division of reactor development, Rickover had managed to get his two hats. This was one reason the odds favored him. If he was thwarted by the Navy, he put on his AEC hat and got his way there. If funding for a particular item was doubtful with AEC money, he would turn to Navy resources. This wasn't just a possibility. It often happened.

Back in Washington, the politically shrewd Navy captain had everything pretty well lined up. By adroit maneuvering, he was able to get his nuclear submarine program rolling even though nobody was volunteering to help. He prepared letters for senior officials and persuaded them to sign, urging yet other officials to make favorable decisions. But it was tough going.

He now enjoyed the backing, at least theoretically, of both the Navy and the AEC. ANL was authorized to start developing submarine nuclear propulsion. Rickover fought and cajoled, connived and pleaded almost singlehandedly to bring the naval nuclear propulsion program into existence. Without him, the nuclear submarine certainly would not have existed for many years. He might have been able to get another contractor and gotten the development and design done without Westinghouse, but he didn't.

Bettis managers had direct-line supervision over all research, development, design, testing, procurement, and construction. This work was done mostly at Bettis and in Idaho by Bettis personnel. Subcontracted work was also under Bettis direction.

In judging performance Rickover demanded three things: excellence, speed, and total dedication to the job. Westinghouse gave him all three.

The Navy nuclear program was one of the few successful AEC reactor development programs. There had been a long string of failures or partial successes—the sodium graphite, organic-moderated, homogeneous, and gas-cooled reactors and the nuclear airplane. The only other successes during this period were materials testing and nuclear materials production reactors that operated not much above room temperature and the small experimental breeder reactor (EBR-1). The Navy program's aura of success was a key factor in Rickover's good relations with the Joint Committee on Atomic Energy (JCAE) and other government officials.

This is what happened, but the real story is how it was accomplished.

The Historic Contract and Bettis

The Wizard contract did not call for Westinghouse to do nuclear development, but it enabled the corporation to send people to ANL to gain experience. Soon it became irrelevant when the basic contract put Westinghouse into the nuclear business "big time." However, Price had already moved to form the Atomic Power Division.

Because the company had had little project management responsibility even during World War II, having built defense products largely as a subcontractor or a components supplier, Price had no obvious candidate available with experience in large, high-technology projects. But he had good reports on a man who had been manager of the marine marketing department during the war and had started an innovative program called "This Time Let's Keep Our Merchant Marine." Clearly no run-of-the-mill manager, this man brought with him the added advantage of having interfaced successfully with Rickover.

Although his degree was in engineering, Charles H. Weaver worked in marketing and had considerable experience in contractual negotiations with the Navy. An articulate redhead, Weaver didn't hesitate when making hard decisions and Price knew this. So he skipped down several layers of management to give Weaver the assignment that years later would propel him to top levels of the corporate hierarchy.

The new organization was given division status and Weaver, the general manager of the Atomic Power Division (later called the Bettis Atomic Power Laboratory), reported directly to Price. This unusual direct reporting showed the emphasis Price gave this activity—a clear signal to the rest of the corporation of the importance being placed on this activity by top management. Price was trusting Weaver with the most important job at Westinghouse and—as it turned out—one vital to the country.

William Shoupp was Weaver's first employee. He was the company's top nuclear scientist and had been operating industry's first Van de Graaff generator or atom smasher at the research laboratories. He was not knowledgeable about reactors, but he was competent in basic nuclear physics and in managing a scientific staff. He assembled a cadre of highly competent people who soon were able to recruit the other talent needed. He held that post until he left to manage the new Westinghouse commercial nuclear venture. P.N. Ross was assistant director.

Price and Weaver next moved to provide facilities for the division. As a site for the laboratory-to-be, Westinghouse purchased in early 1949 the original Allegheny County Airport site in the Pittsburgh suburb of West Mifflin, which still was being operated for private aircraft as Bettis Field. When preparing the old airport hangers for office and laboratory space, a number of birds were found occupying the hangers and we were not able to get them out. Bill Shoupp solved the problem with an air rifle. That wasn't his only experience with fire arms. He sometimes would conduct meetings while sitting at his desk with a .45-caliber pistol in front of him, as he engaged in his hobby of repairing watches.

Figure 2.2 Atom smasher at Westinghouse Research Laboratories.

So fast did the work move ahead that early construction on the Bettis site was several times interrupted as private pilots ignored the big X marks and attempted to land on the partially existing runways. To their embarrassment, they found trenches dug across their path, which lead to the word quickly getting around that Bettis was no longer an airport.

This historic Bettis contract (AT-11-1-Gen-14), initially for $6,081,000, was the foundation for the Westinghouse role in nuclear power. It was signed on December 10, 1948, and continued through 22 supplements and several decades. The contract was drafted by James T. Ramey of the Chicago Operations Office of the AEC, who later became executive director of the JCAE and then an AEC commissioner.

The contract called for Westinghouse to design a prototype nuclear engine called Mark I and a subsequent shipboard plant (Mark II) "which will meet the specifications of the Navy Department for installation in a submarine . . . in the shortest practicable time." However, there were no specifications at that time. They were to evolve over the next few years.

Supplemental funding was to be as agreed on with the AEC each year by March 31 for the ensuing fiscal year, which started July 1. The estimated funding for fiscal year 1950 was only $2,431,430, with a fee of $121,430. Fees for operating the laboratory were never very large, but the work of the laboratory formed the basis for the very financially rewarding commercial nuclear program covered later in this book.

To Westinghouse, the AEC assigned detailed engineering design, construction, and operation of the STR Mark I prototype plant and design and construction of the subsequent shipboard plant, Mark II. The contract also provided for Westinghouse to do such research and development work on Mark I as agreed on by ANL and Westinghouse.

The contract further stated, "It is the intent of the Commission and Westinghouse that this agreement shall be carried on in a spirit of partnership and friendly cooperation with a maximum of effort and common sense in achieving their common objectives.

"Every effort will be made to design Mark I so that it will meet the specifications of the Navy Department for installation in a submarine. It is recognized, however, that this first plant may not meet all such requirements. It is desired, under this contract, that Westinghouse will be in a position to assume responsibility for all work which may be necessary in the transition from a land-based reactor to a submarine reactor, including the responsibility for any subsequent designs modified as a result of experience with Mark I to accomplish this purpose."

To do this, Westinghouse was authorized to build and operate a government-owned facility that became the Bettis Atomic Power Laboratory. Westinghouse now had its foot in the door, but could it force its way in as a major participant? Now was the time to perform, because if nuclear power was to be the energy source of the future, the company must succeed in this first venture. It had to show a solid commitment to the nuclear propulsion field. The final contract with all the details was signed July 15, 1949. But Price didn't wait for that. He moved fast to get Westinghouse into action.

The contract allowed much leeway for changing the scope of Westinghouse's responsibility. While originally we were to perform only engineering design and production, Rickover also let us have a small research staff that essentially paralleled ANL's work. Of course, we saw this as an opportunity to take control of the naval reactor program, which in the end we did. Rickover didn't object to this because he really didn't think a national laboratory could build anything successfully. Our cause was aided by the fact that Walter Zinn considered that ANL had a broader mission than to serve Rickover's single-minded objectives, an attitude that did not sit well with Rickover.

The Westinghouse CEO put all the technical assets of the corporation solidly behind the new nuclear boss and his fledgling atomic power division. This meant giving him carte blanche to handpick people throughout the vast Westinghouse organization. For the first couple of weeks after his appointment in October 1948, Weaver's division consisted of himself and his secretary—period. Then he began cherry-picking the rest of the corporation with remarkable speed.

Simpson Goes to Bettis

In November 1949, Rickover wanted Westinghouse to transfer me to Bettis from the switchgear division at East Pittsburgh. But division manager John MacNeil objected and offered me a two-level promotion as an inducement to stay.

I was uncertain about what to do. I knew Westinghouse had the nuclear submarine development contract, but I also knew that major funding was required and it was far from a sure thing.

Learning of my indecision, Latham Osborne, corporate executive vice president, called me into his office. Osborne was a soft-spoken, kindly gentleman who commanded great respect from his peers.

"John," he said. "I don't want to put pressure on you, but I hope you will go to Bettis. We need you out there."

That made my decision easy, of course. Top management wanted me to go to Bettis and, when put that way, it was an offer I couldn't refuse. Furthermore, the move obviously would be in my best interest from a career standpoint if, that was, *if* we were successful. So in December 1949, the transfer was made.

At Bettis, my job was assistant engineering manager—a staff job, but approximately equivalent to the one I was offered at the switchgear division. The difference was that switchgear was mainline Westinghouse and one of the most profitable divisions. Bettis was a developmental gamble with uncertain funding. There was a strong effort by some in the Navy Department to kill the program, even after a successful operation of the prototype in the Idaho desert.

Magnitude of the Task

To appreciate the task set before Bettis, one must realize that nobody had ever before built a high-temperature nuclear power plant. We were being asked to harness the tremendous power of nuclear fission and turn it into a steady, well-regulated release of energy to run an engine—safely.

Technical data were scarce and we didn't even know what the problems were. Few people in our organization had ever worked together as a team.

The laboratory lacked procedures for administration, scheduling, budgeting, and reporting. We started with a clean sheet of paper.

We faced the task of performing research and development in unused airplane hangars while the new laboratory was being constructed. It was occupied in the spring of 1950, more than a year after the contract was signed. A major development contract had been given to a Bettis organization that didn't yet exist and that was to require thousands of technical people. Rickover remembered how Westinghouse had performed during the war and believed we could do it again. Talk about having faith!

The job of pulling together thousands of technical people to handle a major development contract required Westinghouse to depart from the normal behavior of postwar companies. Other firms at the war's end pulled their best people off government work. Westinghouse believed in the future of nuclear power, and Price gave Weaver top priority to pick the best talent in the corporation to run that division.

Although there were only about two dozen people brought in from other Westinghouse locations out of several thousand eventually needed, this enabled Bettis to get a good start. I don't believe we caused any serious harm to any of the other Westinghouse divisions.

The emphasis was not on just hiring good technical people; we wanted and got the best. Had the rest of the company followed this practice Westinghouse would have been in far better shape today.

Weaver's job posed a big, individual challenge for him. He had the responsibility for motivating a large number of the most talented technical people in the company to accomplish the "impossible" job. He was managing a group of highly talented and headstrong technical people who were constantly at odds with Rickover and NRB. That he held this team together is nothing short of a miracle.

It was inevitable that not all the senior team picked to do a job with no precedent would meet the test, and so it happened. Weaver's job was to make the necessary changes, find better replacements, keep the team together, and maintain the very tough schedule. This he did exceptionally well. All the while, he had to take the brunt of dealing with Rickover—undoubtedly the most demanding individual any of us had ever known. Rickover made Simon Legree look like a scoutmaster.

Bettis Management Organization

Organization charts are deadly dull things unless your name is on them. But it is important to give recognition to the outstanding leaders at Bettis who performed what I still view as the nuclear miracle. Figures 2.3 and 2.4 show charts representing the start of the project and the organization as it evolved in 1952.

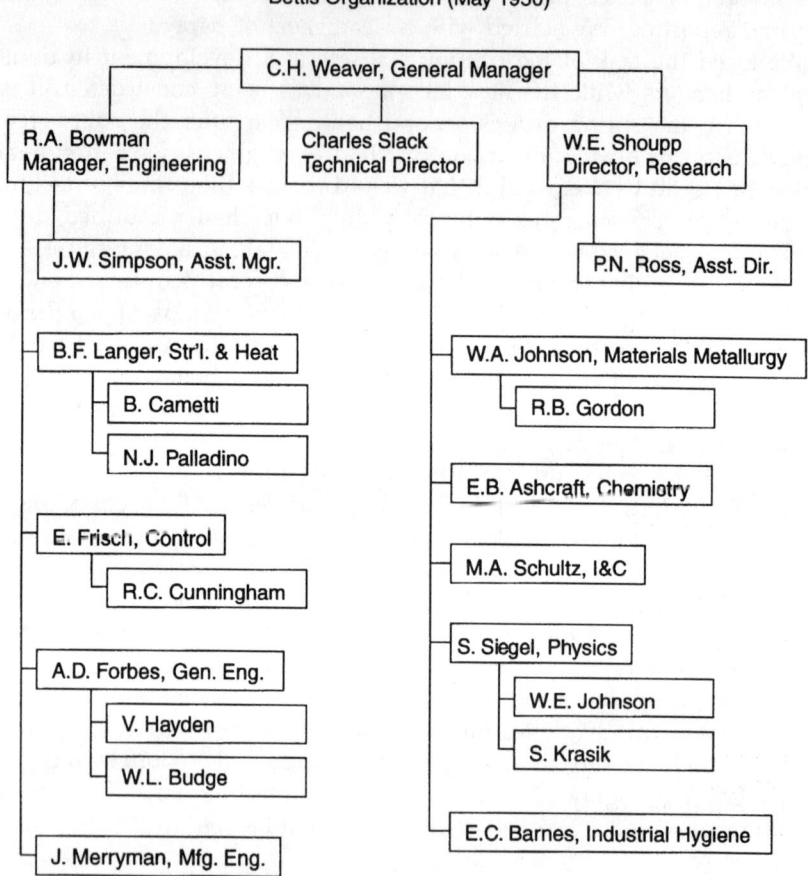

Figure 2.3 Bettis organization chart, June 1950.

Charles Slack, the first technical director, was given no line responsibility. He had been director of research for the Westinghouse lamp division, which did the early work refining uranium for the Fermi pile at Stagg Field, a real piece of atomic pioneering. He was a competent, low-key manager, who did much to help Weaver recognize technical problems and deal with the technical staff. Slack was relieved of his responsibilities as technical director and made a consultant in 1951 because he couldn't get along with Rickover.

He was followed by Leon Ludwig, an excellent engineer and former manager of the motor and control division of Westinghouse in Buffalo, who was made director of engineering and research, with line responsibility for

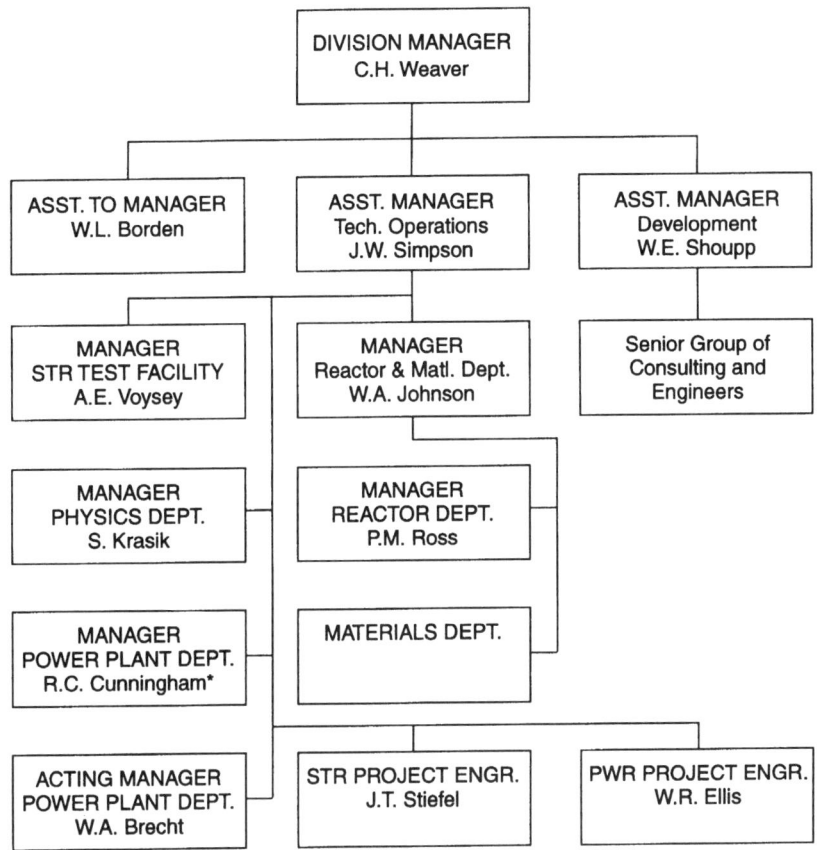

*On loan to General Dynamics Corporation for startup of the *Nautilus*

Figure 2.4 Bettis organization chart, June 1952.

all technical activities. But Leon was at Bettis only a day or so before he became ill and died.

Frank Benedict was brought on board for his quality control capability, but he knew nothing about reactors or system design engineering. He was unable to meet the challenge and did not stay long.

Robert Bowman, from the Westinghouse heat transfer division, was doing a good job as engineering manager, but he ran afoul of Rickover and left to join the Bechtel Corporation, where he became a senior vice president.

Bowman was replaced by Clarence Lynn, who had designed motors and generators for Rickover during the war. However, he had little concept of how to run a complicated major development program, and he compensated for this by becoming involved in many minor engineering problems, paying little attention to the overall picture and was forced to leave. More about this later. I was assistant to Bowman and later to Lynn.

In 1951, a project department was split from engineering with Phil Ross as manager. This department was responsible for liaison with the shipbuilder on overall Mark I construction. There were two project engineers: John Stiefel for Mark I and Robert Lusk for Mark II. In June 1952 this department was renamed the reactor department and was given the added responsibility for the nuclear core design (thermal and mechanical but not physics).

Bettis Technical Organization

Weaver fared much better in the selection of other technical people than he did with the top technical managers. None of the senior technical staff failed to measure up to the job.

Weaver used his priority well in selecting the senior technical people. He had help from A.C. Monteith, the electric utility group vice president, and J.A. Hutcheson, vice president for engineering and research. The people selected had great respect for each other's competence and soon were a smoothly working team, albeit with some problems.

You can see the names of the other key people on the accompanying organization charts. A more talented and dedicated group of engineers and scientists I had never encountered. Every one an "All-American" in my book.

Technical Problems Faced

To understand the technical challenge facing the Bettis engineers and scientists, here in as simple language as possible is a brief review of what we thought it would take to construct the nuclear propulsion plant for the *Nautilus*.

We started with a set of "givens"—Navy assumptions and requirements that set up our target. The Navy wanted a certain speed for the submarine and insisted it have two propellers, and preferably two complete sets of main propulsion equipment.

The concept of a nuclear propulsion plant was disarmingly simple. Just put enough uranium, enriched to the proper amount of the uranium-235 isotope, into fuel elements; the fissioning of the uranium will produce heat.

Then flow a coolant over these hot fuel elements to generate steam that will then drive a turbine. The turbine turns the propeller shaft.

This reactor will be controlled by the insertion of rods between the fuel elements, which will serve to slow down or speed up the fission process.

Sounds easy, doesn't it? The trouble was, none of this theory was well enough advanced to know precisely how much or how many, or how big or how small, or how hard or corrosion resistant everything had to be. Most of the hardware we needed didn't exist. Some of the materials we needed didn't exist either. They had to be improved or developed from scratch. They had to be tested.

The theory of nuclear reactors was not adequate, for example, to determine the amount of enrichment or just how the heat generated would be distributed. We had to develop the theory and validate it by means of what are called "critical" experiments. And, for building the first prototype reactor, Mark I, we were forced to do the calculations using hand calculators, then develop computer codes for far more sophisticated calculations later.

What would be the best coolant to use? Even today there is no agreement as to which coolant is the best for various reactor types. Etherington, although he favored a pressurized water reactor, had three different groups studying alternative concepts. Each of the three Level 1 studies examined a different coolant—pressurized water, helium gas, and liquid metal.

These ANL studies were reviewed at both Bettis and GE's Knolls Atomic Power Laboratory (KAPL). Neither group found gas cooling to be very attractive, mainly because of the low specific heat of helium, the difficulty of containing the gas, and its availability. KAPL argued strongly for the higher efficiency of superheated steam, but this required a liquid metal coolant. Bowman, on the other hand, was willing to accept dry and saturated steam and pressurized water. In the end KAPL elected to use the liquid metal approach for their submarine program.

When these studies were finished, the pressurized water reactor (PWR) was deemed the most likely to be completed successfully in a reasonable period of time. Even before the studies were completed, Rickover was certain that the PWR concept was the correct one, and he succeeded in forcing a decision in the spring of 1949. From then on, work at ANL and Westinghouse concentrated on the PWR.

This decision lead to dealing with the problem that water is highly corrosive at high temperatures. The fuel elements had to be clad in a durable corrosion resistant material. But what? No one at that time realized just how corrosive high-temperature water was. The possibilities were very limited. We soon learned that zirconium in pure form, but not too pure, resisted corrosion quite well. But even small amounts of certain impurities made it unacceptable. Pure zirconium was unavailable, so we had to develop a method of producing it and then build the necessary facility.

Some impurities, we found, actually improved corrosion resistance in zirconium. So although we had to live with pure zirconium for Mark I, we undertook a major program to develop a good zirconium alloy, which was used later in the naval program and throughout the nuclear industry.

Figure 2.5 Al Squire and early zirconium crystal bar production.

Figure 2.6 Canned motor main coolant pumps.

How to insert and move the control rods through the thick steel head of the vessel that surrounds the reactor posed another difficult problem. The method proposed by ANL was judged to be unsuitable. It was necessary to develop "canned motors" to move the control rods. We couldn't tolerate the leaks that normal motors with seals would allow.

To know how and when to move the control rods, neutron detection instruments had to be developed that would be suitable for use in the reactor environment. The control rod alloy (Cd-In-Ag) that was the reference design for a long time didn't meet the requirements, but we found that the rare metal hafnium would do the trick. However, it wasn't available in commercial quantities. Again, that meant we had to produce it ourselves.

Ordinary motors couldn't be used to drive the main coolant pumps either, so we developed large motors with a sheet of metal covering the stator and another the rotor, thus protecting the electrical parts from contact with the coolant water. The rotor was inside the reactor envelope, thus preventing any leakage of coolant.

We had to have instrumentation that would determine the temperatures, pressures, and flows within the reactor. There were none suitable; so, you guessed it, we had to develop and make our own instruments.

Figure 2.7 Bettis system test loop.

The extreme corrosiveness of high-temperature water required the use of many unusual metals and alloys. All of the equipment had to be suitable not only for the nuclear environment, but also for vibration and battle shocks encountered in submarine service. It was difficult to get equipment suppliers and the shipyards doing the installation to meet our necessarily strict standards. Bettis took the lead in educating and monitoring both the suppliers and the shipyards.

After components had been tested individually they were combined into a sytem and tested. A nonradioactive loop was installed at Bettis for these system tests.

Such were the challenges and such was the nature of pioneering a new technology. Some of these problems and their solution are covered later in this chapter. All of them are covered extensively in the Naval Reactors Technical Story chapter.

Periodic Technical Reviews

Periodic reviews at ANL included Rickover, Cyril Smith of the University of Chicago, and Al Weinberg, director of ORNL. While this review group discussed such things as the uranium inventory, zirconium metallurgy, control rod requirements for shutdown and xenon override, and multi-cycle control rod response, they really gave just broad-brush guidelines. Implementation was still left with the technical people in Bettis and ANL.

Rickover and his NRB organization provided competent contractual management and guidance and demanded excellence and speed of performance. However, essentially all of the research, development, design, procurement, construction, and testing—for both the prototype and the *Nautilus* propulsion plants—was done by Bettis or under its supervision, after the early work at ANL.

Many contracts were given to outside firms for components. For example, main coolant pump development contracts went to Byron Jackson, Allis-Chalmers, and Westinghouse. However, Westinghouse pumps were eventually chosen and were manufactured at the Westinghouse Electro-Mechanical Division in Cheswick, Pennsylvania. Eager to characterize this new industry, the press called this the "world's first atomic manufacturing plant."

Later on, when the naval program expanded, five vendors provided nuclear fuel elements including Westinghouse at Cheswick; Metals and Controls of Attleboro, Massachusetts; Combustion Engineering in Windsor, Connecticut; Babcock & Wilcox in Lynchburg, Virginia; and Nuclear Materials of White Plains, New York.

What Happened to GE?

So where was General Electric?

GE's situation was complicated by a number of factors. First, the AEC had prevailed on that company to take over operation of the Hanford, Washington, nuclear materials production facilities that made plutonium for bombs. In return, the AEC provided a nuclear development laboratory—the Knolls Atomic Power Laboratory (KAPL)—for GE at Schenectady, New York. At KAPL GE was developing an intermediate-energy spectrum breeder reactor.

But the AEC obviously gave the Hanford assignment greater priority where GE was concerned. The Cold War was at its height and it had no intention of jeopardizing the plutonium production job because of Rickover's questionable nuclear submarine program.

Rickover believed GE was the best qualified company to develop the submarine propulsion plant, but he also saw the advantage in having the two major electrical manufacturers compete. Getting GE more involved at this time did not seem to be in the cards. However, GE would continue work on the development of auxiliary systems for a liquid metal reactor, which paralleled the Westinghouse water reactor system. That was as far as GE was allowed or prepared to go at that time.

GE, at the Knolls Atomic Power Laboratory, had been working on a central station liquid-metal-cooled intermediate-energy spectrum breeder reactor. When that program appeared to be stalled, Rickover persuaded the company to undertake instead development of a liquid metal, submarine-propulsion plant.

Sodium was clearly more suitable for a central station reactor than for a submarine reactor, but it was a possible alternative to the PWR type. At that time there was no certainty that a PWR submarine reactor could be developed. A failure in any one of many areas, such as cladding for the fuel, might have doomed the PWR.

Eventually GE completed a submarine intermediate breeder (SIR) prototype at West Milton, New York, and ultimately a submarine, the *Seawolf*, went to sea in 1957. Although it technically was a success, Rickover decided that the liquid metal reactor was not as suitable as the PWR for submarine or other naval vessel purposes. And on December 16, 1958, the *Seawolf* entered a shipyard to have its center section containing the reactor cut out and a PWR designed by Bettis installed.

The problem was leaking heat exchanger tubes. As T. R. Rockwell points out, "Heat-exchanger tubes of any type sooner or later develop small leaks, which in a pressurized-water reactor results in water leaking into water. But in SIR, a leak would result in high-pressure steam and water leaking into sodium, which would cause a disastrous explosion. Even if the leak were a tiny one, the reaction of sodium and water at the leak would create sodium hydroxide—lye—which would quickly corrode the tiny leak into a large one."[2] This problem was overcome by using a double-walled heat exchanger with mercury as the intermediate liquid. This introduced the problem of mercury on a submarine and was a difficult design to fabricate.

The heat exchanger tubes in the SIR prototype did leak as did those in the *Seawolf*. Had the PWR not been a success, undoubtedly the development of liquid metal submarine reactors would have continued.

GE engineers came to Bettis to learn about PWRs. They learned fast and handled the propulsion for many later naval vessels, all essentially copies of the Westinghouse design, but they never caught up with Westinghouse in the number of reactors for the Navy.

---

[2]T.R. Rockwell, *The Rickover Effect*, Naval Institute Press, Annapolis, MD, 1992.

Building a Submarine in the Desert

The traditional approach to a project of this magnitude was to build a small pilot plant first. This would be purely experimental and would provide experience before attempting to build the full-size plant. Bettis and Rickover made a courageous decision to skip that step and immediately build a full-size prototype. As it turned out, this saved much time in getting the *Nautilus* to sea.

So it was that we set out to build a submarine in the desert, almost 600 miles from the nearest ocean, in the middle of the AEC's 400,000 acre National Reactor Testing Station (NRTS) in Idaho. There, the Mark I nuclear power plant and the associated propulsion equipment were to be installed inside two hull sections of a submarine built to the exact dimensions of the *Nautilus*.

In simple terms, it would be an underwater steam engine burning nuclear fuel. The engine's vital parts would be fashioned from relatively unknown materials. Intense radioactivity would require ingenious shielding for safety. The engine must run continuously for periods of time unheard of for nuclear reactors and operate without requiring oxygen. And it all must fit inside a submarine hull. No wonder some people called it the "impossible project." It was our job to make it possible.

Rickover permitted no exceptions to the rule that every component fit into the exact space it would occupy in the hull of the *Nautilus*. There would be no cutting corners, such as putting parts together breadboard style with some equipment too big to fit in, while promising to downsize it in time for the actual submarine construction. Every thing was going to fit before actual construction began.

The Mark I propulsion plant was mounted in a steel hull, the actual size of the engine room and reactor compartment of the *Nautilus*. Lacking an ocean, the submarine builders brought their own—a 50-foot-diameter, 40-foot-deep "sea tank" that surrounded the reactor compartment to make it possible to study back-scattering of radiation for the purpose of designing shielding.

As it turned out, it would have been technically possible to build the Mark I in an actual submarine hull. However, this would have been an unwise risk for something so revolutionary.

Needed—An Overall Plan

Bettis was a completely new experience for me, and I'm sure for the other Westinghouse managers as well. In the past, our experience had been concerned almost completely with improvements and extension of existing product lines. Now we were creating something new, real new. And it took time to realize the magnitude of the job ahead. Developing every component and system from scratch was going to take several times the number of technical people we first estimated.

I soon realized we were not making proper progress and the reason was clear. We had no master plan. Sure each section knew, more or less, what it was supposed to do, but everything being developed depended on how something else was to be designed. So there was a lot of groping around.

There was one person who could be a great help to me on this problem. He was Commander Bill Turnbaugh, Rickover's man at Bettis. Bill had been in the production end of shipbuilding. Though he lacked experience with nuclear power, he knew of the need for an overall plan.

Working together, we prepared an outline that showed a schedule for every major component of each system. There was a line for each system into which the component lines converged. These system lines converged into an overall schedule for the entire project. This gave us a road map on which to set priorities and judge progress. John Stiefel was made project manager and was responsible for making sure performance stayed on schedule. Although things improved, there were still many problems.

Rickover became concerned that his organization and Bettis were not working well together. So, in the spring of 1951, he decided to set up a group of six people—three from NRB and three from Westinghouse—to look into the problem. It was officially called the Project Review Board. No one except those involved were told the purpose of this group. For NRB they were Lieutenant Ray Dick, R. Panoff, and E.B. Roth, and for Westinghouse P.N. Ross was the senior man.

The meetings of this group were conducted as "Quaker meetings." There was no agenda. Each was free to speak at any time. They just sat around trying to establish mutual respect and understanding while attempting to determine the cause of problems between the two organizations. Bettis dubbed this group "the Secret Six."

In the beginning the group focused on the problem of removing the antagonisms and lack of trust that had developed in the working relations between Bettis and NRB. This led inevitably to criticisms and appraisals of the performance capabilities of specific individuals in both organizations. Because of the sensitive nature, this was not written up but was reported verbally to Weaver and Rickover. There were many heated exchanges between Ross and two of the NRB members, Dick and Panoff, during which Roth acted as moderator and conciliator.

Lieutenant Dick was considered Rickover's hatchet man and no one at Bettis thought highly of him. Of course, through the grapevine, Bettis technical people knew that secret meetings were going on and that Dick was a part of it. They assumed that Rickover would take drastic action against those not meeting his standards and this caused considerable consternation.

Phil Ross believed that in the end, though it is difficult to point to specific improvements, all the thrashing about did result in some improvements in the rapport between the two organizations. The board did play a significant role in the selection of the canned motor roller nut control drive mechanism to replace the ANL "three gang drive." The final decision was made at a meeting at ANL of Bettis and ANL people called by Rickover. This is discussed further in the Naval Reactors Technical Story chapter.

From my point of view I could see little or no tangible good that resulted from the exercise. They produced no report that ever was distributed, and we were unable to see that much was changed as a result of their actions.

There was another "inquisition" that took place later. It concerned nuclear design codes, shielding calculations, and thermal design and involved both Bettis and KAPL. Ross was the chairman of the group, which included John Taylor and one other person from Bettis. Mandil, Rockwell, and Radkowsky represented NRB. Harry Stevens and Henry Hurwitz came from KAPL for these meetings. Round-the-clock sessions were held at the DuPont Circle Hotel in Washington for a couple of weeks. An exhaustive and comprehensive review of the validity of various calculations, basic physics data, and analytical methods was made and incorporated into a voluminous report, which promptly sank in the ocean of paper without a trace. The net result, however, was renewed faith on the part of Rickover in the competence of Bettis. Also Radkowsky and Rockwell gave us somewhat less trouble afterward.

Rickover's Management Style

To say that Rickover had a unique management style is a serious understatement. You might liken it to the full-court press in basketball. Keep the pressure on, use unorthodox tactics, and never let up.

A favorite Rickover ploy was to call meetings on nights, weekends, and holidays. Whether his purpose was to demonstrate the importance of accomplishing something, or to stress the urgency of the project, or simply to harass the troops, I cannot say. But it happened too frequently not to be an intentional strategy.

Rickover broke every management rule in the book, but for him it worked. It is hard to see how we could have worked any harder or done any better. The same I believe goes for all of our subcontractors. It has often been said that Admiral Raborn did a good job on the Polaris missile program in a more normal manner and why couldn't Rickover have been more like him. In my opinion, to each his own. Rickover couldn't have gotten the job done the Raborn way nor could Raborn have succeeded in the Rickover manner.

General review meetings covering the entire program usually lasted all day. At these meetings, Rickover wasn't much interested in successes. He wanted to know about troubles. Woe betide anyone who gave an optimistic picture, only to have a problem surface shortly thereafter. These meetings were virtual "wailing-wall" experiences—every real or imaginary problem was described in gruesome detail, making it difficult even for Bettis management to determine how bad the situation really was.

When key decisions were to be made, Rickover would come to Bettis, accompanied by one or more of his NRB staff and someone from the AEC's Pittsburgh area office. The Bettis technical people would make a presentation and recommend a course of action for approval. The air was tense because you never knew what Rickover's reaction would be. It was seldom calm and often resulted in a tirade, sometimes with serious consequences for the poor devils involved.

The day before an evening meeting with Rickover, several of his staff would come to Bettis and grill our people on the technical subject scheduled for the evening discussion. Their job was to dig out every possible problem and report those problems to Rickover before the meeting. Sometimes the topics for the meeting changed, because they had not uncovered enough problems on the main topic chosen.

They seldom found a problem we were not aware of, but Rickover took delight in catching people off guard in order to obtain other than the "party line" information. Sometimes such meetings did not produce a decision that was completely clear and we had to straighten it out the next day by discussions with the NRB staff. The Bettis staff would not be sure, from the premeeting grilling, what issues would be highlighted, and thus it was hard for our top management to prepare properly.

Strangely enough, this decision-making process was both speedy and effective, but it also was traumatic and produced a lot of ulcers.

Certainly no one could be sure about the subject of a meeting with Rickover. A meeting would be set up in advance by Rickover's staff with a well-defined agenda. On the appointed day, Rickover would show up for, say, a meeting on fuel rods. After a brief premeeting conference with the division general manager, he would then make his grand entrance to the conference room saying: "I want to talk about control rods." Of course, nobody had prepared for a discussion of control rods and the meeting didn't turn out well. Then Westinghouse would be chastised for its poor performance. Not fair, you say? What does fairness have to do with it? That was a Rickover way of getting information that he thought would be free of management bias.

Engineers from NRB frequently visited Bettis to discuss design details. When Bettis completed the design of a system or component, the necessary drawings and specifications were submitted to NRB for approval.

When technical presentations were made, Rickover was quite good at asking searching questions to bring out the facts and then deciding who and what to believe and support—although he frequently didn't fully understand the technical facts involved.

One sure way to get into trouble was to try to answer a question when you weren't sure of the answer, rather than just saying, "I don't know." You soon found yourself backed into a corner by the inevitable second question and further probing. Rickover's extraordinary demand for certitude in all matters great and small and however complex caught everyone off guard at one time or another. One of the secrets of his phenomenal success was his temerity and tenacity to ask the second question whatever the circumstances. He took absolutely nothing for granted and no one on faith.

Confronted with an impressive array of experimental apparatus on a laboratory tour of inspection, he would cast doubt on the competence of the scientist merely by discovering the presence of a dead storage battery under the bench, for example. He was the ultimate inquisitor as well as a masterfully Machiavellian politician.

Perhaps Rickover's worst trait was to make snap judgments about people—judgments based on a single incident rather than on overall performance. This led to many highly qualified people being transferred out of Bettis. Such people formed the cadre of highly competent employees who enabled Westinghouse to enter the commercial nuclear power field. Although this was a traumatic experience for the people involved, in the end few were badly hurt, and I doubt that it adversely damaged the project.

NRB Representatives

In every discipline, NRB had very competent people who were the basic interface between the AEC/Navy and Bettis. The NRB representatives located at Bettis in Pittsburgh and at NRTS in Idaho played a key role in interpreting NRB positions to Bettis and vice versa. They frequently had a more thorough understanding of the circumstances surrounding an issue than the Washington people could acquire after a few visits.

Rickover's local representatives, as well as the Washington NRB staff, had free access to all parts of Bettis and to all personnel. This occasionally caused problems when Rickover had more up-to-the-minute information than Bettis management. Often, our people in the laboratory didn't consider a minor failure all that important, and I would not be notified immediately, only to have Rickover call and say, "Did you know that . . . ?"

Lawton Geiger, manager of the AEC's Pittsburgh area office, was responsible for monitoring the details of the contract and did a superb job, so did Commander Turnbaugh in the technical liaison role.

Government Relations Got a High Priority

Frequently Rickover brought groups of government officials to Bettis. This usually was just before the government was preparing the next year's budget or before a decision on proceeding with a new project. We prepared for these visits the way Eisenhower prepared for D-Day.

Rickover would evaluate what part each visitor might play in the upcoming decision. Was he for or against us? Did he directly control the decision? Was he in a position to greatly influence the decision?

The day before such a visit, Rickover held a briefing session. Nothing was left to chance, from the moment they were met at the airport until the moment they departed. There was an assignment as to who would be in each car and what subjects were to be raised or dodged. At the laboratory, we followed the same procedure: who would talk, what points would be made, and what questions we would avoid. The lab tour was arranged so certain of the visitors would not be with key people. We were told to make sure each visitor went where he was supposed to go—stopping just short of using force, if necessary. Anyone who would play no role in whatever decision was to be made would be ignored, no matter what his rank or title might be. Thought control at its best!

Of course, Rickover was a master at political maneuvering. He was keenly aware who had control in every situation—who made the decisions and who controlled the money. He knew exactly when to work through the AEC and when through the Navy. He switched hats frequently and adroitly.

He also was a master at understanding what power the administration had and what power Congress had. When he needed help, he would persuade Congress to put pressure on the Navy or the AEC. He courted key members of Congress by taking them to laboratories and shipyards and for rides in submarines.

He was at his theatrical best at hearings and worked hard to persuade the JCAE to hold them. His presentations were always well prepared. Part of his success in receiving congressional backing was the fact that he was running the most successful, if not the only successful, nuclear power program in the world.

Rickover Wants Changes

Once we were trying to explain a method for inserting control rods into the reactor core that was designed by ANL. Rickover thought that the way we were presenting the situation was very poor and he let us know it. The meeting was floundering. He had our engineers so flustered that their explanations weren't altogether clear. Our problem was we really weren't sure it would work, and at that time we didn't have an alternative. How we solved this problem with canned electric drive mechanisms is explained in the technical chapter, Chapter 12.

Finally, in desperation, I tried my hand at explaining the situation. Rickover fortunately began to understand the pros and cons and the meeting gained its momentum. Sometimes you have to ignore the warning "never volunteer." My stock went up, but it might go down 24 hours later.

On one occasion, during a visit to Washington, I asked one of the Rickover senior staff what my problem was. He replied, "One frequently doesn't know. Some day Rickover will need a left-handed monkey wrench and you'll be back in."

In general, my standing must have been fairly good. The secret of whatever success I had in dealing with Rickover over a period of 20 years was quite simple and straightforward: I sincerely believed in what he was trying to accomplish and I was convinced he could do it better than anyone else. I also never put the company's commercial interests ahead of his programs, and was always frank in giving him my opinion, whether it was what he wanted to hear or not.

However, working with him was always a problem. Maybe the toughest situation I had to handle came about when we were coming back on the train from a meeting at Argonne in Chicago. Shortly after the train pulled out, Rickover called me into his roomette. He wanted to talk and at once I knew there was trouble in River City.

"John," he said. "I want you to do something about Clarence Lynn." Lynn, of course, was my boss, the engineering manager.

"What do you want me to do?" I asked.

"I want you to get rid of him," said Rickover. "I don't want him as engineering manager."

I was dumbfounded. "Admiral," I protested, "he's my boss. I couldn't do that even if I wanted to."

"Listen," he said, "I know you are in very good standing at corporate headquarters. If you threaten to quit unless they remove Lynn, it will be done."

"C'mon Admiral," I protested again. "You can't ask me to do that."

"Look," Rickover said grimly. "If you don't do it, I'll have you fired. And I can do it."

Somehow I got out of the roomette without committing myself and spent a sleepless night between Chicago and Pittsburgh trying to decide what to do.

When I got back to Bettis the next morning, I told Charlie Slack about the conversation and asked his advice. At his suggestion I went to Weaver—Lynn's boss. I told Weaver the story and did nothing else. I believe Weaver felt that Lynn was not performing, but in any event, he soon was transferred to another division.

It was only a few months after that incident that Rickover told Monteith he was disappointed in my performance. And by the following spring, Rickover was pushing for an organizational change, and opposed my being given a key role.

Much later, I acquired a memo from Slack to Weaver dated May 5, 1952. The following quote from the memo depicts Slack's evaluation of the situation:

> As you well know, Lynn and Simpson have always disagreed in many rather basic issues in regard to the operation of the engineering department. Lynn does not like to spend his time in regard to organization, budgets, manpower, schedules, delegating responsibility, and digging into those phases of engineering where his past experience has not led him.
>
> Simpson, on the other hand, attempts to understand all phases of the operation and struggles to keep them in balance. While he has not agreed with the methods of operation, he has attempted to adjust himself to the present circumstances under the assurance of both of us of a sympathetic understanding of the overall conditions. You had mentioned to me at the time of the reorganization of the project department last fall that you had told Simpson he was next in line should Lynn leave for any reason.
>
> I suspect strongly that following the reorganization of Lynn's old division, Simpson believes Lynn plans to stay with the atomic power division indefinitely. Simpson has exerted every effort to make things work under the present conditions, he has considered employment elsewhere should he become convinced that the situation will not change for the better. It is now my considered opinion, contrary perhaps to that expressed earlier, that this situation would be aggravated rather than relieved under any of the proposed schemes.

Most of the time, Rickover got his way concerning organizational matters. Fortunately for me, Weaver and senior Westinghouse officials stood up for me at this critical time. Without that, my career would have taken a completely different path. I had been seriously considering an Atomics International job offer. A year or so later, I learned that Rickover had gotten Weaver to admit I had told him of Rickover's threat to have me fired.

The result of this was my promotion in June 1952 to assistant division manager, with an office next to Weaver's. I was given staff responsibility for all technical work in the division but had no real authority. Rickover told me that was my problem as far as he was concerned. I was in charge of all technical work. He made clear that he was holding me responsible. I persuaded Weaver to give me control of the budget process. This, plus a little Simpson arrogance, succeeded in generating the needed authority.

Westinghouse Senior Management Involvement

One of the problems in any corporation is keeping top management people fully aware of what is going on below them. The nuclear program was of such importance to Westinghouse that Bettis and the NRB felt obliged to keep in touch with top management to assure their understanding and support.

So, early in the program, a series of classes was put on at Bettis for senior Westinghouse executives, from Price on down to key public relations executives. They attended a series of afternoon classes designed to give them a real appreciation of what was involved in the project. It also made the Bettis team very visible at headquarters, which was clearly an advantage for later career advancement.

Latham Osborne and other senior corporate officers frequently visited Bettis to judge progress. When they came, Weaver would ask me to join them in his office for the meeting. Lynn was seldom asked. I would give an overall status report, so they could view the project with perspective. Like Rickover, they were often far from pleased. This caused many management changes.

Mark Cresap, who was to become the Westinghouse CEO in 1958, visited Bettis frequently. Almost every time, we had meetings with senior AEC, Navy, or congressional people. Top Westinghouse management people also attended.

The AEC contract called for payment of a percentage of the direct expenses for overhead. This included assistance from the corporate staff people—law, industrial relations, purchasing, public relations, etc.—and these people were always available to help.

Early Performance

Despite all the problems, by 1952 we had completed much of the design, construction of the prototype (STR Mark I) was under way, and the Electric Boat Division of General Dynamics was well on the road toward constructing the USS *Nautilus*—the world's first atomic-powered submarine. The keel was laid and Harry Truman was there to see his initials welded on the keel plate.

In a development program of the magnitude of the submarine thermal reactor program, many dozens of items, completely new in concept and design, are combined to form a final power plant. The principle of individual tests for each item and then each subsystem was followed to the maximum extent possible, so the final assembly represented a test of the complete power plant, rather than the test of individual items.

The most complete system test erected at Bettis was a full-scale duplicate of the primary coolant system. Of course, there was no radiation shielding, because it was not intended that this unit would ever be operated radioactively. The purpose of this test was to permit flow and water treatment problems to be studied in much greater detail than is possible in the shielded and highly radioactive Mark I prototype plant itself. It permitted a study of certain very difficult water treatment problems, with resulting simplifications.

The concept of extensive testing of individual parts, then of the parts combined into individual systems, and then of the propulsion system combined as a whole, before installation in an actual ship, was quite new to shipbuilding and marine industries, though extensively followed in other industries.

Bettis Management Responsibilities

Rickover and NRB leaned on Bettis almost unmercifully for performance and speed. But it is also fair to say that Bettis management pushed its own organization and also NRB. We were a dedicated bunch who believed strongly that what we were doing was of the utmost importance to our country, then involved in the Cold War and the Korean War. This caused us to accept our relationship with Rickover with good grace—most of the time. More than once, we could have cheerfully murdered him. We worked long hours, late into the evening, many days a week, many Saturdays and some Sundays—and seemed always to be on the road.

It was easy to forget that Rickover and his people were the customer and to keep fighting with them. To a great extent the attitude of our top management as expressed by Price kept the fighting down. At a strategy meeting one day, Price laid it on the line for us.

"Gentlemen, Rickover is the customer," he said. "I know the man. I've had to deal with him a number of times. And I know how arrogant and unreasonable he can be. But you have to put up with that. I know it will be hell for all of you from time to time. But that's the price of the project. Swallow your pride and get the job done."

CHAPTER 3

Building and Operation of the Mark I in Idaho

There was as much drama in the building of Submarine Thermal Reactor (STR) Mark I—the world's first atomic engine—as in the launching and building of the *Nautilus* herself. For the scientist and engineer there was even more, I believe. Engineering feats were being achieved for the first time in history. In a way, the *Nautilus* propulsion plant was a rerun because the prototype plant was constructed exactly to the shipyard standards of the submarine. Each piece of equipment, down to the most minute detail, was exactly what would go into the *Nautilus*.

During one of his early visits to the Idaho site, Rickover stopped short. "What's that equipment over there by the bulkhead?" he asked, although he obviously knew what it was.

"That's a coffee maker we use during work," a supervisor assured him.

"Get it out of here," the Admiral insisted. "You know the rules. Move it outside the hull."

So it was that Mark I took shape as an exact copy of what would be in two hull sections of the *Nautilus*—nothing more, nothing less.

Timing was important. The prototype was to be finished and tested ahead of the *Nautilus* construction, but not far ahead. Actually construction of the Mark X at the National Reactor Testing Station (NRTS) began on almost the same day in August 1950 on which President Truman signed Public Law 674. That law authorized construction of the first nuclear-powered submarine SSN-571, later named USS *Nautilus*. The submarine would be powered by STR Mark II.

Bill Budge, who had been a steam division service engineer installing large turbogenerators, was given the assignment of site manager for the construction of Mark I. Woodrow (Woody) Johnson, who was a solid-state physicist by training, but had demonstrated a knowledge of reactor physics and engineering, was selected as Budge's assistant for technical matters. Woody came to me for advice as to whether he should take the assignment. He told me that he wanted to eventually get into general management. I told him that he might not get another chance at a job that was as likely to lead to general management and suggested that he accept. It would give him a chance to see if he liked that kind of work or was good at it. He could always go back to physics.

Since Electric Boat (EB) had been selected by the Navy to build the *Nautilus*, it was logical that EB should also build the prototype. Its contract to construct the Mark I plant at the NRTS was signed February 25, 1950. The firm was to handle detailed design of the hull, plant arrangements, and piping and electrical systems. They put their best people on the job. Thomas W. Dunn was responsible for the reactor compartment design and Frank T. Horan for the engine room. Joe Milligan, one of Electric Boat's most experienced shipbuilders, and his crew of skilled craftsmen came over from Groton to build the submarine hull section, which included both compartments. The reactor compartment was surrounded by the big sea tank to allow for tests of back-scattered radiation from the water.

Despite the skilled personnel on hand from both EB and Westinghouse, Rickover was very uneasy when he was in Washington and couldn't see what was happening. When he visited during construction or a test and something went wrong, he took it in stride. But when he wasn't there, he always imagined the worst. So he set up a three-man group to act as his alter ego.

Bob Panoff, Harry Mandil, two of Rickover's most senior people, and I had to approve every departure from the program or procedures and the corrective action to be taken and whether it was satisfactorily accomplished. No ad-libbing by anybody. As you would imagine, this three-way approval was required frequently in such a developmental project. With the time difference between Idaho and Pittsburgh, and with construction and testing going on 24 hours a day, the three of us were on the phone with one another and with our people at all hours—most of which seemed to be well after midnight.

Mark I Design Frozen

The design and development of the Mark I plant proceeded at top pace until mid-1952. With Mark I under construction in Idaho, we placed a freeze on design changes. This was our opportunity to release the designs for the

BUILDING AND OPERATION OF THE MARK I 45

Figure 3.1 The Mark I in the sea tank at NRTS.

Mark II *Nautilus* plant, but not without a thorough review. With reliability foremost in our mind, simplification became the rallying cry.

Dick Cunningham had responsibility for the *Nautilus* plant design at that time. We initiated the systems approach. Functional requirements for each system and component were carefully defined. This task fell to Walt Esselman and his plant systems department managers, Bill Hamilton, Vern Hayden, Owen Woodruff, and Doug Spencer. They tackled the task with enthusiasm. Up to this time, fluid systems and plant controls were being designed in different parts of the Bettis organization. Now the engineers were working with back-to-back desks and the synergism resulted in a remarkable number of ideas.

Each plant system was reviewed to reduce complexity. No stone was unturned. We knew changes could be checked on Mark I. The need for each valve, shutdown signal, and instrument and control function was questioned and a significant number were eliminated.

Esselman recalls the elimination of more than half of the remotely controlled valves in the shielded reactor compartment. Another result of this effort was the simplified reactor shutdown system described later in this chapter.

Rickover formed a simplification committee to review the plant systems. Bob Panoff had the key Naval Reactors Branch (NRB) role on this effort. While some may argue that these committees did not uncover many problems, Bob certainly provided strong motivation to the Bettis engineers not to be caught with any design problems.

At the same time the equipment design managers under Al Voysey were reviewing their designs for Mark II. This included Ben Cametti on pumps, Karl Schwanekamp on the pressurizer and heat exchangers, Ed Kreh on valves, Basil Lide on instrumentation, and Bill Elmendorf on process instrumentation. Jim Hunter put the final touches on the welding procedures.

Testing of the design changes became a major part of the prototype test program. As the systems were checked in Idaho, deficiencies—no matter how small—were relayed to Bettis, corrected, and forwarded to Electric Boat. The goal was to unearth and correct all problems in Idaho rather than at sea. The value of Rickover's decision to make Mark I a true prototype of the *Nautilus* was proven again and again.

Dick Cunningham's key responsibility for the *Nautilus* plant continued to the *Nautilus* testing and sea trials.

First Criticality

By late March 1953, the main-coolant piping was installed, but not tested. All of the components needed for criticality were in place and tested. The nuclear core had been loaded and we were ready for the first major milestone—criticality.

We had made the theoretical calculations and run critical experiments to validate them. But had we missed something? Would everything work out correctly? The night before, as we checked and rechecked, was a nervous one for us.

New babies arrive and new reactors start up in the middle of the night it seems. At least that's been my experience. The Mark I was ready to be taken to criticality on March 29, 1953, but Rickover postponed it because he felt that there were people in the control room who should not have been there and some were not there who should have been. So history had to wait until all the right hands were present or accounted for. So it was that STR Mark I went critical just before midnight on March 30, 1953.

Final Phase of Construction

The next step was to finish construction of the plant and testing required for power operation. Naturally there were a lot of problems. For a few weeks the three-man procedure, while cumbersome, worked well enough, but nothing was fast enough for Rickover. He was pushing for operation at power by June. Finally Weaver and I agreed that I should go to Idaho and stay there until construction was completed and the prototype was in full operation.

I went out in early April and stayed until late June. But my going was more dramatic than I would have preferred. For reasons of his own, Rickover was dissatisfied with the performance of Bill Budge, the Westinghouse site manager, and Woody Johnson, his technical assistant. Upon my arrival, Budge was to be relieved of his responsibilities, and I was to become the acting site manager. To prevent any premature leaks, Rickover ordered all telephone and other communication between Bettis and NRTS cut until I arrived. I'm not sure how the telephone operators explained the mysterious blackout.

When I arrived I went to Budge's office to tell him that he was being relieved and transferred, only to find that he had gone hunting and fishing somewhere on the Salmon River and could not be located. Budge had been doing a satisfactory job, but wasn't going all out the way Rickover wanted.

I then went to talk to Johnson, who was in charge of technical matters including construction. Johnson was very dissatisfied with the way Rickover and NRB were handling the monitoring and said he would prefer to be transferred. Later Weaver came out and tried to persuade Johnson to stay but to no avail. Johnson remained at the site in the same position until June 1 when Al Voysey assumed the role of site manager with Walt Esselman as his technical assistant.

When Budge returned a few days later, I called all of the senior managers in and told them that I was taking over as acting site manager and that Budge and Johnson were being transferred to other locations. Budge went to the commercial atomic power division and Johnson went to Pittsburgh for a short while and then to Princeton to be a Westinghouse representative on the Stellarator fusion project. This was another case of Westinghouse getting good, experienced people out of Bettis.

Fortunately, Rickover allowed his man at the site, Ed Kintner, and me to make on-the-spot decisions. We only had to confer with the NRB when we felt it was necessary. Because important decisions were to be made, however, we had to have the most competent people present. Joining me from Bettis were four senior managers—Sid Krasik (physics), Dick Cunningham (plant engineering), Paul Cohen (chemistry), and Bernard Langer (mechanical). Frankie Frisch and John Taylor from Bettis were also on hand for special tests.

To prevent getting crossed up with Rickover, Ed Kintner and I had an agreement. He would not call Rickover and I would not call Weaver unless we were together. We didn't stick to this procedure 100% of the time, but almost. It was a big help in keeping our stories straight.

We had other means of keeping Rickover's blood pressure down. He had a rule when he was in Washington, that every time the reactor "scrammed" (automatically shut down), there had to be a full-dress investigation. During this one particular test, when he was on the site, we had one scram and then another, but he was taking it pretty calmly. We were sure, however, his serenity wouldn't last. So Krasik suggested I take Rickover out for a cup of coffee. He guaranteed there wouldn't be any more scrams.

There weren't. Krasik saw to that. Since he knew that the pressure could not instantaneously go from 2000 psi to 2500, where the reactor would scram, he put a clip lead across the overpressure signal. He kept his eye on the pressure gauge and his hand on the scram button. This was probably a gross violation of some rule or other, but we were knowledgeable people and we were sure it was perfectly safe. There was no regulatory commission at that time from which to get approval. From a practical standpoint, if Kintner and I said go, we went.

We took extreme precautions to prevent trouble, of course. Whenever there was an important operation to be performed, such as loading the nuclear core or putting on the reactor vessel head, a detailed plan was prepared listing each step to be taken and what every individual was to do. All cranes were given special tests.

When anyone was working above the reactor vessel, with the head off, each tool was secured to the man by a lanyard so nothing could drop in. And after the operation was completed, we made sure all tools were accounted for.

BUILDING AND OPERATION OF THE MARK I 49

Figure 3.2 The NRTS complex.

The reactor compartment was very crowded, particularly in the area below the shield deck. This lack of space was a constant handicap. Take welding, for example. It sometimes required the welders to make welds lying on their backs or in other strange positions. Repair of equipment that required removal was particularly troublesome. So to help during both installation and maintenance, we built a complete large-scale model of this area.

The equipment in the prototype was so tightly packed together that welding the piping to our strict standards produced many defective welds. After the coolant water was in the plant, repairing a weld was next to impossible and draining the plant was very time consuming. A method was developed that used liquid nitrogen to freeze the water in the piping so that repairs could be made without draining the system.

Incidents During Construction

There were removable fuel elements in the core, so we could determine how the fuel elements were performing. These removable fuel elements had very little clearance between them and surrounding fixed elements. On one occasion we were removing one of them and it stuck part of the way out. This caused consternation, because if we could not completely remove it, the plant would have to be shut down, drained, the head weld cut, and the head removed. This would require possibly several weeks. If the fuel element had started to gall, continued attempts to remove it would have made the situation worse. We finally decided to take the gamble and just apply a lot more force. Fortunately the fuel element came free.

As the June deadline approached, everyone was being pushed to the limit. Tempers sometimes flared, but we all were working toward the same goal and understood the urgency.

The dedication of those Idaho people was amazing. Once, we needed to get some data to Pittsburgh by the next morning. Remember, this was before fax machines. The last plane for Salt Lake City had already left Idaho Falls, so we sent the data to Salt Lake City by a driver, who could still make the connection with the midnight plane for Pittsburgh. Unfortunately, he ran out of gas while still in Idaho. But he was undaunted.

The state police came by, and he persuaded them to drive him to the state line and to radio ahead for the Utah state police to meet him and take him on to the airport. He reached the airport just in time and found the Westinghouse courier. The pony express had nothing on such guys.

On one occasion, Roger Keyes, Deputy Secretary of Defense, and Admiral H.N. Wallin, head of the Bureau of Ships, were to visit the site. They were to be given the red carpet treatment, because government authorization was needed soon for the SFR (submarines fleet reactor). Rickover had just started us on preliminary design of this new class of submarine reactors and we also were working on reactors for an aircraft carrier. So these two officials were particularly important to us at this point. We got a call saying they were ahead of schedule and would be arriving at Idaho Falls within the hour.

This was while Bill Budge was still site manager. Bill started for the airport and his office called AEC security asking them please not to stop a green Buick that would be breaking the speed limit. Budge kept the accelerator on the floor all the way into Idaho Falls and arrived just as Keyes, the former General Motors executive, descended from the plane. Keyes ceremoniously took his hat off and bowed to the Buick before getting in. But then Budge tried to start the car. It was vapor locked. The car had to be pushed to get it started—one of life's embarrassing moments.

As it turned out, authorization for the aircraft carrier wasn't forthcoming then anyway. On the drive to the site, Keyes told Admiral Wallin that if he got a nuclear-powered carrier, he would have to give up almost all the rest of the Navy construction program. Wallin knew then there would be no nuclear carrier, for some time to come.

The flight between Salt Lake and Idaho Falls wasn't my favorite trip. I'm a white knuckle flyer anyway and something always seemed to happen to keep me on edge. One time Rickover had flown up to the site in a Colorado Navy Air National Guard plane. Since my schedule out of Salt Lake was the same as his, we planned to fly from Idaho Falls together. The pilot wanted to know my rank to see if I should be allowed on the plane. Rickover just brushed him off with some remark and we got on board. As it says in the song, "the weather outside was frightful," but I didn't feel delightful because the wings of the plane were icing up as the snow fell. Two Waves got out, threw a rope over the wing, and proceeded to saw back and forth—removing most of the snow and ice. Then they went to the other wing. After they cleared that one, they had to give the first wing another once-over lightly. And when we finally got into the air, the DC3 couldn't fly over the mountains. We bounced through them. I really can't remember what Rickover talked about on that ride. My mind wasn't on business very much.

Training and Maintenance

Rickover was a bear for good maintenance and training, which is why the American nuclear fleet in the years to come would be in readiness far more of the time than would the Soviet fleet. (This opinion was reinforced in 1963 when I visited the Soviet Union.) He insisted we prepare a complete maintenance manual on every piece of equipment not just the major components. We also set up a comprehensive program to train the personnel who would operate this first-of-a-kind propulsion plant.

From the start we had designed the STR plant to be operated by regular submarine engineering personnel. However, we decided that the operating personnel, both officers and enlisted, should be so thoroughly trained that they would have a full understanding of this completely new type of plant. This program set a standard for nuclear submarine training that is still followed today.

The Navy ordered specially selected officers and enlisted personnel, qualified in submarine operation, to Pittsburgh first, then to the Mark I in Idaho. The first group arrived at Bettis in 1951. Their training ranged from basic nuclear physics to the welding of stainless steel. Each individual had to pass a regular welder's qualification test. We encouraged these operating personnel to recommend changes in the detailed design when they believed improvements could be made. Their recommendations would be reviewed by a formal board. Many *Nautilus* improvements resulted from recommendations of the subs engineering personnel.

Strict compliance with the operating procedures that Bettis developed was required, in both the reactor and engine control rooms. Techniques were developed for auditing performance of personnel as they operated the prototypes (STR Mark I and later A1W) like ships at sea. By 1973, Bettis had trained 40,000 Navy personnel to operate submarines and surface ships.[1] This training program continued until 1989 with many thousands being trained each year.

Testing and Operation

No machinery has ever been tested more than the components and systems for the Mark I and *Nautilus*. As each component was installed it was tested again. When we finished a section of the plant and sealed it off, a hydrostatic test was carried out at 3750 psi to determine if there were any leaks. After the plant was completed, a final hydrostatic test of the entire nuclear plant was performed.

Before operating the plant with nuclear energy, a hot functional test was performed. When the main-coolant pumps circulate water, its temperature rises to normal operating levels and any adverse effect on performance of equipment is revealed.

While en route to Idaho one day, I was informed that during a hot functional test a leak had developed. Budge was going into the lower reactor compartment to investigate. What he was doing was potentially very dangerous. If the pipe had ruptured, release of the energy stored in the steam would have wrecked the entire reactor building. Then when Budge met me on my arrival, I saw he was wearing several bandages. I thought, "My God, he was injured checking on the leak." I wondered what else had happened. Was I glad to have Bill tell me that his injuries were from a skiing accident.

Mark I Produces Nuclear Power

As June approached the plant construction was completed. Now was the real test. Were we really sure that we were ready? Kintner, Krasik, Cunningham, Langer, Cohen, and I got together and reviewed everything. Yes, we knew the reactor could go critical, but was there anything about operating at power and at high temperature that we had overlooked or was there something completely unknown that might cause a problem? This was no time to make a mistake and jeopardize the whole program. We

---

[1]"Underway on Nuclear Power," Westinghouse Electric Corporation publication, p. 45, December 10, 1973.

finally convinced ourselves that we were ready to go to power. I decided that the next day was the day.

History was made on the afternoon of May 31, 1953. Rickover had come out with AEC Commissioner Thomas Murray, the only commissioner with a technical background, for a routine visit. When Rickover arrived, Ed Kintner and I told him we were ready to go to power. We expected more than a few searching questions, but Rickover had only one.

"Are you sure?" he asked. We assured him we were.

"OK. Let's go."

Inside the large building erected on the desert 60 miles from Idaho Falls, the atmosphere was tense. Beside a control valve in a room filled with meters, instruments, and panel boards, a group of men stood waiting. Then, on the spur of the moment, Rickover asked Murray to open the valve that would cause the reactor to go up to power. Rickover always found ways to impress important people. The commissioner was thrilled at the chance to play an active role in making history. Murray stepped forward, grasped the valve handle, and slowly turned it.

In the adjoining area, inside the hull of this submarine-on-land, hot steam hissed against the turbine blades. A propeller shaft began to turn. For the first time in history, man had the means to produce a substantial quantity of controlled power from the atom.

This was the Kitty Hawk of the Atomic Age. Half a century after the practicality of powered flight had been proved at Kitty Hawk, North Carolina, man now was proving the practicality of atomic power.

Yes, historic photographs were duly taken. But Rickover was at odds with the AEC Idaho operations manager and, mysteriously, the photos with his picture did not turn out.

Mark I Goes to Full Power

Less than a month after the first power operation, on June 23, 1953, we were ready for full power. Mark I was brought to full power and we got ready to put the revolutionary propulsion plant to its first strenuous test—a 24-hour endurance run to establish reliability and get information on some nuclear physics aspects.

We had checked and rechecked everything to get ready for this endurance run. We carefully reviewed all operating procedures with the shift supervisors who would run the endurance test and were sure that everyone in the operating crew was letter perfect in what they were to do.

For the test, Al Voysey took the day shift and Walt Esselman took over at night. My station was in the control room observation booth, observing but not taking an active part. Al Sanderson was in charge of the operating crews. We had decided that Kintner and I would not interfere or try to second guess the trained operating personnel.

After a few hours of smooth running, the phone rang. My wife was on the line, calling from the hospital in Pittsburgh. The birth of our third child was imminent. My first decision of the day was where I ought to be for the rest of the test run. Since all was going so well and I had no active part in the test, I decided it would be all right to return to Pittsburgh and observe the launching there—in which I also would have no active role. I didn't figure on the Mark I test going so well that Rickover would get ideas.

Well, he did. During the scheduled 24-hour test, Rickover decided to simulate a submerged run across the Atlantic. This would require about 100 hours of operation. So while I was in the air heading for Pittsburgh, the Mark I headed across the Atlantic, so to speak.

Actually, I missed both events. My daughter Patty was born shortly before I arrived at the hospital. After checking everybody there and finding all was well, I got on the phone to Idaho and got the same positive report on Mark I's progress. The life of an observer, I decided, was easy if not always predictable.

For the next two-and-a-half days, Mark I proceeded at top speed and experienced no difficulties. I kept in close touch with Idaho from my office at Bettis and from a hospital phone. Yes, there was some change in the pump noise level, but Esselman did not believe it was a problem. And there was some arcing of the generator brushes, but the power level instrumentation was steady and the other instrumentation, being pneumatic, was not affected. A capacitor was added to prevent any instrument problem due to arcing.

Between the 60th and 70th hour of the run, however, a condenser tube leak occurred. When Esselman reported at 8:00 p.m. the third day, although the power level was steady and the plant output was 100%, condensate water was overflowing because of the tube leak. However, it was being watched and was not causing any trouble. There were some steam generator leaks by this time and there was some minor radiation on the secondary side.

But then some people began to get uneasy. The morning of the fourth day, I got a call from Jack Kyger, Rickover's chief technical assistant, who worried about the condenser tube leak and the detection of some radiation. In the opinion of Rickover's staff people, the reactor ought to be shut down. Kyger said they had given this opinion to Rickover and his reply was, "Call John Simpson. He's in charge."

When I was elected president of the American Nuclear Society and this incident was printed later in *Nuclear News,* I was told Rickover objected to that version. However, he was told by one of his senior staff, "Admiral, that's what happened."

When Rickover thus passed the ball to me, I called my personal staff at the site—Krasik, Cunningham, Langer, and Cohen—and we reviewed the problems. We agreed that it was OK to continue the run. The next dissenter was Voysey, who called to say that he now thought it was unsafe to continue and recommended shutting down. Again I talked with my staff and they held firm that the problems were being exaggerated and there was no real danger. Voysey then agreed and the run continued. I was on the phone almost constantly from the hospital.

A little later, Voysey called once more to say he was now convinced the reactor should be shut down. Again I called my four experts and again we decided it was safe to continue. But this time, Voysey did not agree. He announced that he was shutting down, unless I gave him a direct order to continue the run. I did so.

At stake here was the future of the nuclear submarine program, certainly, but not public safety. Since there was only a small amount of fission products even a meltdown would not have been as serious as for a power reactor that had operated for a long time. We felt sure we were correct in our decisions. In any event we were not gambling with the health or safety of the public. Our gamble, if it was one, was the future of nuclear submarines as a part of the U.S. Navy. An accident on this test run could have made the project politically impossible. On the other hand, since there were many in the Navy who thought that the nuclear submarine was just too complex to be practical, not being able to make the submerged crossing could have jeopardized the program.

In the early morning of the fourth day,[2] the feedwater control on the port steam generator failed, and the water level dropped rapidly. As a result, steam pressure dropped and the reactor power level varied. Esselman ordered a reduction in power level to 50% to give some margin for reactor power level changes without risking a power level scram (an automatic shutdown).

Since this was the first time an event of this type had occurred the crew had to improvise the corrective action. Feedwater control was switched to manual and the crew took action to fill the steam generator.

Because Esselman was concerned the reactor might be shut down by an overpower trip, he called Sid Krasik, who was asleep in the Quonset hut at the site. Sid came to the control room to provide welcome support. In time, the level in the steam generator was brought back to normal, and

---

[2]Private communication from W. H. Esselman.

power was brought back to 100%. By 8:00 a.m. the plant was again running normally at full power. Although the leak in the condenser continued, this did not pose a safety problem. And there were no more feedwater control difficulties.

The bottom line was this. It was never necessary to shut down the reactor or even to "surface" during this simulated trans-Atlantic run of Mark I. Only three times during the entire "crossing" did the Mark I slow down. Once the crew throttled back to two-thirds power for 7 minutes and twice to half power for a total of less than 90 minutes. Each time the crew made the necessary adjustments to permit the run to continue.

We ran for 96 hours at full power, except for short periods. This demonstration of reliability did much to assure continuance of the Navy program.

It was with great anticipation that Krasik and the whole physics department had looked at the first power run of the prototype as a way to gain experimental data they so badly needed. One of the fission products is an isotope of iodine that decays to an isotope of xenon that has a high capture cross section. An equilibrium is reached during operation between production of xenon and its capture of neutrons. However, when the reactor is shut down the xenon is not removed by capture of neutrons and builds up. Quantitative data were needed. This test had originally been scheduled for the end of the first 24-hour full-power run, but because Rickover decided to continue that run for almost 100 hours, the xenon override test had to be postponed until the run was completed. The test confirmed, as was expected for such a small reactor, that a xenon-caused spatial power oscillation at full power would not be a problem. The test also confirmed that xenon override after shutting down from full power would not be a problem either.

I must admit that my decision to return to Pittsburgh during the early part of the test run to be with my wife has bothered me ever since. If I had known the run was to be prolonged by Rickover, I believe staying at the site would have been the thing to do. On the other hand, perhaps decisions were made with less emotion from Pittsburgh. After all, you can't see what's going on in the reactor compartment, even if you are there.

It's interesting to ponder over this episode from Rickover's point of view. He never mentioned the run to me at any time afterwards. Why did he place the burden of command decisions on me at this critical time? Was he, uncharacteristically, just following the straight line of command? Was he buying himself some personal career insurance? He was eager to have the run completed, of course, but perhaps he feared an accident for which he would need someone to blame. Did he see me as a possible "fall guy" or did he believe I was in a better position technically to make the judgment calls. We'll never know. And fortunately there was no need to know. As they say, success has a hundred fathers, but failure is an orphan.

Later, Mark I was operated at various powers with relative ease, demonstrating the known conservatism of design of the reactor and primary coolant system. There was no deliberate attempt to operate the plant above its rated power. Later a large and extensive testing program was to be run to obtain detailed design information on all parts of the plant to permit better design of future power plants of this general type. During the end-of-life tests on the first reactor core, the reactor was operated for significantly longer than expected design life. Experiments were conducted with fuel elements with heat fluxes significantly higher than design values. When practical, off-normal conditions were tested.

At the time of the 96-hour endurance run, the future of the naval nuclear program was anything but certain. Bettis had been working on the design of a reactor for an aircraft carrier since late 1951, but this project had been canceled at the suggestion of Admiral Strauss, chairman of the AEC, during the National Security Council meeting on March 31, 1953.[3] The Shippingport project had not yet been authorized, and the SFR project also was not authorized until September 1953. So you can see why Rickover attached such importance to a demonstration of a Mark I success. It was the only show in town.

We in the Mark I project knew the program was remarkably successful, but that success was far from being common knowledge among all the important people who influenced future program decisions. Even such prestigious naval officers as Slade Cutter were opposed and leaked adverse information on the program.

The Significance of Extended Test and Training[4]

For the next few years, the STR prototype had the dual objectives of testing the plant and training the operating crew of the *Nautilus* and later submarines.

Following the 96-hour run a complete review was made at the site of all operating procedures. By Rickover's orders Esselman, Bud Simmons, Max Johnson, and key Navy officers could not leave the site until a complete set of revised procedures was completed. Some eight weeks of around-the-clock work was required to complete this task.

Esselman was in the often humbling position of revising procedures he had approved only a few months before at Bettis. He now had to criticize the procedures from an operator's viewpoint.

---

[3]R.G. Hewlett and F. Duncan, *The Nuclear Navy*, p. 198, University of Chicago, Chicago, 1974.
[4]Private communication from W. H. Esselman, October 16, 1975.

We were constantly seeking ways of improving reliability of plant operations. New methods of heating the primary system and of operating the plant and responding to system malfunctions were all part of the learning process.

A comprehensive series of tests was developed to examine the characteristics of the plant under normal and abnormal conditions. The ability to perform both routine and major maintenance was evaluated.

The thermal performance of the core was examined at all power levels and off-normal conditions. For example, tests were conducted to measure reactor response to loss of main coolant flow. Other experiments examined effects of unbalanced core flow distributions resulting from disabling one of the main coolant loops. Effects of restarting a loop after it had been disabled for some time showed the effect of cooler water entering the reactor.

At that time considerable concern existed over possible buildup of corrosion products, called crud, that would affect heat transfer properties of the core. To address this point, careful measurements were made to detect any changes, such as increased pressure drop, that occurred during core lifetime. None were observed.

Control and transient experiments were always preceded by analog computer predictions of what should be expected. Initial experiments were made by moving control rods to obtain small positive and negative reactivity changes. Reactor transient response compared well with predictions. Following these positive results, increasing rates of power demands were attempted until they were only limited by inherent characteristics of the power plant. Full ahead to full astern was achieved in surprisingly short times, eight seconds to be exact. Power could be increased from 20% to 100% in four seconds.

Maximum use was made of inherent properties of the reactor to respond to load demands. The ability of the plant to operate for long periods without operator attention was tested to simulate a possible emergency operating requirement. The plant was operated for many hours without any automatic or manual adjustments of the control rods. (Control rod positions were fixed.) As the demands of the system were varied, inherent reactivity effects ensured that the reactor met demands without exceeding any scram settings.

Reliability and simplification were major motivations during this period. Each plant shutdown raised Rickover's wrath and had to be thoroughly explained. As a result, the plant shutdown system was the focus of much attention. During the previous year, Walt Esselman and Bill Hamilton had developed a simplified shutdown system for the *Nautilus* plant. This system eliminated more than half of shutdown signals and where possible increased margins to the set points. This simplified system was installed during the summer of 1953. These changes resulted in the number of reactor scrams being reduced from 52 in 1953 to only 2 in 1954. This improvement was achieved even though a demanding series of tests was being conducted every day during the year.

The program of simplification extended to development of more straightforward operating procedures.

There were many tests to determine factors affecting the achievement of the required chemistry conditions of the primary water. Different methods of oxygen removal were tried, and a simple method of controlling the oxygen was developed. There were many experiments on the primary water purification system. Careful observations were made to detect crud buildup and changes in heat transfer conditions in both the reactor and the steam generators.

During the extended test period a sense of urgency existed at the site. The goal was to discover and correct all potential problems on the STR prototype rather than on the *Nautilus*, which was being built at that time.

Any needed changes were immediately forwarded to the *Nautilus* design team at Bettis. We were learning fast about problems that caused plant shutdowns. For several months after the 96-hour run, we tried to run a week without having a scram or failure of any type. Gradually as the root causes of the malfunctions were identified and corrected the operation improved. Simplification and reliability were the key themes.

Rickover received daily reports on the progress and problems on the prototype. Each deficiency was reported by the operators as an incident and corrective action was taken. No problem was too small to receive his attention. These corrective actions were reviewed with Bud Barker who represented the NRB at that time.

The following example illustrates the intensity of the emotions during this period. Following a main coolant pump incident, Rickover called to learn what happened. As Esselman explained the event, Rickover interrupted with "Who is in charge there? Is Simpson there?" As Esselman said he was in charge, Rickover replied, "Who the hell is there that knows what he is doing?" and hung up. A day later Rickover and I were at the site and found the problem was being handled capably. Rickover patiently waited for two days while site staff worked on the problem and then asked to have it explained to him. In the end he was convinced the site people knew what they were doing.

The performance of the crew during this period was impressive. By this time, they had received about three years of instruction on nuclear reactors and power plants. Operating a steam plant in a submarine was a challenge to submariners whose previous experience had been on diesel-powered boats. Dennis Wilkinson, first skipper of the *Nautilus*, spent time at the prototype, as did each of his senior officers—Les Kelly, Jack Nicholson, Bus Cobean, Bill Layman, and Dean Axene.

Even during the early tests, operations were performed by the Navy crew from stations within the hull. The wisdom of the Admiral's insistence on making an exact prototype of the submarine became ever more evident as the test program proceeded. Maintenance and repairs had to be performed under the same conditions as would exist in the *Nautilus*.

As plant operation improved, each system was tested beyond its design conditions—including some that would be considered unusual at a landlocked test site in the Idaho desert. Particularly unusual were the snorkeling tests performed to simulate the operation effects of the associated hull pressure changes. These tests made certain that plant pneumatic instrumentation systems were not adversely affected by this mode of operation. Tests were made to determine the effects of failure of each auxiliary system, such as the electrical and the plant cooling systems. Nothing was left to chance—every mode of operation was methodically checked. George Conley contributed much to the planning of these tests.

Not only were we trying improve the existing plant and its operating procedures, we were thinking ahead as to more basic improvements. As an example, Esselman was flying with Rickover in a Navy plane and was asked what could be done to improve the plant reliability. This was all Walt needed as he rattled off a number of ideas that had been on his mind. Perhaps the most significant of these suggestions was that NRB should support development of semiconductor devices to replace magnetic amplifiers and pneumatic instrumentation. Rickover liked the idea and immediately started a development program.

Ireland Committee

Of great importance to the people at the site was the visit by a committee chaired by Mark Ireland, chief engineer of Newport News Shipbuilding and Dry Dock Company. The Ireland Committee included some of the most experienced men in ship design and operation in the United States. They came to critically review STR operating performance and maneuverability and to compare them with other vessels.

We prepared thoroughly for this visit and planned a series of rapid maneuvers that would put the plant through its paces. I mean we didn't miss a trick—fast startups, full speed ahead and astern, shutdown and restart, fast reactor power changes—every rapid maneuver in the book. Mark I performed like a thoroughbred. It didn't shy away from anything. Its operating capabilities exceeded all of the committee's expectations. This demonstration was an important step and Rickover was pleased.

End-of-Core-Life Tests

One of the most exciting periods at the test site was during the time we called the "end-of-core-life" tests. We had put the plant through all of its paces and it had met all performance criteria. Now we would explore its operating limits—how much more could we get out of the plant with its initial core loading?

The questions now became:

What can we learn from the initial core during its last hours of operation?

How much longer than design lifetime can be achieved?

How close is the core thermal design to its limit?

What can we learn about plant operation with a nearly depleted core?

The plant operated with the first core several times longer than the design life of the initial core. The neutron source level, which was designed for the design life of the core, became almost nonexistent and core excess reactivity decreased so that it became almost impossible to override the effects of xenon. Startups under these conditions had to be made very shortly after a shutdown.

Esselman remembers one plant scram when they had less than 15 minutes to get power or they could not restart until the xenon had decayed. This would have taken more than 24 hours. This would not have been a major problem for the prototype, but being unable to restart could be crucial for a submarine, so learning as much as possible about xenon buildup was necessary. They analyzed the problem. Bud Barker, Rickover's man at the site, agreed and called Radkowsky in Washington for the approval to restart, which was required after every scram. He called back with the Admiral's approval. By that time they had only a few minutes in which to get to power. They were motivated to prove that it could be done.

In experiments with specially designed fuel elements, we deliberately inserted hot spots, with about 10 times normal heat flux. These hot spots were well instrumented, so much was learned about temperature conditions during steady-state and transient test conditions. Experiments were conducted to determine the maximum power at which the fuel with the hot spot could safely operate. This information was coordinated with the extensive heat transfer testing being done at Bettis to guide the design of later reactor cores.

An important result of this program was that the operating crews became aware of what to expect under each operating condition. There were some surprises in this series of tests, and it was good they occurred on the prototype rather than on the *Nautilus*.

A typical example is the test in which we deliberately allowed the primary system chemistry to go "off normal." To everyone's surprise, the water samples quickly became bright yellow. The yellow color was the result of formation of nitric acid, which was made in the reactor cooling water, reacting with the chromium that was present in the stainless steel reactor coolant system.

Years later Bill Layman, as engineering officer in the first *Nautilus* crew, told Esselman that it was these types of lessons that were of greatest value to operators at sea. An off-chemistry incident occurred when they were at sea, and they knew immediately what had to be done.

The ultimate goal of these tests was to inform the design team back at Bettis and the crew of the *Nautilus* of various design problems and of deficiencies in operating procedures. The many unusual incidents and tests on the prototype did much to build an invaluable experience base for the submarine operators.

In the early years (1953 and 1954), a continual stream of design changes resulted from testing Mark I. By the time the *Nautilus* went to sea in January 1955, a vast majority of the surprises had occurred in the prototype. Design problems had been corrected, and appropriate operating procedures had been established. I am convinced that the Idaho prototype tests were a key factor in the outstanding operating performance of the *Nautilus*.

As might be expected in a new plant as complex as the STR Mark I, many bugs and maintenance problems were uncovered. Many simplifications were made in the Mark II plant and many more were possible, but time and the basic design did not permit the incorporation of all in the *Nautilus*, though they were incorporated in later plants of this type.

People were tested too. I was one of them.

When I was in Idaho, trying to bludgeon the job to completion, my tactics probably were too rough. Soon after I was back at Bettis, Weaver told me that Mel Yadon, one of our top men in Idaho, who was being transferred to Bettis, had said he would not work for me. Weaver persuaded him that my actions in Idaho were under the stress of circumstances.

The fact that Yadon took the job is a tribute to Weaver's powers of persuasion.

Regulatory Approval

There were very few regulatory approvals required from anyone not connected with the project. No one knew what to look for, and if we had to answer the questions asked today, we could not have done so. It is hard to see how the submarine development could have been completed in today's environment. The only review for both naval reactors and Shippingport was that of the Advisory Committee on Reactor Safety.

For that approval Bob Creagan and Al Henry met with the committee at NRTS. Edward Teller was chairman of the committee and did most of the talking. The various design features were discussed, but there was very little discussion of technicalities of safety.

The hypothetical accident that Creagan and Henry had analyzed considered what would have happened if the Mark I core, without control rods, and hanging on the crane cable waiting to be installed, suddenly fell into the sea tank. Instantaneous criticality would pulse the core out of the water like a rocket. The "fix" was to put boric acid in the sea tank.

Control stability, negative temperature coefficient of reactivity, automatic scrams, nuclear understanding, radiation shielding, and design considerations all were topics of conversations that were conducted like a classroom

Figure 3.3 The *Nautilus* at sea.

question and answer session. Teller then canvassed the committee for comments and all went home.

Nautilus Goes to Sea

Eighteen months after the startup of the prototype, the installation of the propulsion plant for the *Nautilus* was completed and final preparation for startup was under way. Dick Cunningham of Westinghouse and Frank Horan of Electric Boat were the key engineers in the startup and testing in the shipyard and on the sea trials.

Figure 3.4 Walt Disney cartoon of the *Nautilus*.

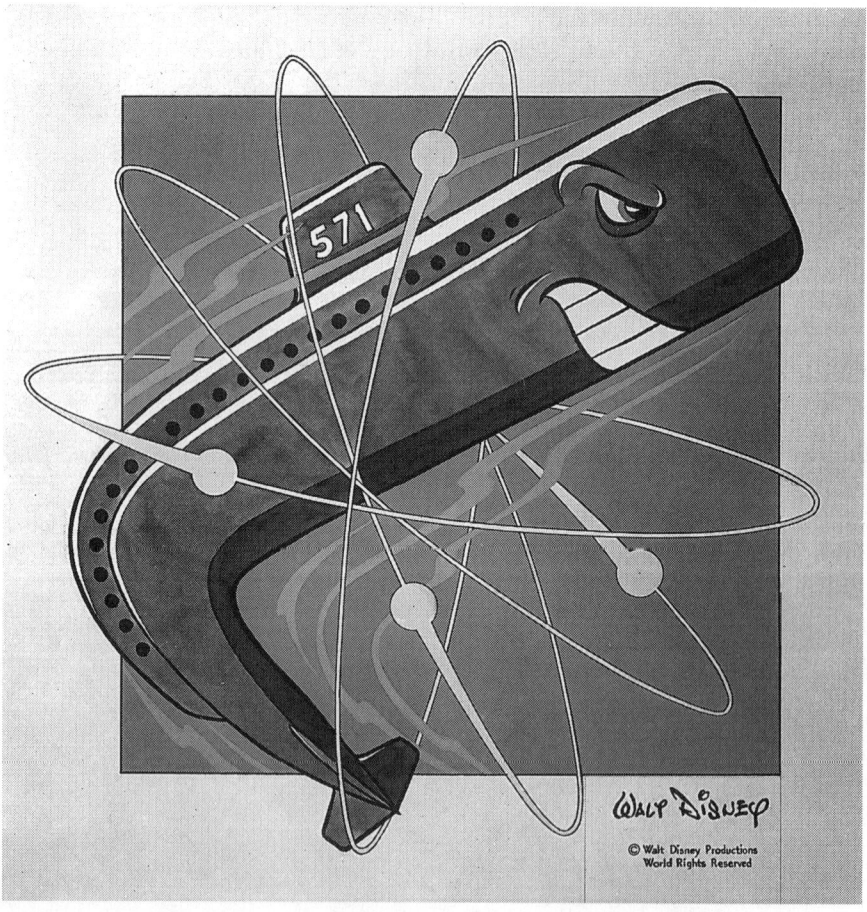

The *Nautilus* power plant first operated on December 30, 1954, and on January 3, 1955, at full power. On January 17, 1955, the message "Underway on nuclear power" was flashed to Admiral F.T. Watkins. In May, the *Nautilus* steamed submerged 1300 miles in 84 hours. From the moment the *Nautilus* put to sea, she was the most powerful fighting vessel in the world and naval warfare was dramatically changed. The change was comparable for its suddenness to the arrival of ironclads during the Civil War, but was far more fundamental in naval warfare.

It is not generally understood that pre-*Nautilus* submarines were in reality just surface ships that could operate under water for only brief periods—as little as 30 to 40 miles at full power. The *Nautilus*, on the other hand, could circumnavigate the world submerged.

Westinghouse got more good publicity out of its participation in the *Nautilus* development than from any other project in its history. Most Westinghouse advertisements carried the slogan "Westinghouse, First in Atomic Power."

Awarding Orders of Merit

Price once told me that the only thing that allowed him to hold his head high on visits to Washington, when he had to go there in connection with the company's jet engine problems, was the Bettis success. By contrast, Westinghouse's early leadership in jet aircraft engines had gotten into serious trouble with the J-40 engine trouble, which later lead to the company getting out of that business. It is reassuring to me to know that our nuclear concentration and priorities had nothing to do with that failure. No one had been transferred from the aviation gas turbine division to any nuclear activity until after the J-40 episode and they were cutting back on personnel.

Price felt that development of the *Nautilus* prototype was one of the most successful accomplishments in the history of Westinghouse. So he recommended to the board of directors that the Order of Merit, the highest award the corporation provides, be awarded to Bill Johnson, Krasik, Lustman,[5] Shoupp, and me. Weaver had already received the award. This was the first time Westinghouse had made multiple awards to a group of individuals for their performance. Quite a tribute!

---

[5]Lustman headed up the metallurgy development.

CHAPTER 4

Nuclear Propulsion Moves Forward

Planning America's nuclear Navy really didn't begin until the Mark I prototype successfully completed its 96-hour full-power run in June 1953. At that time, the only contract for a reactor plant on the Bettis order book was for Mark II—the actual *Nautilus* propulsion plant.

Was the Bureau of Ships waiting to see if Rickover's submarine would actually work before taking the big plunge to nuclear power? Maybe so, but in any case neither the Navy nor the Atomic Energy Commission (AEC) had a plan for the next step. And the Navy had not yet developed a mission for either nuclear submarines or surface ships. Rickover realized this and wanted as broad a development program as possible to meet whatever needs the Navy foresaw. He was going to be ready for any mission the admirals specified.

The only thing he was sure of was that the Navy would want more nuclear subs, but not necessarily ones of advanced design. Fingering both of his hats, Rickover knew that to keep AEC money coming, any further programs would have to involve advanced nuclear design. So he worked to obtain a compromise. He talked the Navy into stating its needs not just for a second submarine but also for one of advanced design.

SFR (*Skate*) Class Submarines

For the 1955 shipbuilding program the Navy included one small, nuclear-powered fleet submarine. Bettis received a contract for development of the nuclear propulsion plant for this submarine, the fleet submarine *Skate*. This sub was much smaller than the *Nautilus*. Its SFR reactor later carried the

Westinghouse designation S3W. Later versions of this nuclear plant were designated S4W.

Rickover asked his staff to list objectives to guide Bettis in the development and design of the new propulsion plant. For the nuclear design, they asked for longer core lifetime and simplified control and refueling. For the plant equipment, the need was for improved accessibility for maintenance. In other words, we would strive to make the plant run longer on each loading of fuel and make it simpler to operate, maintain, and refuel. When our new design was finished, most of these objectives were met and the AEC was satisfied that advances would be achieved.

Rickover wanted the first submarine of this type built at Electric Boat where the *Nautilus* was under construction, but the Navy wanted to broaden its own capability and assigned it to Portsmouth Naval Shipyard. The Joint Committee on Atomic Energy (JCAE) put pressure on the Navy, probably at Rickover's urging, and a second fleet submarine to be built at Electric Boat was added to the 1955 program.

Soon, three other fleet submarines were added to the list, one in the 1955 shipbuilding program and two for 1956. One of these was to be built at Electric Boat and two at Portsmouth. Then another fleet sub was added for the Mare Island Naval Shipyard, followed by a guided-missile submarine, also for Mare Island.

As the SFR plant design progressed, a strong difference of opinion developed within Rickover's shop on one aspect of these propulsion plants. It concerned whether horizontal steam generators, like those on the original SFR, or vertical ones should be used. Not a big deal, actually. Milt Shaw had proposed the vertical design for the upcoming S5W class. Incidentally this design made it easier to redesign the S5W class missile subs. The vertical design allowed greater accessibility in the reactor compartment, where space was at a premium. The horizontal design would provide more room above the reactor compartment. Panoff, Rickover's project officer, favored the vertical design. Kintner, head of the development group, was for the horizontal design. Surprisingly, both men felt strongly about their preference.

Electric Boat built full-scale wooden mockups of both designs. After viewing the mockups Rickover then made the nondecision. A vertical steam generator would be used for the Electric Boat and Mare Island submarines, while a horizontal layout would be used for the subs authorized for Portsmouth.

Rickover assigned Commander Marshall (Bill) Turnbaugh, who had been his representative at Bettis, to be the nuclear power plant superintendent at Portsmouth. Rickover took special delight in assigning Kintner to the same position at Mare Island. There Kintner would have responsibility for building two plants with the steam generator design he had opposed. I'm sure that made Rickover's day.

Success breeds success. Rickover was getting increased support from the JCAE as the naval program kept succeeding. The Navy, too, was increasing its support for nuclear propulsion, although not always in the way Rickover wanted.

We carried on work for the *Skate* with the same discipline-oriented organization as that for the *Nautilus* but, in 1954, Rickover decided he wanted to set up a project-type organization. He was looking forward to the possibility of building many nuclear-powered ships and thought that each project should be essentially self sustaining—with its own research, development, engineering, and procurement capabilities.

I felt the project approach would be a mistake and the idea also provoked mixed reactions from the senior Bettis managers. Coordination and transfer of technology between projects would be lacking, and that was something in which I could provide leadership from my position as *de facto* technical director. The ability of top men in each technical discipline to have a say in each project could outweigh the value of single-purpose concentration and undivided responsibility for each project.

I told Weaver of my objections to the project approach. Was I giving an unbiased opinion? I wasn't sure. As top technical man, I was responsible for the technical aspects of all projects but had no production responsibility. My position was clearly next in line for the top job. Charlie Slack, by then a senior consultant and advisor, advised me to go along with the Rickover-Weaver proposal. This I did, but I don't think it would have changed their minds if I hadn't.

I became PWR (Shippingport) project manager, Joe Rengel took the CVR (the aircraft carrier) project, John Stiefel the STR (Mark I and Mark II) project, and Al Squire the SFR (the fleet submarines) project.

The Skipjack-class Provides More Speed

The *Nautilus* trials with the Atlantic fleet had demonstrated the advantages of more speed. Even though the *Nautilus* was faster than any previous submarine, more speed was wanted.

It did not seem possible to increase the power enough with just a modified version of the *Nautilus* reactor, and a complete new reactor design would take too long. Fortunately, the Navy had recently tested a new hull design in the experimental submarine *Albacore*. The essential feature of the *Albacore* was a more rounded hull, with a length-to-diameter ratio of approximately 7.6 to 1, rather than the conventional 11 to 1. This hull design created less drag and permitted more speed for the same power. This project was given the highest priority by the Navy, and this new reactor was to become the major nuclear plant for future submarines.

Douglas C. Spencer was put in charge of this project and the Navy requirements were given to Bettis on September 20, 1955. The contract for the S5W reactor, first used in the Skipjack, was awarded in December. Combining the most advanced hull design with nuclear propulsion produced the best performing submarine in the world. There were six more Skipjacks and then 53 more attack submarines, which used a version of the S5W reactor.

The principal changes from earlier reactors were in the core, control system, and refueling capabilities. The core design was sufficiently new that a critical experiment was performed at Bettis.

The next great advance in the capability of nuclear submarines had to do not with speed or stamina, but with firepower. Nuclear subs could run and hide from any warship in the world. What if they also could launch missiles? Wouldn't that provide a greater deterrent capability?

The development of a submarine that could launch the Polaris missile without surfacing brought the U.S. Navy what many considered the unstoppable weapon. A contract for Polaris-class ballistic missile submarine reactors was awarded to Bettis on March 14, 1956. There were 41 of these submarines, all using the S5W reactor.

George Mechlin, manager of the SFR physics department at Bettis, went to the Westinghouse plant in Sunnyvale, California, where he was put in charge of developing the launching system for Polaris missiles.

Naturally, I was enthusiastic about the capabilities of the Polaris submarines. Sometime in the mid-1950s, I gave a speech to the National Science Writers Association in Cleveland. Among other things, I said that Polaris submarines were more important to the defense of the United States than the Strategic Air Command. It was true, but a dumb thing for me to say.

That night, when I got home, I received a telegram from Rear Admiral Raborn, head of the Polaris program, complimenting me on the speech. I felt good about it. And the next morning there was a newsletter on the desk of every flag officer in the Pentagon, quoting me. How nice.

But unbeknownst to me, Westinghouse, at that time, was in final negotiations for the electronics for the B-70 bomber—which would be the key weapon of the Strategic Air Command. Our electronics people decided Simpson was just another four-letter word!

Within a few days, a directive came from Westinghouse corporate headquarters stating that henceforth corporate officers would not take part in interservice rivalry. Fortunately, my head didn't roll. I had ducked another bullet. A missile actually!

It is interesting to note that a recent report of the Congressional General Accounting Office[1] contains this statement: "Evidence from eight classified GAO reports assessing weapon system upgrades suggests that the sea leg is the strongest, most cost-effective component of the U.S. strategic triad."

I really don't think it was any of my doing, but a contract for Polaris-class ballistic missile submarine reactors was awarded to Bettis on March 14, 1956. The components for these reactors were procured by the plant apparatus department, which is described shortly.

Aircraft Carrier Program[2]

In July 1951, the large ship reactor (LSR) project was started with William Ellis as the project manager. We all believed it was for a carrier, but the government did not want to say so. A group was formed to recommend a reactor type for a surface ship and they considered five reactor types by assigning an advocate for each type.

The advocates and reactor types were PWR—Walt Roman and others; BWR—Willie Comtois, who had just come from Combustion Engineering at Windsor, Connecticut; and Homogeneous—Stuart Mims, who had been at the Oak Ridge School of Reactor Technology and had acquired that laboratory's liking for the homogeneous concept. Heavy water and sodium-cooled reactors were also considered.

Owen Woodruff was consultant on heat transfer and Alan Henry analyst for reactivity considerations for all types. Creagan served as chairman for the studies.

These groups were second-guessed by almost everyone at Bettis. Everybody, it seemed, had an opinion. Louis Roddis, of Rickover's staff, flew up from Washington every day and spent time with the various groups. Then he would fly back each night and brief Rickover.

Apparently Roddis's reports worried the Admiral, because he called Weaver complaining that "a bunch of long hairs" were doing an academic study of what the reactor should be. Rickover obviously didn't want academic studies. He wanted the studies to be done by experienced engineers, who knew the right questions to ask and the right problems to probe. Weaver called Creagan and Roman into his office and asked what they were doing. Roman said he had gotten the ship blueprints of the latest carrier and they were designing the layout of the reactor system consistent with

---

[1]Eleanor Chelimsky, Assistant Comptroller General for Program Evaluation and Methodology before the Senate Committee on Governmental Affairs, GAO/T-PEMD-95-5, June 10, 1993.
[2]Private communications from P.N. Ross, R.J. Creagan, D.E. Thomas, G.W. Hardigg, and A.P. Zechella.

the constraints of that hull, the watertight compartments, and the number of screws. That made sense to Weaver and, as he frequently had to do, Charlie calmed Rickover down.

Of the other possible reactor types, the homogeneous reactor was ruled out due to a lack of development experience, plus maintenance difficulties and possible battle vulnerability. Heat-exchanger tubes inevitably leak and leaking of the radioactive slurry would have posed major problems. This decision was objected to by the Oak Ridge advocates. Boiling water was considered risky because of possible sloshing around of the water-steam mixture at sea, and possible instabilities associated with that motion. The sodium-cooled reactor was ruled out because of enrichment considerations, maintenance, and shielding. The maintenance problem was somewhat similar to that of the homogeneous reactor. It was realized that in the case of an intermediate neutron energy spectrum reactor, such as SIR, the number of uranium-235 atoms destroyed per shaft horsepower hour were greater than in the STR. This is because of parasitic captures of neutrons in the intermediate (or resonance) energy region, in spite of the greater thermal efficiency of the sodium reactor. This, at the time, was an important consideration, more so for a carrier than for a submarine, because of the greater energy requirements and the perceived scarcity of enriched uranium. However, it later turned out that this was not an important consideration.

The heavy water reactor had a lot of people on its side—including Roddis of Rickover's group and Walter Zinn of Argonne—this made it difficult to eliminate it in the design study. The design study group did eventually rule it out, however, because of its greater size and weight, plus the fact that its greatest alleged advantage—use of natural uranium—made it subject to not being able to be restarted at times due to the buildup of the fission product xenon.

Later Creagan found out that Rickover's group had done (in-house or subcontracted) an evaluation of reactors for a carrier, and had selected the heavy water reactor so they could use natural uranium. This would get around the concern about enriched uranium. Rickover also was aware that there were no facilities for reprocessing slightly enriched used fuel elements. Zinn said that it was a lot easier to enrich heavy water than uranium. Also he had built CP-3, which was a heavy water reactor.

This is why Creagan spent so much time evaluating the technology and economics of separating isotopes of uranium and heavy water. He concluded that the practicality and economics of drawing slightly enriched uranium from the diffusion plant cascades were excellent, and that heavy water isotope separation was costly, as indicated by experiences trying to get heavy water for the Canadian nuclear plants. Also, it was costly to design for extreme leak-tightness, which was necessary to facilitate maintenance and refueling and to reenrich and replace leakage of heavy water.

The group selected the PWR as the preferred reactor type for the surface ship, and spent most of the time making a realistic design of a slightly enriched PWR, with considerable effort spent on the economics of uranium enrichment versus heavy water enrichment. Enriched uranium was considered in short supply, so a highly enriched PWR was not favored.

After much discussion as to the practicability of a slightly enriched uranium reactor, Creagan and Henry concluded that additional reactivity experimental data were desirable for slightly enriched uranium reactors. They agreed that curves for three independent variables would be useful for reactor experimental and theoretical physics. The three variables were uranium enrichment, rod diameter, and metal-to-water ratio (moderation).

They decided on three enrichments and a rod size and determined the amount of material required for critical experiments. To change rod diameter, they had the rods swaged down in diameter until they were twice as long, cut them in two, and thus had a new diameter with no loss of material. This process was repeated to provide rods of a third diameter. Later, Creagan found it interesting to read how some unknowing analysts were reaching far-out reasons for the odd dimensions of these and later rods.

These plans were implemented as critical experiments at Bettis, but to convince the AEC to fund these experiments, they also visited Herb Kouts at the Brookhaven National Laboratory (BNL). They got his agreement to use the same rods in exponential experiments on top of the BNL large graphite reactor. Both Bettis and BNL ended up with curves for three different enrichments, three rod diameters, and many metal-to-water ratios. And, of course, it gave them the evidence they needed to support the use of a slightly enriched uranium reactor.

These data were used by many theorists who were trying to predict loadings and associated economics for slightly enriched uranium reactors. It was timely, because there was much interest in slightly enriched reactor economics at that time. The fact that the results were confirmed by BNL was helpful, and BNL was able to explore some metal-to-water ratios that were beyond critical experiment range. Information on specific aspects of reactor physics, such as the fast effect, resonance escape, and thermal utilization, were useful to theorists, who made their analyses agree with the experimental results.

It seemed we had been spinning our wheels, however, when at this time the surface ship contract was canceled. This was a traumatic time at Bettis because the large ship project was an exciting one and there was nothing to take its place. But our time was not wasted. The contract was resurrected the following year as the A1W project. In the meantime the Shippingport project was begun.

Carrier Contract Awarded

The contract for the A1W project, covering design, construction, and operation of the aircraft carrier prototype, was awarded on October 15, 1954. The initial plant design basis was 70,000 shaft horsepower (SHP) per propulsion plant, with two reactors per screw and four coolant loops per reactor (with an objective of providing full power from any three loops). Coolant flow was provided by a canned motor pump for each loop. There were U-tube steam generators and a reactor vessel with the inlet nozzles for core cooling to be below the level of the core. Containment was provided by a sealed, rectangular parallelepiped reactor compartment capable of containing the pressure resulting from the maximum credible leak in the primary system without discharging radioactivity into the environment. The features of bottom reactor vessel inlets and compartment containment against leakage were to cause major problems in the safety analysis.

The organization of the project after several fits and starts was as shown in Figure 4.1. After several months of shakedown, the A1W project settled down to the task of crystallizing the technical and strategic objectives that were to occupy the efforts of hundreds of people, mostly technical, for the period from 1954 to 1960. Briefly stated the technical objectives were as follows:

- Design, construct, and operate a prototype of one of four propulsion plants for the USS *Enterprise*, CVN 65, and for the cruiser USS *Long Beach*, CGN 9.
- Test two different types of nuclear cores, one slightly enriched and one a seed-and-blanket.
- Develop metallic fuel material for the slightly enriched core and for the blanket of the seed-and-blanket core.
- One of the reactors was to have stainless steel piping and components and the other stainless clad carbon steel.
- Complete the prototype plant sufficiently earlier than the mid-1960 carrier launch date to provide a training facility for the operating crews.

Achievement of these objectives in a timely and cost-effective manner was far more important to the naval reactors program than were the corresponding aims of the submarine program. The technical superiority of nuclear-powered submarines was unarguable, thus cost and speed of

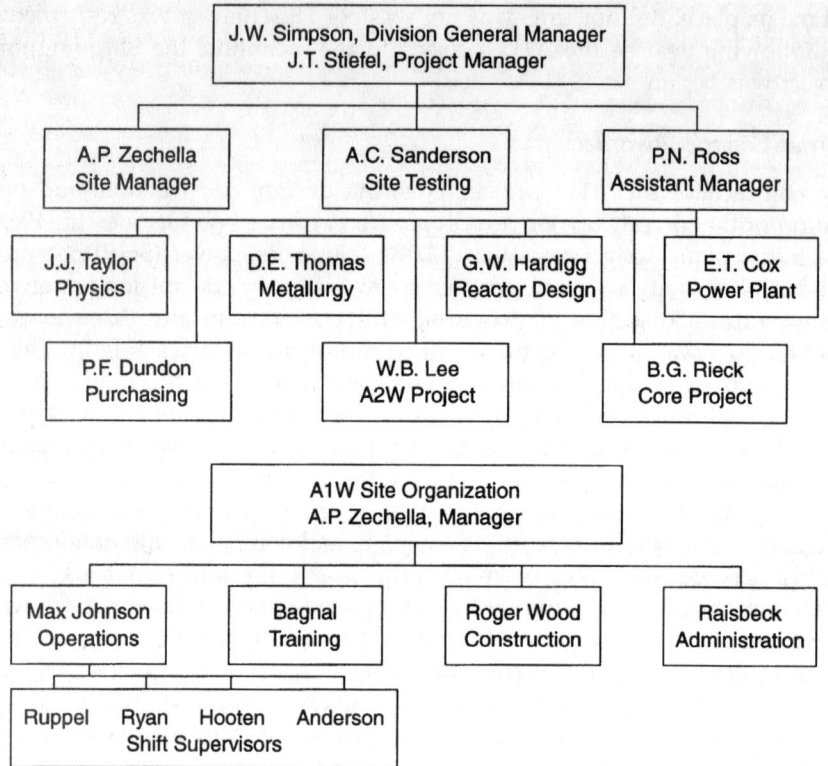

Figure 4.1 Organization of the A1W project.

completion were not as crucial. Rickover's enemies, and they were legion, were using cost as their principal antinuclear argument for surface vessels. Any significant failure to meet the goals of the carrier and cruiser program would have been seized as a way to get the programs canceled. Nuclear power for carriers was justified on a strategic basis by demonstration of the effectiveness of the *Enterprise* in Vietnam. She was able to remain on station in the Gulf of Tonkin for months on end, setting record after record in sorties flown.

The schedular constraints were particularly binding on the fuel element development. It takes more than a decade to develop and completely prove out a new fuel, which is expected to last years in the actual ship's core. If, during testing, problems are encountered that require changes in design or composition, then one must start over on a new multiyear program.

The initial selection of fuel material and fuel element design was based on Shippingport experience (covered in Chapter 5), modified to accommodate the special requirements of a naval application. In the latter context it was argued, based on nothing more than intuition, that rods would be unable to withstand the stresses imposed by the shock loads resulting from explosions in battle. Although extensive study later concluded that this was not the case, such results came too late to be factored into the A1W cores. The choice of plate geometry was based also on the belief that plates were better able to accommodate fuel material swelling resulting from fissioning.

The Shippingport project was proceeding concurrently, but a little ahead of the carrier program. Although there were firm orders from Rickover to establish a "zirconium curtain" between Shippingport and the naval projects for reason of military security, as well as AEC versus Navy politics, each design category profited from its proximity to the other. The problem of radiation damage to reactor vessels was attacked concurrently by the two projects. Had both groups not worked together, it is doubtful that either would have been able to hold to the schedule, since roll-bonded plate was a limiting item for both. In an effort to reduce costs, we were attempting to develop a method of roll-bonding a thin sheet of stainless steel to a thick piece of carbon steel and then forming the reactor vessel. In this one area, despite the zirconium curtain, there was a joint development effort.

The Shippingport project had chosen, as one optional fuel type, a gamma phase alloy of U-Mo rods clad with Zircaloy and produced by coextrusion. Reactor tests showed that under the design operating conditions the metastable gamma phase alloy would begin to decompose quickly to alpha uranium and a molybdenum-rich phase, and its corrosion resistance would quickly degrade. The Shippingport project dropped this effort in favor of the UO^2 approach.

As the deadline to cut metal for A1W approached, people became uneasy about the behavior of the gamma phase alloy, and began to incorporate a fully enriched core approach. Just in time the A1W metallurgists came up with the alloy U-10.6 Nb-4Zr, which had good corrosion resistance out of reactor. It was also determined that this alloy had a sufficiently slow gamma decomposition rate that thermal spiking due to irradiation caused the gamma phase to be stable under all plant conditions. The development program included alloy studies, gamma decomposition kinetics, and irradiation behavior, while the fabrication people were also heavily involved.

Thus at rug-cutting time, one core for the land-based A1W was fully enriched and one was the seed-and-blanket type. The blanket used plates of Zircaloy-2 clad U-10.6 Nb-4Zr fuel alloy. The seed was similar mechanically and metallurgically to that of the *Nautilus*.

The seed-and-blanket concept was a counterproposal by Alvin Radkowsky of the Nuclear Reactors Branch (NRB) to the original Bettis slightly enriched design. It prevailed from the outset regardless of merit, just as it had in the case of Shippingport. Both of these designs required a new uranium-rich alloy and it was the risk inherent in this that ultimately led to retention of the fully enriched *Nautilus* type design for one of the cores.

Just as in the case of Shippingport, the added complexity of the seed-and-blanket design required critical experiments and a considerable amount of reactor theory development.

The great reduction in concern about the supply of uranium that occurred in the late 1950s effectively eliminated further interest in the slightly enriched concept for naval reactors. The emphasis shifted entirely to achieving designs with long intervals between refuelings.

Both cores performed successfully for their design lifetimes. Also specimens of gamma phase alloys continued in the Canadian Chalk River reactor loops with superb results.

With this background in mind, it is easy to imagine the consternation that developed during 1956 and 1957 when a few in-pile test specimens began to leak and corrode. Rickover and his reactor design manager, Harry Mandil, made frequent visits to Bettis, during which they bore down unmercifully on the cognizant Bettis scientists, as if the failure were their fault. On one such occasion Robert Gordon, a highly competent senior supervisor, bore the brunt of the admiral's attack. He left the program shortly thereafter for a senior job with another company.

The outcome of these discussions was the decision to continue the seed-and-blanket design and to abandon the slightly enriched design in favor of one based on the U-Zr highly enriched fuel material. The cores for all reactors on the *Enterprise* were highly enriched. Considering the enormous consequences of failure and the probabilities of success of the U-Nb-Zr fuel, this decision appears in retrospect to have been appropriate. Further testing and operation of the seed-and-blanket core in the prototype later, however, proved the adequacy of the U-Nb-Zr fuel.

Other design decisions made by Rickover were less defensible in the light of later experience. One example that stands out is the requirement to be able to find and remove a fuel cluster containing a defective fuel element (that is, one in which the cladding has been penetrated) without removing the massive reactor closure head. This requirement imposed major compromises on the nuclear and thermal design, but even more so on the mechanical design of the reactor for the small benefit of reducing the outage time to correct an extremely low probability accident. All Bettis project managers strenuously objected to this requirement. In one evening meeting the admiral and I shouted at each other for what seemed like hours, but Rickover stood fast. This design requirement carried over to later designs. When applied to the reactors of the Nimitz-class ships, it resulted in the GE designed "skewed" cores, adding enormously to the cost and compromising the ship design.

The change in fuel material for the non-seed-and-blanket core came very late, particularly since this design was soon designated as the reference for the ship. Specifically, the reactor core and closure head were due at the Idaho site eight months after the fuel material decision! To meet this all but impossible schedule required the imposition of even more design constraints. The steel forgings, with very long lead time items, that had been ordered for the closure head of the slightly enriched core had to be converted to the later design, because the penetrations were different. That this was accomplished in the short time available is a credit to the engineers at Bettis and Combustion Engineering, the fabrication contractor.

Another potentially delaying category of equipment was the core structurals. Large and massive, these parts were required to be built to watchmaker tolerances from hard to machine corrosion-resistant alloys. Moreover, the forgings and other raw material that had been ordered for the earlier core were for the most part unsuitable.

The design of the reactor vessel had been frozen for about two years when the core type decision was made, therefore, it was out of the question to consider any change. This was the only fixed boundary condition to the core structurals when the contract for construction of these parts was let to the Baldwin Lima Hamilton company in Lester, Pennsylvania. To meet the schedule, raw material sizes were much larger than necessary, allowing for design changes right up to the last minute. Drawings were literally snatched off the drawing boards and flown to the supplier, often without approvals from NRB. Had there not been the utmost of cooperation among the design, procurement, quality control, and government representatives, it would not have been possible to deliver the core by the required date. In particular, the manufacturing oversight provided by the nuclear core department under Ike Eisenschmidt was extraordinary; at one time there were 80 production coordinators at the vendor's plant.

It is a known fact that any group of people, however compatible, working under extreme pressure for months at a time tend to get on each others' nerves. One of the prime tasks of management under such conditions is to keep the troops soothed and working toward the common goal. In the case of A1W this was aggravated by the fact that the groups of people involved were often separated by thousands of miles.

A good example of the type of problem that arises was the writing of the Safety Analysis Report. This was necessary for the prototype and the carrier before the prototype core could be taken to its initial criticality. Such a report cannot be written until all design, experimental physics tests, and analyses were completed. This report was really a first of its kind and there were a lot of new concepts, technical problems, and "what if" accidents to be dealt with. In the case of A1W these tasks were finished only after the core was shipped.

It befell the lot of Phil Ross and his key designers to gather the scraps of information on these subjects and stitch them together into as convincing a story of public safety as possible. Time was extremely short; for a week, none of the key personnel went home. Sleep was snatched in short naps stretched out on desks.

During this time John Stiefel was at the site seeing to the completion of the prototype. While the design people were beating their brains out, John kept sending wires complaining and cajoling Phil to complete the report. Finally Ross called Stiefel and blew up in front of all the report writers. This action proved to be the stimulus needed to motivate the troops to complete the job. When the dust cleared, the humor of the situation was apparent; a standing joke among A1W people was to quote a Stiefel telegram "gotta get on with the lagging." The thermal insulation (lagging) of the piping and some components was the almost final item of construction and could not be completed until the emergency core cooling system was finalized and installed.

After the learning experience of A1W the construction of the plants for the *Long Beach* and the *Enterprise* proceeded in a more straightforward manner. All elements of the ship plants were ready when required. Had this not been the case, you can imagine the problems involved with constructing and testing eight reactor plants for the *Enterprise* at one time. The first of these eight went critical on December 2, 1960. Prototype operation provided trained contractor test personnel and experienced naval operators.

This is a good point at which to digress concerning Rickover's philosophy on pursuing a complex project with a challenging schedule. Each development feature was required to be verified by prototype testing; research, development, confirmation tests, design, and construction must, of necessity, proceed concurrently or with great overlap between phases. In the case of A1W, as we have seen, no failure could be brooked; therefore, each development feature was backed up by a more nearly proven one. The backup for the U-10.6 Nb-4Zr fuel was the *Nautilus* U-Zr alloy fuel. Rickover went about high-risk development in the most conservative manner possible.

Although the *Enterprise* program was successful in all of its major aspects, the question of nuclear power for carriers continued to engross Navy planners. The *Kennedy*, which followed the *Enterprise*, was powered by oil-fired boilers. Thus, the question of the follow-on ships was hotly contested. When then-Secretary of Defense Robert McNamara complained about the cost of nuclear carriers, Rickover publicly suggested that the Navy go back to sail as the cheapest form of propulsion. Finally, in a meeting at Bettis in May 1964, Rickover decided on two reactors per ship as the reference for the later ships to be designated the Nimitz class. Bettis had recommended that there be four reactors. The factor that decided the issue was the size of the operating crew for a carrier. Four reactors would have dictated an operating staff more than 50% larger than for a two-reactor carrier.

The *Enterprise* did not need to have eight reactors. Four could have done the job, and many people thought so at the time, but Rickover insisted on eight. Phil Ross recalls vividly going to Washington at the outset of the project to present and plead the case for one reactor per shaft—but in vain. Rickover was adamant that two were necessary, not (Ross thinks) for reason of reliability or battle damage concerns, but because he believed one reactor would be too great an extrapolation in power from the *Nautilus*.

Ross, by then general manager of Bettis, further recalls that he and Ellis Cox, then LSR project manager, made a presentation to Robert McNamara describing a design with one long-life reactor per shaft. McNamara maintained his usual inscrutable, enigmatic, noncommittal attitude throughout the discussion. Nothing resulted since McNamara did nothing about it and Rickover was opposed.

The fight over whether to even build more nuclear carriers and how many reactors they should have and how to make cost comparisons was a bitter one. This controversy is covered in a report by John Conway, then executive director of the JCAE.

A1W had its share of Rickover incidents. He liked to put people down and make them feel their lack of expertise. At one meeting on reactor metallurgy, John Stiefel, the project manager, was the victim. The meeting convened late in the afternoon, broke for dinner, then continued into the evening. All during this time Rickover told Stiefel to sit in one corner of the room and not interrupt. Phil Ross, Stiefel's right-hand man on technical matters, was told to do the talking—even at dinner. Rickover wanted to give the impression that he considered Stiefel a technical idiot. Of course, he was not. (John was a good engineer, although obviously not an expert in every field.)

On one occasion, Don Thomas presented what amounted to the final piece of information that made the U-Nb-Zr alloy (gamma phase) acceptable for the A1W prototype reactor core. He was able to explain carefully to Rickover that all our previous evidence on the behavior of this type of alloy had been wrapped up in a unifying concept. This alloy, he explained, remains corrosion resistant when the neutron level is high enough and swells in a predictable manner. Further, he said, the neutron level in the A1W core would be well above this critical value.

Rickover had been hearing ever since he first entered the nuclear field what a terrible effect irradiation wreaked on materials. This new idea that radiation could have a beneficial effect flabbergasted him. Others at the meeting said afterward that they had never seen Rickover so engrossed nor had they seen his facial expression run through such emotions of surprise, incredulity, and finally satisfaction.

There was one series of meetings with the admiral that involved several consultants whom we engaged at his insistence. In the critical stages of the PWR and A1W fuel programs, he apparently wasn't sure we were competent enough to solve the problems, so he wanted a panel of consultants to review these programs periodically. The consultants selected and engaged were Morris Cohen, Fred Seitz, and Cyril Smith, all outstanding figures in materials science.

Although I generally am leery of committees, this was one that worked. The discipline of preparing for meetings with these consultants, as well as the helpful contributions received from them, made the effort worthwhile. Our consultants were getting paid and were contributing. They also gave us useful insight into problems and frequently backed up our people.

Strangely, at these meetings Rickover's behavior was entirely different from usual. He stood in awe of these men, apparently believing anything they said. He paid rapt attention, launched no tirades, and remained civil and in good humor. I hardly recognized him.

Zechella Named A1W Site Manager

Zeke Zechella wound up in charge of construction of the A1W reactor after bouncing around—the Navy, then Bettis—with first the carrier, then the submarine program and Shippingport, before landing in the Idaho sage brush.

He had just been selected for promotion to commander in 1953 when he was offered a job at Bettis on the carrier project. The Chief of the Bureau of Yards and Docks tried to persuade him to change his mind and remain in the Navy, but Zeke said no and arrived at Bettis on August 10. Almost immediately, the carrier program for which he had been hired was canceled. He was assigned to the submarine program and then named fluid systems supervisor for the Shippingport project.

About this time, we assigned Al Voysey, our Idaho site A1W manager, to the commercial nuclear activity because he had gotten crossed up with Rickover somehow. So one evening I called Zeke at his home in Pittsburgh and asked him to come to Bettis to talk about a job. I'm sure Zeke wondered "what now?" How many new jobs could a guy have in such a short time?

Rickover was in my office when Zechella arrived, and Rickover proceeded to give him one of his famous 45-minute interviews before agreeing that he was the man to head up construction of the A1W. Within a week, Zechella was at the Idaho site, which was just a plot of sage brush adjacent to the S1W site. He hired Hank Ruppel, his assistant public works officer in his last Navy post, to be one of the four A1W reactor supervisors. The others were Don Anderson, Dick Gardner, and Bill Hooten.

There were only about seven or eight people on the A1W site at the time. Newport News Shipbuilding and Dry Dock Company had been selected as the construction subcontractor and had started sending people out to Idaho, with Charlie Palen as construction project manager. He was retiring and this was his last project. Stu Foster was the operations manager for the Newport News contingent and Felix Bledsoe was the engineering manager. Zechella in his capacity as project manager, reported to John Stiefel, the Bettis A1W manager. Max Johnson came over from S1W as operations manager and Roy Bagnall was in charge of the program for training Navy and Westinghouse people on A1W. The administrative group was under Frank Raisbeck.

If you have never been in Idaho in the winter, you can't understand how tough a job it was to build something out there at that time of year. It was below zero most of the time when we were trying to pour concrete foundations and erect the building. They covered all the freshly poured concrete with tarpaulins and used big gasoline-fired heaters to blow hot air under the canvas so the concrete would cure before it froze. It worked.

And it wasn't easy to get craftworkers with shipbuilding experience out there in the middle of the desert 600 miles from the nearest ocean. Newport News did an excellent job of coordinating with the unions in Idaho and importing shipbuilding crafts from Seattle and San Francisco.

A1W Plant Layout

The A1W reactor was in a hull section, which was inside what appeared to be a rather normal five-story building. Inside that hull, you'd swear you were at sea on a ship. There was a large test instrument room in the building that covered about 25,000 square feet. Most of the instrumentation was for readout and recording of test data from the plant itself. The plant was operated from a compartment just above the reactors.

The A1W plant consisted of two four-loop reactors, each rated at 35,000 SHP. On the *Enterprise* there were four reactor compartments with two four-loop reactors of 35,000 SHP in each providing steam for a single 70,000-SHP turbine, which powered a single shaft.

The A1W shaft did not turn a screw. Rather it turned a slow-speed 70,000-HP generator with the power being absorbed by a large water rheostat. The rheostat plates were raised or lowered to match the heat dissipation required. The water was cooled by a large cooling tower. At full power, the shaft turned at 171 rpm. This unique system for loading the shaft worked very well.

With two reactors furnishing steam for a single turbine in A1W, we were uncertain how well they would share the load without "hunting" (alternating in carrying more than half the load) unless special controls were added. We need not have worried. Due to the large negative temperature coefficient of reactivity, the balance was excellent.

Training on A1W

The A1W prototype took a big training load. It trained crews destined for both carriers and submarines. The STR Mark I couldn't handle the load for all the submariners that had to be trained.

Rickover instituted the chief engineer concept. He installed a lieutenant commander in that role in each of the reactor compartments and also put a sea-experienced lieutenant on each of the operating crews as engineering duty officer. Navy personnel previously qualified as chief operators were qualified as engineering officers of the watch. These men reported to the chief engineer, who in turn reported to the shift supervisor for crew operations.

Rickover put some outstanding Navy men in these positions, and they were later to move to numerous positions in Westinghouse and the nuclear power industry in general. There were Westinghouse reactors, Westinghouse turbines, and Westinghouse reduction gears throughout the prototype and the *Enterprise* herself.

A1W Goes Critical

When the A1W was ready to go critical, a number of Westinghouse physicists and other senior technical people from Bettis were on hand. In addition Zechella wanted as many of his people as possible to see it happen. It was their moment of success. So he had directed his local people into the building containing the readout instrumentation.

Rickover called Zechella over and said, "Do we need all these people for this test?"

"No, Admiral, we don't."

"Well, what the hell are they doing here?"

"Admiral," Zechella countered, "these people came here when it was a pile of sagebrush, and they're eager to see us go critical."

Said Rickover, "If we don't need 'em, then get 'em outta here!"

Zechella replied, "If they go, I go."

"Well, suit yourself," Rickover answered.

With that, Zechella rounded up his people, and they all went over to the office building. I've always admired Zeke for that stand on principle.

Westinghouse supervisors were in charge, but it was a Westinghouse-trained Navy crew that ran the criticality test. Everything went off shipshape, and Rickover came into Zechella's office afterward. "That was the best criticality I have ever witnessed," he said. But those words of praise didn't make Zechella feel any better. He still was steaming because his people hadn't had the opportunity to see it happen.

At a dinner party that night at the country club, Rickover asked Zechella to go to Newport News to head up the Westinghouse team installing reactors in the carrier *Enterprise*. Zeke said he would go, if he could pick his people. Rickover agreed, so Zechella took about 25 people from A1W and they all moved to Newport News. This time he apparently liked what Zeke did, but you could never count on Rickover's reactions.

Installation at Newport News

The Westinghouse role was reversed at Newport News. The shipyard had the prime contract and we were the subcontractor for reactor facilities. By this time, however, we had worked out a very good relationship.

Our organization consisted of a shaft supervisor for each of the four shafts—Hank Ruppel, Pat Ryan, Roger Wood, and Ed Ney. Our total complement varied from 35 to 50 people, primarily supervising the work of the shipyard on the reactor construction. John McLaughlin was our quality control man, Al Sanderson the chief test engineer, and Bill DiPetro assistant chief test engineer. We scheduled and conducted all tests on the reactor compartment with the shipyard. A Bettis man, Bill McKim, was in charge of core loading and control rod drive installation and test. Al Sanderson and DiPetro were in charge of reactor startup tests.

NRB had one man overseeing all Bettis activities at the shipyard plus all construction work on the carrier—an impossibly large assignment, so he didn't play much of a role in any one area.

This was the only time in the history of nuclear power that 8 reactors went on line at the same time. The Idaho crew that went to Newport News has the distinction of being the only group of people ever in charge of building and operating 10 nuclear reactors—and all of them operated successfully.

Sea Trials of the *Enterprise*

When we were on the sea trials of the *Enterprise*, the Navy was operating the four reactor plants, but we had a supervisor in each plant. Because the

Figure 4.2 CVN 65, the USS *Enterprise*.

turbines and gears were Westinghouse design, the company turbine engineer was on board in the control room. We were up to full power, making our four-hour full-power run, and we were about halfway into it.

Full power on the *Enterprise* was 171 rpm and Rickover kept asking for a few more turns on the shaft. We got up to 178 or 179 rpm and still were not using all the steam available. With the *Enterprise* making nearly 34 knots, the destroyers couldn't keep up with the big carrier, because the seas were a little heavy. They left and we finished the full-power run alone.

With the ship cutting the waves at 34 knots with 178 rpm, Rickover said, "Well, we're going to put on a couple more turns." At that the Westinghouse turbine engineer looked him straight in the eye and said, "Admiral, one more turn and all our guarantees are null and void." So we stayed right there, at about 178 rpm for the rest of the time.

Rickover said it was the most successful sea trial he had been on and the shipyard agreed that it was the finest carrier trial they had ever performed. So everybody was happy with the performance of the ship. As we came back into Newport News, the admiral was very gracious, as he could

Figure 4.3 CGN 9, the USS *Long Beach*.

be at times, telling us what a great job we had done. It was an interesting sea trial.

Plant Apparatus Department

With the increase in nuclear Navy construction, Rickover and Weaver took steps in late 1953 to streamline the procurement of non-nuclear components by setting up an organization at Large, Pennsylvania, using this group with Al Squire as manager. Three years later, the Plant Apparatus Division (PAD) was formed at Cheswick, Pennsylvania, with William Borden as manager. Bill had been my executive assistant. PAD reported to Weaver, independently of Bettis. Borden was a lawyer by training and had been staff director of the JCAE before coming to Bettis. The first-of-a-kind components would be developed and procured by Bettis, with later components, except nuclear cores, procured by PAD.

While PAD was procuring only components that duplicated those procured by Bettis, much technical work was involved. The Bettis specifications required a lot of interpretation, particularly for a new manufacturer—and those new manufacturers had first to be found and qualified. Therefore, in addition to having a good legal and administrative staff, PAD needed a strong technical staff. These technical people were mostly recruited from places other than Bettis, but they were sent to Bettis for intensive training.

It was hard to impress new vendors with the quality of work required and the need for every specification to be met completely. Most of them had never had to do this before. So while PAD was not 100% effective in attaining its quality objectives, the quality of product did continue to improve.

PAD soon became a really big business.[3] By 1958, it was dealing with 400 suppliers, 55 of whom had contracts for more than $100,000 and 21 contracts ranged from $1 to $15 million. This growing volume caused the PAD organization to be changed from a project to a product orientation. There were five departments: reactor components, other than cores; heat transfer equipment; pumps; instrumentation; and auxiliary components. To put the above numbers in perspective, the entire research, development, and construction of the Mark I prototype and the entire cost of the *Nautilus* was only $178 million.

Electro-Mechanical Division

One of the early Bettis problems had been how to get leak-proof pumps for reactor systems. This led to the development of the "canned motor" pumps in a program led by Ben Cametti. In 1952, capitalizing on its capability of manufacturing high technology equipment that had been designed at Bettis, Westinghouse formed the Atomic Equipment Department, later called the Electro-Mechanical Division. It was to build these pumps and also control drive mechanisms. Under the direction of William Miller, this division reported directly to Weaver and was located at Cheswick, Pennsylvania. Later, this division developed and manufactured valves and other components used in nuclear plants for Bettis, and still later, for Westinghouse commercial nuclear plants.

Nuclear Fuels Department

Another function that grew to be an important business for Westinghouse was the manufacture of nuclear fuel. This is covered in some detail in a later chapter.

---

[3]R.G. Hewlett and F. Duncan, *The Nuclear Navy, 1946–1962*, p. 286, University of Chicago Press, Chicago, 1974.

Bettis had manufactured all cores up to 1957. Some time before that, it was decided that Bettis would manufacture the first core of each new design, but subsequent cores would be built by others. The problem was to locate and qualify firms that could do this.

Westinghouse was to be one of the commercial suppliers, and the nuclear fuel department was established at Cheswick, Pennsylvania, with Ira Fox as manager. The other fuel manufacturers were Metals & Controls, Attleboro, Massachusetts; Combustion Engineering, Windsor, Connecticut; Babcock and Wilcox, Lynchburg, Virginia; and Nuclear Materials Co., White Plains, New York.

The British Get on Board

The British Admiralty was impressed by the success of the Navy's nuclear submarine program and initiated one of its own. Rickover lost no time in convincing U.S. and U.K. authorities that it would be to their mutual advantage for the British to use American reactor plant and submarine designs. This was formalized in the Agreement for Cooperation on the Uses of Atomic Energy for Mutual Defense Purposes dated July 3, 1958.

Westinghouse was selected by the Navy and the AEC to supply a Skipjack-class (SSN585) nuclear propulsion plant for the first British nuclear submarine, the HMS *Dreadnought*. We also were ordered to transfer to the British Navy and industry the technology and know-how needed to design, build, and operate additional nuclear submarines. Under Rickover's general surveillance, a contract was negotiated between Westinghouse and Rolls Royce Associates (a consortium of Rolls Royce, Vickers-Armstrong, and Babcock & Wilcox). It called for Westinghouse to:

- Manufacture, at a fixed price, those items of reactor and propulsion plant equipment that Westinghouse normally sold to the U.S. Navy, including the reactor core, main coolant pumps, steam generators, reactor control equipment and the steam turbine and gears.
- Purchase the remaining equipment for the reactor plant on a cost-reimbursement basis.
- Train Rolls Royce Associates in the design of the overall reactor plant and to provide detailed drawings and manufacturing know-how to enable them in the future to build the equipment furnished by Westinghouse for the *Dreadnought*. In addition, Westinghouse was to provide the field service needed to install and test the reactor plant and to initiate manufacture of equipment in England.

- Subcontract to Electric Boat the responsibility for procuring, on a cost plus fixed fee basis, those items they furnished U.S. Navy propulsion plants. Electric Boat was to be responsible for providing the British shipbuilder with technical and manufacturing assistance and field service needed in building the hull and installing the plant.

In other words, the United States was giving the British full information on everything about our nuclear submarines except missions and strategy. The operation reported directly to Charlie Weaver who worked closely throughout the project with Sir Denning Pearson, chairman of Rolls Royce.

To carry out this contract Westinghouse established, in August 1958, the Special Atomic Project (WSAP). This organization consisted of about 50 people drawn from Bettis and the PAD and was located away from both those organizations for security reasons. Another reason was to avoid interference with the U.S. Navy nuclear program.

The British Admiralty sent three officers and Rolls Royce Associates sent a manager to be in residence at the WSAP offices to provide the required contract and technical approvals. More than 100 other Admiralty and Rolls Royce Associates personnel were sent to the United States to be trained on reactor and plant design and to obtain knowledge required to manufacture reactor plant equipment. The trainees spent from three weeks to more than a year in the United States. Westinghouse sent technical and manufacturing experts to the United Kingdom to assist in the manufacture of reactor plant equipment.

Some of the key personnel involved were W.R. Baker, Admiralty *Dreadnought* project manager; Captain D.A. Cotman, Admiralty U.S. representative; J. Fawn, manufacturing director of Rolls Royce; and W. Gilligan, U.S. resident representative of Rolls Royce. WSAP personnel included W.R. Ellis, project manager; H. McCreary, engineering manager; M. Penfield, purchasing manager; and M. Berguson, Q.A. manager.

The *Dreadnought* hull was launched on Trafalgar Day 1960 by Her Majesty the Queen at the Barrows Shipyard of Vickers-Armstrong. The U.S.-supplied equipment for the *Dreadnought* was provided in line with schedule requirements and within cost estimates.

Life as Lab Director

My life and my association with Rickover took another turn in 1955, when Charlie Weaver was made vice president of all Westinghouse Atomic Power Activities, i.e., Bettis, the Plant Apparatus Department, the Atomic Equipment and Fuel Departments at Cheswick, and Commercial Atomic

Power Activity. I moved up as division general manager, in effect, laboratory director. Now my association with Rickover became closer and more frequent—but not necessarily any easier.

The admiral was as complex a person as anyone I ever have known. If you put any company business ahead of what he wanted, he was furious. But he understood that sometimes personal problems might have to take precedence. His routine went something like this. Almost every week he would call and announce that he was coming to Pittsburgh for a meeting the next day or even that day. I would pick him up at the airport about five, then we would either eat dinner at a hamburger joint or go on to the Bettis cafeteria, if others were to join us for the discussion. After the meeting, I would take him to the midnight train, and he would be at his desk in Washington the next morning.

On one occasion when he called, I told him that Price, our CEO, had asked me to have dinner with the directors the night before the Westinghouse board meeting. He slammed the phone down and in about five minutes Price called me. He said they had changed the schedule and it would not be necessary for me to attend.

A few weeks later, when he called to set up one of his spur of the moment meetings, I told him it was my daughter's fifth birthday and he said, "OK, I'll come tomorrow." An unpredictable man.

A lot of his irascibility and bad temper was for effect. One evening he was with me in my office discussing a schedule problem with the president of a supplier company. The admiral got so mad he pounded the table and was actually almost jumping out of his chair. He was so upset I thought he might have a stroke. However, when the president had left the room, he turned to me and asked, "How'd I do, John?" I replied, "Admiral, Lionel Barrymore couldn't have done the scene any better." He grinned and took it as a compliment.

Another time, he was coming from a visit to a supplier of zirconium, one of our critical needs. Before he arrived at my office, the president of that company called me. He said they simply would no longer take any orders from Bettis. They were through with Rickover. And that was it.

So when the admiral arrived, I told him that this time he had gone too far. We needed that supplier. He agreed and promised he would call the supplier the next day and make amends. Early the next morning, the president of the company called me.

"Mr. Simpson," he said, "we have reconsidered and we've decided to continue doing business with you. We are not going to let that little S.O.B. get to us." The ball was now in my court. I dreaded having to tell that to the admiral. The next morning I told Rickover about the call. I did not use that president's exact words, but gave him the gist of what he said. Rickover broke every normal rule of doing business with people—but most of the time he got away with it.

On another occasion, he told the president of a supplier company that one of his vice presidents was a liar. It seemed the vice president had assured Rickover that an order had been shipped, while Rickover's man, who was visiting the plant told Rickover that the equipment had not been shipped. The real story, it turned out, was that the order was on a truck but was stopped at the gate because of incorrect paperwork.

Did Rickover wait to get to the bottom of it before calling somebody a liar or taking some drastic action? Not the admiral.

One Christmas Eve, Rickover called me at home about six and told me to call the president of Allis Chalmers and persuade him to speed the delivery of some equipment. My voice grew louder and louder as I tried to explain that even if I called, the president wouldn't do anything until after Christmas. Even if he did relay the order, nothing would happen until then.

This was one time I simply refused Rickover's request and hung up. As I left the den, my wife said the kids had remarked, "Well, this will be a bad Christmas Eve. We can hear Dad talking to the admiral."

Calls from Rickover were usually a traumatic experience. He might ask your opinion on something, and if the answer did not suit him or you didn't have an answer, he would go into a tirade. You often didn't know how bad he considered your performance to be. He might have just been letting off steam.

There was a direct dedicated line from my office to Rickover's office. I had AT&T provide me with a special phone. If you pushed any one of 14 buttons you reached whomever you needed on a dedicated line. I could also push another button, and all 14 phones would ring. I could put Rickover on hold and find out if any of them knew the answer. These calls had to be answered immediately.

Testifying before the JCAE with Rickover was quite an experience. He always had everything arranged beforehand. His staff and the committee staff worked out the questions and answers in advance. If a member, not in on the plan, asked something difficult or embarrassing, the chairman found a way to cut him off. Rickover frequently misspoke, usually on a minor detail, but the two staffs then corrected the record.

Rickover presented a completely different face to different audiences. To secretaries, not those working for him, he was thoughtful and helpful; to people in a social situation, he was charming; to contractors and people who worked for him, he was a tyrant; to government officials, he was a master politician. Even those under his thumb, who most of the time hated him, also respected him and worked hard for him. Why? I guess, as much as anything, it was because we believed in what we were trying to do—and he was a winner.

Several times, Rickover insisted we get rid of one of my people. If I completely agreed and was just putting it off, I acted. If I disagreed, I could usually talk him out of it or refuse. The tough situations were those where I was beginning to think the man wasn't making it, but I still thought it was too soon to act. Another situation, which was pretty tricky, was when I thought the man was being misjudged, but we could use him in the commercial atomic activity. Sometimes I had no choice and had to move him. Sometimes, it was a case of balancing the Westinghouse commercial program against Bettis interests. However, if the man could not work with Rickover, he was useless to me. We moved a lot of people, many of whom later played key roles in the commercial activity.

I attended the Atoms for Peace conference in Geneva in 1955 and presented a paper on Shippingport, coauthored with Rickover, at the plenary session. The Soviets were also there. This was the first time the Soviets had given the West any significant information on their nuclear program.

Before each session in which questions could be asked, the U.S. attendees met to consider what questions should be asked and who should ask them. It was my opinion that the most important information we could get from the Soviets had to do with their industrial capability, e.g., what size reactor vessels they could build. But the scientists had other ideas. They did not consider this important and they prevailed. We clearly had little to learn in reactor physics or other scientific areas. Knowing their industrial capability would have given us a clue as to how competitive they might be worldwide in nuclear plants. However, as it turned out, they were never a competitor on the world scene.

I was lucky I stayed in Geneva only long enough to present my paper. After I left, Westinghouse gave a boat-ride party for our potential customers and many delegates to the conference. The boat went to the far end of Lake Geneva. By that time, the captain was drunk and kept wandering around the lake and not returning to the dock as had been planned. There was an open bar and by the time the boat returned to the dock it was considerably past midnight. Many of the guests, particularly the younger ones, had enjoyed the bar too much. A real bacchanal, I guess, nuclear style.

The captain was the first off at the gang plank as the ship docked. Weaver had some explaining to do to Senator Clinton Anderson at a JCAE hearing the next day in Geneva.

The event was reported in *Time* magazine the following week and was the subject of much kidding the next few times I testified before the JCAE. That I "missed the boat" didn't make much difference to those guys.

A Voyage of Importance

The *Nautilus* was in Hawaii and, after a maintenance stay, was ready again to go to sea. Rickover told me to expect an important event to occur, but he wouldn't tell me what. In just a few days, the event hit the front pages of almost every newspaper in the world. The *Nautilus* left Pearl Harbor at 2:00 a.m., July 23, 1958, and headed north. She cruised 2900 miles submerged to the narrow Bering Strait between Alaska and Siberia. There, she surfaced briefly for a final navigation check. Then she submerged to begin the 2114 mile trip under the polar ice cap. At exactly 11:15 p.m. (EDT), August 3, 1958, the *Nautilus* passed under the North Pole. About 36 hours later, she emerged from under the ice in the North Atlantic. Captain W.R. Anderson, the skipper, sent me a telegram from "Ninety North."

The search for the Northwest Passage, a voyage across the top of North America from one ocean to the other, had been a goal of seagoing adventurers for centuries. At last, the goal was reached.

The *Nautilus* beat the *Skate* to the North Pole. The *Skate* reached it at 1:47 p.m. Greenwich time on August 12, 1958. She was the first to surface at the pole, accomplishing that on March 17, 1959.

These voyages showed that submarines could operate in this area on routine military missions. This became a major part of the U.S. strategic plan, particularly as a hiding place for ballistic-missile submarines close to Soviet territory.

Many had searched for a water route between the oceans, only to be turned back by impenetrable ice. These included Hudson, Frobisher, Baffin, Foxxe, Parr, and Franklin. Amundsen, between 1903 and 1906, in a 70-foot sloop, made his way through the passage from east to west. Several other small vessels had made this passage, but far south of the North Pole.

In the early 1960s, the Reynolds Aluminum Company studied the possibility of building a submarine to transport cargo between the two oceans. That goal has yet to be reached.

New York gave the *Nautilus* a hero's reception when she arrived. I went with Mark Cresap to greet her and was on deck for the Westinghouse commercial that evening on Studio One.

Looking to the Future

Many nations were beginning to consider the possibility of building nuclear plants and most were at the 1958 Atoms for Peace conference. This gave Westinghouse a chance to show what it had done and was capable of doing. By this time, Westinghouse had a materials-testing reactor in operation and had started on the BR-3 reactor for Belgium, Shippingport was operating,

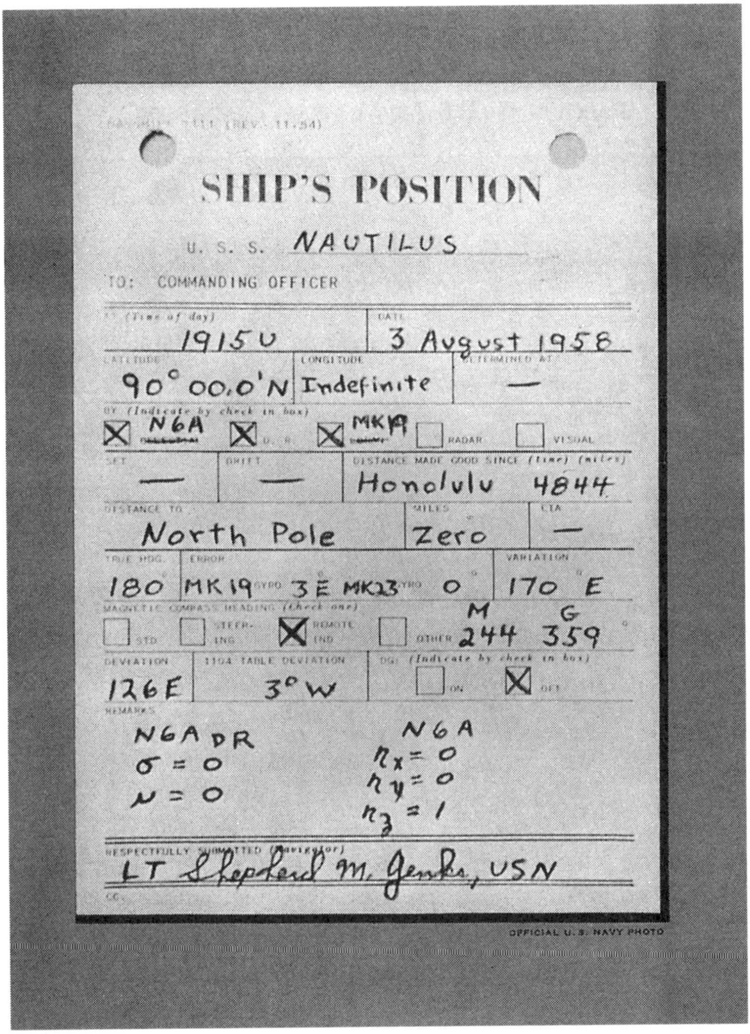
SHIP'S POSITION

U. S. S. NAUTILUS

TO: COMMANDING OFFICER

(Time of day) 1915U

DATE 3 August 1958

LATITUDE 90° 00.0'N

LONGITUDE Indefinite

DETERMINED AT —

BY (Indicate by check in box) [X] N6A [X] D.R. [X] MK19 [ ] RADAR [ ] VISUAL

SET —

DRIFT —

DISTANCE MADE GOOD SINCE (time) (miles) Honolulu 4844

DISTANCE TO North Pole

MILES Zero

ETA —

TRUE HDG. 180

ERROR MK19 GYRO 3E MK23 GYRO 0

VARIATION 170 E

MAGNETIC COMPASS HEADING (check one) [ ] STD [ ] STEERING [X] REMOTE IND [ ] OTHER M 244 G 359

DEVIATION 126E

1104 TABLE DEVIATION 3° W

DGI (Indicate by check in box) [ ] ON [X] OFF

REMARKS
N6A DR
$\sigma = 0$
$\nu = 0$
N6A
$n_x = 0$
$n_y = 0$
$n_z = 1$

RESPECTFULLY SUBMITTED (Navigator)
LT Shepherd M. Jenks, USN

CC:

OFFICIAL U. S. NAVY PHOTO

Figure 4.4 Report of the *Nautilus*'s position at the North Pole.

and Yankee Rowe was under construction. We also had entered into nuclear license agreements with Siemens in Germany, Schneider in France, and Fiat in Italy. There also were contracts with Pennsylvania Power & Light and Carolinas-Virginia for small nuclear plants.

Figure 4.5 Simpson receives Navy Certificate of Merit for Bettis from Assistant Secretary of the Navy Garrison Norton.

In 1958 I prepared a report for Westinghouse management and the government stating my views on the future needs of the division. The conclusions were the following:

"The Bettis Atomic Power Division, as a participant in the Government program for advancement of nuclear power, has been able to make significant contributions in both the fields of nuclear power for naval propulsion and nuclear power for generation of electricity. This was possible because of the efforts of the technical and supporting staffs using the facilities provided by the AEC.

"There is a recognized need for a continuation of research and development activities in the field of nuclear power, particularly in the area of national defense and those areas where nuclear power may be considered as an instrument of our foreign policy. If the Government intends to participate to the fullest extent possible in the furtherance of these policies, maximum utilization must be made of the personnel and facilities available. At Bettis, past accomplishments have shown that the technical and supporting staff is capable of developing, designing and constructing reactor plants with proven performance. Performance of reactors produced as a result of NRB-Bettis participation in the naval propulsion program has been instrumental in the realignment of military and civilian leaders' thinking regarding the role of the Navy in the present allocation of forces. The success of the *Nautilus* not only has stimulated construction of nuclear powered submarines, but has provided a basis for broadening the mission of the Navy and extending its capabilities.

"Although advances to date have been significant, it is recognized that there is much to be done. Of particular importance is the reduction of costs of both military and civilian nuclear plants. Inherent in any program of cost reduction will be development of more efficient nuclear plants and components. These objectives will require the services of personnel experienced in the problems associated with nuclear plants. This experience has been built up at Bettis through long association and contact with the various aspects of development connected with nuclear power.

"To protect the Government's investment in the facilities and to obtain the maximum benefit from the trained staff which is presently available at the Bettis Division, action should be taken to obtain:

1. Two programs equal in magnitude to an S5W effort. Since there is a reduction in the R&D programs for FY 1959, E & S personnel will be available commencing July 1958 to facilitate work on these short range projects.

2. Because of the rapid phaseout of existing programs, work on a long range development program comparable to a PWR or A1W effort should begin by January 1959. As the present projects phase out, the manpower becoming available would be assigned to this project.

3. Even though an effort comparable to the PWR or A1W Program would commence by January 1959, there still will be additional E & S personnel becoming available during the last six months of calendar year 1959. Sufficient engineers and scientists could thus be

utilized on another new short range program equal to the present S5W effort.

"In the interest of advancement, these programs should be virile developmental projects designed to increase knowledge and promote development of nuclear power.

"Failure to undertake these programs will require the Division to reduce operations. Not only will this entail a loss of personnel trained in all aspects of nuclear power, but it might result in a lowering of morale among the remaining staff. The high 'esprit de corps' that has always existed at this Division has in part been due to the urgency associated with the programs undertaken. Reduced operations will also mean idle facilities at a time when the domestic and international situations demand that every effort should be directed toward new advances in the nuclear sciences.

"The Bettis Atomic Power Division will have personnel and facilities available to undertake new programs. If the Division is to continue to participate to the maximum of its potential, consideration should be given immediately to the awarding of new projects."

A Retrospective View

Why was the naval nuclear program so successful? Clearly Admiral Rickover was the *sine qua non*, but he couldn't do the job alone. He wisely chose two large corporations with extensive technical capability, but perhaps more important, corporations who not only traditionally supplied equipment to the Navy, but who could see the potential of nuclear power for generation of electricity. This gave them an added incentive to succeed.

We were also fortunate that there was a JCAE, for this certainly was a big help in dealing with Congress. This one committee had almost complete jurisdiction over authorization of nuclear projects. Today, many committees have some jurisdiction. Any of a myriad of congressional committees can block a project, but none, acting alone, can get anything approved. Also, the JCAE, concerned only with nuclear matters, was much more knowledgeable than is possible for committees today.

Chapter 5

Shippingport

History is made by the decisions of men and the force of circumstance. It was a combination of the two that brought into being the world's first nuclear plant built exclusively for the generation of commercial electric power.

Bettis had been at work for only a short time on the propulsion plant for a large naval vessel, presumably an aircraft carrier, when in mid-1953 the Department of Defense (DOD) decided to cancel the project. Budget restrictions and the anticipated high cost of a nuclear carrier made the move politically necessary.

The cancellation put a road block in the path of Hyman Rickover's drive to develop large reactors. If the Navy's capital ships were someday to be powered by nuclear energy, they would require reactors far bigger than the ones that were being built to drive submarines.

A change of strategy was called for. It had been clear to the admiral ever since President Eisenhower had launched the Atoms for Peace program that nuclear energy someday would be harnessed to generate electricity for civilian use. Why not now? What better source of energy than the pressurized water reactor that already had proved its merit in the desert in Idaho.

He lost no time in meeting with officials of the Atomic Energy Commission (AEC) and the Joint Committee on Atomic Energy (JCAE) to sell them on the idea of building a pressurized water central station power plant.

"We have the technology in hand and the organization to handle the project," he told them, "and timing is politically perfect."

Timing is everything in politics and government. And what could be more timely and farsighted than launching this peacetime initiative? The AEC and JCAE saw the logic of Rickover's argument.

AEC Chairman Lewis Strauss was undoubtedly influenced by the fact that President Eisenhower had launched the Atoms for Peace goal—"to strip the atom of its military casing and adapt it to the arts of peace." He also was uncomfortably aware that important members of Congress were already ridiculing the spending of billions on atomic weapons and warships and essentially nothing on atomic energy for peaceful purposes.

The members of the JCAE also saw the political value in beating the British to the punch. A gas-cooled reactor at Calder Hall in England was being built to generate electricity in addition to producing weapons material. Why not make the United States the leader in harnessing the atom solely for peace?

There were, of course, some doubters. They thought it premature to choose the type of reactor, without lengthy exploration of all possibilities. Others were not happy about having the Navy, and in particular a gadfly like Rickover, in charge of essentially a commercial reactor prototype. They feared he would design naval features into it and lessen its value as a commercial reactor prototype. They also worried that Rickover would not be sufficiently cost conscious. After all, the atom eventually would have to compete with coal and oil.

There is a rather complete story on the events leading up to this decision to develop a pressurized water central station reactor in *Nuclear Navy, 1946–1962* by Hewlett and Duncan (Chicago: University of Chicago Press, 1974).

Rickover knew political strategy as well as he knew naval strategy—maybe better. He pointed out to the various government officials that the purpose of developing this first plant would not be to compete in energy cost with coal or even oil. That would have been impossible anyway. Rather, he pointed out, the purpose of this first plant would be to advance the technology of pressurized water, and prove its potential for electric power generation.

There was a meeting at Argonne National Laboratory (ANL) in 1954 of all the key people in the nuclear program about what to do to reduce the cost of nuclear power. None of the opinions expressed were very helpful. When Rickover was asked for his opinion, he replied with tongue in cheek that the best way to get the cost of nuclear power down would be to reduce the price of coal. It was his way of criticizing those who always were able to predict that their design would result in a power cost as low as that from a coal-fired plant.

Rickover was convincing. With the JCAE backing, the AEC authorized preliminary studies of a central station PWR on July 1, 1953, and the contract was awarded to Westinghouse on October 9. The public announcement that the United States would build a full-scale civilian atomic power plant was made by Commissioner Murray—the same commissioner who had turned the valve to produce the first substantial power from the Mark I reactor in Idaho about a month before. Speaking in Chicago on October 22, he said this historic civilian power plant would produce a minimum of 60,000 kilowatts of electric power. The story made headlines worldwide.

The choice of Westinghouse to develop the pressurized water central station plant was a natural in view of our ongoing progress with the PWR for the Navy, while General Electric (GE) had embarked on a program to design an intermediate-energy spectrum liquid metal (sodium) fast-breeder central station reactor. This program was not showing much promise and was converted to development of a liquid metal submarine propulsion plant, the *Seawolf*.

Organization and Contracts

The Naval Reactors Branch (NRB) of the AEC, with Rickover in charge, was given responsibility for this civilian program of historic proportions—a program that eventually paved the way for some 234 commercial pressurized water reactors worldwide. PWRs represent almost 60% of all nuclear power reactors in the world.

But to conclude that these plants were merely upgraded naval reactors just because a Navy admiral was their godfather would be a great mistake. It is true, some people were saying Rickover's carrier development was simply converted to a utility reactor, and to a degree that was the AEC's intent. But those of us who worked on both projects know differently.

The technology for the central station plant was so different, actually, that very little of the carrier design directly applied. Yes, much was learned in the naval program that could be applied and it provided a good starting point for commercial nuclear power. In particular, all subsequent reactor programs benefited from the basic materials, heat transfer, and reactor physics of the naval program. But a great many new developments were necessary for the civilian plant. Such things as uranium dioxide (UO^2) fuel pellets, the whole idea of containment—which the Soviets failed to copy, to their sorrow—the high-capacity sealed pumps, fuel management, and control drive mechanisms were unique to the commercial atomic power program. This was the first use of UO^2 fuel pellets, which are now basically standard for water reactors throughout the world. Many improvements were required in reactor physics to handle the Shippingport core configuration.

Our initial contract was very vague. It stated that a central station nuclear power plant was wanted. It was to be about as big as seemed reasonable, but no rating was specified. We thought it should be 60 MW(e), plus or minus. Higher capacity, say, 150 or 200 MW(e), would yield little additional information, but would greatly increase the cost of the total project. Because we realized that advances in reactor design technology might well demonstrate that reactor power could be increased, a 100-MW(e) turbine-generator was installed.

Because this was to be a civilian power plant and run as part of a commercial utility business, the AEC wanted a utility partner to build the conventional turbine-generator part of the plant. That utility would buy the steam from the AEC's nuclear portion of the plant. It was a unique partnership.

The AEC had nine proposals to choose from, and in March 1954, it selected Duquesne Light Company of Pittsburgh. With a strong push from Gwilym Price, Philip Fleger, Duquesne Light's CEO, saw this as a great opportunity to participate in the first nuclear central station plant and learn about nuclear power with only a limited financial commitment. He also saw it as a great way for his company to participate in the "Pittsburgh Renaissance," which that city was undergoing. It was a major public relations coup for Duquesne Light as well as a demonstration of technical leadership in the utility industry nationally.

The arrangement was that Duquesne Light would contribute the turbine-generator systems, plus $5 million toward reactor plant structures and systems, and would agree to buy steam from the reactor plant at a price somewhat above its average cost of steam at its other plants.

Duquesne Light brought in John Gray as its project manager. Gray had prior experience in Rickover's organization as well as with Savannah River and both GE and Westinghouse. He formed a team consisting of Albert Stanojev, Edward Gue, William Conwell, Carl Kutschbach, Melvin Oldham, and Charles Jones. This team provided for Duquesne's many contributions to planning, design, construction, training, startup, operation, and maintenance. The Duquesne Light team interacted effectively with Westinghouse and NRB personnel. This made for decisive integrated decision-making on the project as a whole.

Project officer for NRB was Captain Joseph H. Barker, Jr. His man at Bettis was Commander Donald G. Iselin and at the construction site it was S.W.W. Shor.

NRB was charged with the responsibility of giving technical approval to all nuclear plant parameters, performance requirements, details of design, and development on recommendation by Westinghouse. All research, development, and design of the nuclear systems were done at Bettis. Fuel elements for the nuclear core were manufactured at Bettis, and all manufacturing of components and all construction were performed under Bettis supervision.

At this time, Weaver was general manager of Bettis, and I was his technical assistant with staff responsibility for all technical activities, which were organized along functional lines. William E. Shoupp was director of research, and reporting to him were William A. Johnson, materials; Sydney Krasik, physics; Donald M. Wroughton, chemistry; and M. A. Schultz, instrument and controls. The materials department consisted of metallurgy under Benjamin Lustman, and process development under Robert Gordon. Engineering had been divided into two parts—an Engineering Department under Albert Brecht, and a Reactor Department under Philip Ross. Simpson, Shoupp, Brecht and Ross reported to Weaver.

During the summer of 1954, Bettis was reorganized as a project-type organization, and I became Shippingport project manager. In July 1955, I succeeded Charles Weaver, who was elected vice president for the atomic activities, including Bettis, and I became general manager of Bettis. My position was the equivalent of director at other AEC labs.

Joseph C. Rengel then became Shippingport project manager. Reporting to him were Sydney Krasik for physics, Nunzio J. Palladino for reactor design, Benjamin Lustman for metallurgy, Robert Gordon for process development, Dick Cunningham for instrumentation and control, and Bill Ellis for plant equipment. Al Bethel was site manager and Eric Welner was construction manager.

When the site was selected on which to build the plant, it spoke volumes about meeting our energy needs in the future. The new power plant would be built at a little town called Shippingport, down the Ohio River a short distance from Pittsburgh. It would stand on top of a coal mine.

Technical Challenges at Shippingport

Volumes could be written on the Shippingport technology, and some of the detail can be found in the technical chapters at the end of this book. Let me briefly summarize the task that faced Bettis on this project.

As in the case of the submarine program, the concept was simple, and all that was needed was to build a larger plant—about five times as large. The objectives, however, were different and that brought a host of new problems.

The economics of the power production was a major consideration. True, in the beginning we didn't even hope to be competitive with the cost of power from oil or coal, but we did need to have a program that was a step in that direction and could give some hope of later achieving economic power in future plants.

This required a type of nuclear core different from the highly enriched ones in the naval program—but what kind would it be? For Shippingport we finally decided on a core with an annulus of highly enriched fuel elements with natural uranium elements inside and outside of this annulus. The annular seed was an extension of the submarine program in many aspects, but the natural uranium blanket presented a completely different problem. How was this to be done? After considering a clad uranium alloy, we decided on using uranium dioxide pellets, but this required a major development program. We were the first to use uranium dioxide pellets, which are now fairly standard throughout the world.

Zircaloy metallurgy had been developed, but we needed it in the form of tubes to contain the pellets—another development.

We still had to determine how well these fuel elements withstood the radiation environment in the reactor. What would happen if there was a leak in the tube and water entered? How would these elements stand up under many cycles of heating and cooling? What problems would the fission products cause?

The buildup of these fission products is responsible for one of the major problems nuclear reactors face. They continue to produce heat, even after the reactor has been shut down, so there must be a way to remove this heat under any possible circumstance. Emergency core cooling and containment would be required.

The reactor theory for the submarine program was inadequate for this reactor. There were sharp changes in the neutron flux intensity and energy spectrum at the interfaces of the seed and the blanket and a buildup of plutonium in the blanket, as a result of capture of neutrons by the uranium-238. These required the development of a much more sophisticated set of calculations, so complicated that they no longer could be handled by hand calculators. Analog computers had to be developed and later codes for digital computers. The buildup of the fission products was an extra problem for these calculations.

To get the best efficiency, it was necessary to have the temperature of the coolant as high as possible without causing boiling of the coolant water or undue corrosion of the fuel elements. This called for much more detailed knowledge of the reactor flux and the heat transfer characteristics of the reactor.

Chapter 13 on the Shippingport technical story will cover much of the development needed for the reactor vessel, the method of its closure, the new instrumentation, main coolant pumps, and control rod drive systems.

There had been some doubt about the steel industry's ability to make a big enough reactor vessel for Shippingport until it was learned that during World War II, Babcock & Wilcox had fabricated a pressure vessel 12 feet, 8 inches in diameter, 28 feet long, and weighing 214 tons at its shop in Barberton, Ohio, and transported it to Alamogordo, New Mexico. Building and moving that vessel had cost $12 million. That vessel was to contain the first bomb explosion and save the uranium-235 in case it was a dud.

Fortunately the cost of such ventures had come down, but there was still cause for concern because our total budget was only $70 million. This vessel was the largest and heaviest object ever shipped by rail, and a precedent for moving ahead with the Shippingport vessel. Now the Shippingport vessel, weighing 264 tons, would hold the new record.

We knew that even heavier and larger vessels would be required, if nuclear reactors with significantly greater capacity were to be built. That was one of the reasons for trying to learn at the Atoms for Peace conference in Geneva the Soviet capacity for manufacture of large vessels. We realized that if their capacity were limited, they would not be a threat on the world market.

All the while, we were conscious that this plant was to be in a populated area, operated by an electric utility and integrated with their system. And Rickover insisted that the plant was to be operating in four and a half years.

The Project Heard 'Round the World

This nuclear first—a central station power plant run by nuclear fission—quickly became a familiar name throughout the world. We knew we were being watched by the governments, the scientists and engineers—and news media—of every developed nation. We determined that this pioneering power plant's two most important characteristics had to be reliability and safety. Reliability if nuclear energy was to take its place as a major source of energy; safety if it was to be accepted by an uneasy public, which had been introduced to this technology by a bomb falling on Hiroshima.

The construction of Shippingport had sparked worldwide interest in nuclear power. Many countries were attempting to determine what role nuclear power would play, and Shippingport and Calder Hall were the only two nuclear plants being built.

A consortium of European countries decided to pool at least part of their interest in nuclear power and formed an organization called Euratom. Three distinguished men were chosen to conduct a study of what Euratom's role in nuclear power should be. These men—Franz Etzell of Germany, Louis Armand of France, and Francesco Giordani of Italy—became known as the Three Wise Men. They visited Shippingport in 1956 even before the plant was finished.

A Soviet delegation headed by V.S. Emelyanov, chairman of the Soviet Main Administration for Atomic Energy, also visited. They had heard the presentation on Shippingport at the Atoms for Peace Conference the year before.

Figure 5.1 The Three Wise Men visit Shippingport. *Left to right:* Rickover, State Department representative, Etzel, Armand, Giordani, Roddis, Simpson, Geiger, Gray, and Laney.

I hosted Queen Frederica of Greece and her daughter Princess Sophie, now Queen of Spain. Queen Frederica was to fly in a helicopter to Bettis, but her security chief objected. Rickover shamed her into disregarding the advice of her security official. She came to my office, and then we toured the laboratory.

Many American utility executives now were beginning to show interest and came to Pittsburgh to visit Shippingport, which was valuable in illustrating to them our leadership in nuclear power. Of course, Rickover induced many members of Congress to visit. He wanted to be sure they knew that progress was being made in nuclear power.

And we didn't overlook our own people. The wives and families of the Bettis people directly connected with the building of the Shippingport plant were given a special tour of the site.

The tremendous emphasis on reliability and safety was partly responsible, of course, for the high cost of the Shippingport plant. No expense was spared to assure reliability and safety of operation. Backup systems were provided for every key function. No gambles were taken. Nothing was left to chance. No expense was spared to solve the many technical

Figure 5.2 Queen Frederica of Greece visits Shippingport.

problems encountered along the way, and in the end, the plant worked and worked safely. The other reason for Shippingport's high cost—every component was first of its kind. High cost was the plant's only characteristic vulnerable to attack by critics, many of whom claimed they could do it much cheaper.

Getting the Job Done

With Rickover in the lead, our projects always had a way of starting off with a patriotic flair. President Harry Truman had attended the laying of the keel for the *Nautilus*. Mamie Eisenhower christened the ship at the launching. Now President Eisenhower was enlisted to break ground for the Shippingport plant.

But this groundbreaking had to be different. It had to "be nuclear" somehow. And there was a second complication. Ike was in Denver recuperating from a heart attack. He couldn't even be on hand.

Both problems were solved when somebody got the brilliant idea of providing the President with a slightly radioactive "wand," which he

Figure 5.3 President Eisenhower waves the atomic wand that initiated groundbreaking at Shippingport.

would wave over a Geiger counter. The signal from the counter would be sent by telephone line to Shippingport, where it would activate an unmanned bulldozer to break ground.

I was put in charge of the operation and enlisted Dick Cunningham to engineer the remote mechanism. We arranged for a dedicated phone line from Denver which, to save money, was to go into operation only one hour before the ceremony. That made AT&T uneasy; something might go wrong. So it opened the line early in the morning for tests, and every switch from Denver to the site was wired across to be sure there was no open connection.

In case the phone signal didn't work, I secretly had somebody ready to trigger the bulldozer from the site by hand. You can't take chances, you know.

The rest of the deal was also scary. I was afraid the bulldozer might dig too deep and turn over instead of lifting the dirt, so we had steel rails installed under the ground, and the ground to be "broken" was pretty thoroughly broken up before it was ever touched by the bulldozer. A real piece of civil engineering.

The possibility of an embarrassing foul-up was made even more worrisome by the fact that we filmed a rehearsal to be used on the nationwide Westinghouse Studio One national TV program that evening, with the slogan "You can be sure if it's Westinghouse." No matter how the groundbreaking actually went during the day, that's what would be in the TV commercial.

A few weren't so sure as the slogan would have you believe. Mr. Price told me, "John, if the real thing goes wrong, just leave and keep going."

I think the day of the groundbreaking was the hottest of the year—it hit 100 degrees Fahrenheit at Shippingport and the big crowd of VIPs, who ate a luncheon catered by Pittsburgh's Duquesne Club under a big canvas tent, sweltered. But nobody complained, the show went on and everything worked beautifully.

Construction

The contractual setup was a nightmare. Westinghouse was responsible for the nuclear part and retained Stone & Webster for the engineering, the Dravo Corporation for most of the construction, and Blaw Knox for the radioactive waste disposal system. Duquesne Light retained Burns and Roe to construct its part. There were people on the site from so many organizations that they were color coded so you would know who was who.

Westinghouse sold the turbine-generator to Duquesne Light—and I'm sure we made the price "right." We had to be sure the GE nameplate didn't appear in that plant, if we could help it, so our sales people made Philip Fleger an offer he couldn't refuse. We did overlook one minor item. There was a GE drinking fountain in the reception room—for a very short time.

At the government's request, Bechtel was retained as a construction consultant. With all those cooks stirring the broth, you can imagine the buck passing as schedule and interface problems inevitably arose.

Rickover had big ideas about how long it would take to build this first commercial nuclear plant. In 1955, he set the objective of completing Shippingport by the end of 1957. We didn't think such an early date was possible. Neither did a lot of other people. No matter, that was Rickover's goal.

A schedule to meet that end date was put together by Stone & Webster, Westinghouse (Joe Rengel and Eric Welner), Dravo, and NRB (J.H. Barker and Don Iselin). And everybody worked their tails off to meet that schedule. Coordinating meetings were held monthly, then biweekly, weekly, and finally every day.

Figure 5.4 Congressional delegation gets a briefing on Shippingport. *Left to right:* Simpson, Congressmen Vanzandt, Holifield, and Price, then Rickover. In the background are John Conway, Frank Cotter, Jim Ramey, and Geiger.

For the turbine-generator part of the plant there was another coordinating committee consisting of Duquesne Light (Gray, Kutschbach, and Oldham) and Burns & Roe (Grady).

In many such meetings, major subcontractors were present as well as senior management of Westinghouse, Dravo, and Duquesne Light. Once in a while, Rickover was there, cracking the whip. At one such meeting in the office of Carl Jansen, the president of Dravo, Rickover pushed him hard for a schedule improvement. When Jansen finally agreed, Rickover wrote the promise with a crayon on the president's wall. "I want it to stay there until you have met your commitment," he said.

I don't know how Jansen explained the graffiti on his office wall to his other visitors for the next few months. He probably just asked if they had heard of a man named Rickover.

When it appeared that Dravo would not be able to complete construction on schedule, an additional contractor was brought in: Blaw Knox, who had done piping for Duquesne Light, and agreed to the tightest schedule. This helped, but they had a lot of difficulty qualifying their welders to our high standards. With Bettis help, however, they succeeded. I'm sure Blaw Knox regretted the day they took that job, because it cost them far more than the contract price.

The British were making progress on their Calder Hall plant, and Admiral Lewis Strauss, AEC chairman, wanted Shippingport completed ahead of them. He asked how much it would cost to speed up the schedule enough to be first. We were only about a year away from completion, and we had to say that no amount of money could get us to the finish line first.

During the final six months, Joe Rengel essentially lived at the site. I would visit the site about once a week with Rickover and Lawton Geiger, the AEC Pittsburgh area office manager. We would arrive at the site early in the afternoon and go over the schedule and special problems with site personnel. These meetings were always traumatic. Nothing was ever being done well enough or fast enough to suit Rickover. Then we would either head back to Bettis for another meeting or take Rickover to the train. We usually stopped at a roadside restaurant for dinner. By that time, Geiger and I were ready for a martini, despite Rickover's disapproval.

Core Loading, October 1957

The night of October 6, 1957, was one of exhilaration. The nuclear core—a multiple-ton assembly of uranium, Zircaloy, stainless steel, and other exotic materials packaged into a cylinder with the precision of a fine watch—was to be loaded into the 32-foot-high reactor vessel, as big a unit as the steel industry could provide. All the key Bettis managers were there and stayed all night watching this gleaming cylinder being lowered very, very slowly by the overhead crane into the reactor vessel.

The photographs of this historic event appeared in the newspapers and magazines throughout the world and still appear from time to time more than 35 years later.

As we went home that morning, we knew that a major milestone had been passed in the development of nuclear power. The morning paper headlines, however, announced that Sputnik had been launched by the Soviet Union. The seed had been sown for what later would become the Astronuclear Laboratory and our next challenge.

Criticality and Full Power

Then came that tense moment in nuclear power—the time when the reactor goes critical. That's when self-sustaining nuclear fission occurs. That's when you find out if all of your calculations and measurements were correct. You *thought* for sure that they were correct—but were they? Now you *knew* for sure.

Again, as at the Idaho Mark I site, on December 2, 1957, the control room at Shippingport was full. However, this time everything proceeded without any major confrontation as had happened at earlier first criticalities.

Preparing to go to full power, we still had a large margin of uncertainty as to what power level we could actually achieve. The highest power rating is determined by that power at which the hottest spot on any fuel element is below the design limit. The determination of the hot spot temperature involves many different calculations, each of which had a large margin of error in those early days. If all were to turn out unfavorably, the power would be well below the rated value. The plant rating had been set rather arbitrarily at 60 MW(e) at the outset. As indicated earlier, Duquesne Light had installed a Westinghouse 100-MW turbine-generator.

The core was instrumented, so that when the plant was operating, we could more accurately determine actual flux distribution and flow characteristics and thus determine the safe operating power level. The 100-MW rating was eventually achieved.

On December 23, 1957, less than four and a half years after the award of the contract to Bettis, the Shippingport nuclear power plant reached full power and was synchronized with the Duquesne Light Company system furnishing power to Pittsburgh.

Perched at the top of Mt. Washington, just across the Ohio River from Pittsburgh's Golden Triangle, Walter Cronkite of CBS-TV watched and described to the nation the lighting of the city through the use of atomic power. A technological milestone for mankind.

Figure 5.5 Shippingport lights up Pittsburgh, December 23, 1957.

Shippingport Costs

At an Atomic Industrial Forum meeting in New York in 1958, John Gray presented a paper on Shippingport and reported that the bus-bar cost of power was 64.4 mills per kilowatt-hour.

Asked a questioner, "Mr. Gray, why is your cost so high when those planning new plants predict they will produce power at 5 to 10 mills per kilowatt hour?"

Replied Gray, "Shippingport was built and we kept books." The audience roared.

What was the cost to build the Shippingport plant? One day early in the project, Louis Roddis from Rickover's office visited me to discuss the budget for construction.

"John, what is your estimate of the total cost?" he asked.

"It will be about $100 million."

"Well, we have only $70 million available in the budget," he said.

"In that case," I replied, "you have just gotten a new cost estimate, $70 million."

The plant eventually cost about $100 million, but it was enough of a success that the overrun was no big deal.

The Attempt at a Gas-Cooled Reactor Project

It was not likely that the AEC was going to assign any other development to Bettis for central station reactors, and we believed Bettis needed another challenge—something other than turning out more naval reactors.

We decided, therefore, in 1958 to shoot for a gas-cooled reactor contract. Our advanced development department, managed by Esselman, made a conceptual design of such a reactor, and we persuaded Rickover to go along with the idea. There was a bill being prepared by Congress authorizing development of a gas-cooled reactor. Frank Cotter, my executive assistant, was formerly staff director of the JCAE. Through his connections with the JCAE and the legislative assistant to Clarence Cannon, he got it written so as to almost force the project to be done at Bettis. Admiral Strauss, however, had other ideas and was adamant that it not go to Bettis. He prevailed.

Bettis, however, could take credit for pioneering accomplishment in two major areas of nuclear technology: nuclear propulsion and central station electric power generation.

Summary

Despite Shippingport's high cost, its purpose was fulfilled by paving the way for the dominance of the pressurized water reactors, which account for almost 60% of all of the nuclear power reactors in the world.

Shippingport did provide a sufficient technical base for moving on to larger commercial reactors. This had to be done in several steps of increasingly larger size: Yankee at 175 MW(e), Conn Yankee and San Onofre at 450 MW(e), several at about 800 MW(e), then 1000 MW(e) and later 1300 MW(e) plus.

Despite having little technical base from which to start and a need to develop new theory, components, and materials, the Shippingport plant was completed in just four and a half years. The time for construction of nuclear plants has continued to increase. Today, due to the very unfavorable sociopolitical climate, it takes about ten years at best in the United States; other nations can build nuclear plants in about half that time.

Admiral Rickover saw to it that the technology developed and used in the Shippingport program was widely distributed. The naval reactors program was cloaked in security. All of us working on the program had to have AEC "Q" security clearances. This was also extended to the Shippingport project, but there was a difference in that most of the technical information was not classified. Rickover extended every effort to have as much information as possible unclassified or declassified. This enabled not only Westinghouse, but also other companies, to begin development of commercial nuclear power plants.

Rickover believed that a light water breeder (LWBR) was feasible and initiated a development program to convert Shippingport to being such a reactor. This program was widely criticized and was only able to be continued because of Rickover. The final tests to determine whether or not there was breeding were inconclusive. It appeared that at least it was close and some maintained that there was marginal breeding. Had a breeder been needed and for some reason other breeder programs were unsuccessful, the LWBR might have been of value. However, the design required a thorium cycle with the attendant problem of difficult and expensive reprocessing as well as the high cost of the fissionable material required.

During the mid-1980s Shippingport was decommissioned and served as a prototype development and demonstration project for decommissionings. The site has now been returned to a "green field" condition and much of the material from the plant has been shipped to Hanford for burial.

CHAPTER 6

Astronuclear Years

Space Offers a New Challenge

If there was one element that characterized the nuclear power efforts of Westinghouse from the day of its first contract, it was the fierce eagerness to take on new challenges.

The *Nautilus* team put up with frustrations and unbelievable demands cheerfully and determinedly to develop and build the world's first nuclear propulsion plant and then the propulsion plant for the first nuclear carrier. These first challenges were met with great success. The Shippingport team faced a new challenge in building the United States' first commercial nuclear power reactor. Again we had to conquer obstacles, both technical and political, that tested the will of the organization.

Now, on the day the nuclear core was being inserted into the reactor at Shippingport, we all recognized that a second milestone in nuclear age technology was being reached. As we watched, we felt a sense of triumph. Another nuclear "first" was being achieved. But, once again, we looked for a new challenge.

We found it this time, figuratively speaking, by simply looking up.

That very night the Soviet Union launched *Sputnik*, the first man-made orbiting satellite. Space! Where else should we look for a new challenge? But was there a role for nuclear power in space?

Many people in industry, government, and education began at once to reevaluate their thinking about space. Sputnik shook the world. It galvanized the U.S. government into embarking on a major space program.

In the Bettis organization, with my encouragement, Walt Esselman and Walt Roman turned their attention to the potential of nuclear power in space. They were joined by Sid Krasik, manager of the PWR Physics Department. Conceptual studies by Esselman and others made in early 1958 compared the performance of chemical and nuclear rockets.

The potential for chemical rocket missions had been described in a paper entitled "A New Supply System for Satellite Orbits" by Krafft A. Ehricke in the September–October 1954 issue of *Jet Propulsion*. This paper included an analysis of the number of chemically propelled supply vehicles needed to achieve various missions. For example, a three-person vehicle circumnavigating the moon required 55 earth-to-orbit supply flights. An eight-person vehicle circumnavigating Mars required 365 supply flights and a Mars landing 680 supply flights.

Our study indicated the same lunar mission could be accomplished with a single launch of a two-stage nuclear rocket. Circling and landing on Mars would require about 6 and 13 launch flights for a nuclear system. A module would go to the surface, explore it, and return to orbit. Another module would return to earth, leaving behind parts of a vehicle on Mars and in Mars orbit.

We also made studies in which a three-stage nuclear rocket could go to Mars orbit, carrying a payload of the following: a chemical stage for landing on the surface, an explorer, a chemical stage for returning to orbit, and another nuclear stage for returning from Mars orbit to Earth orbit.

These studies made nuclear rockets look like good bets for space. Yes, the technical demands were formidable, even greater than the demands we had faced just a few short years before in designing the first nuclear plant to drive a submarine. But our organization thrived on formidable challenges—and these were formidable challenges indeed.

Here's how Glenn Seaborg, AEC chairman, once put it: "What we are attempting to make is a flyable compact reactor, not bigger than an office desk, which will produce the power of Hoover Dam from a cold start in a matter of minutes."

But at that early moment in 1958, we weren't even aware of those parameters. Our first presentation on nuclear rockets was made by Esselman at a Westinghouse symposium on advanced energy systems in February of that year. He had been asked to speak on nuclear propulsion for aircraft, but his analyses showed that such propulsion had little promise. This assessment was borne out by history: The Nuclear Engine for Propulsion of Aircraft (NEPA) program never was successful. Therefore, he devoted part of his presentation to nuclear rockets, reporting there existed considerable potential.

We were not then aware in any detail of the efforts at the Los Alamos Scientific Laboratory (LASL) on nuclear rockets. Later, Don Thomas, a Bettis metallurgy supervisor made a presentation on space materials to a similar symposium.

U.S. Enters the Space Race

When the launch of *Sputnik* propelled the United States into its major space program, Lyndon Johnson, majority leader of the Senate, set up a space committee with himself as chairman. Of course this caused the House to form a parallel committee and Speaker John McCormack established himself as chairman. He realized that neither he nor his staff was knowledgeable in this area, so he asked Chet Holifield, chairman of the Joint Committee on Atomic Energy (JCAE), for advice.

Holifield suggested that he call Frank Cotter, my executive assistant at Bettis. Cotter had been executive director of the JCAE and was knowledgeable about both space matters and congressional legislation. Frank had worked on the Atomic Energy Act of 1954 and also had become somewhat informed about space matters.

Cotter welcomed new frontiers and enjoyed nothing better than organizational strategy and tactics, particularly in Washington circles. He took a leave of absence from Westinghouse and went on the congressional payroll. In that capacity he helped the speaker write the Space Act. This act specified civilian control, thus giving the AEC a clear role in space.

After the passage of this act in 1958, Keith Glennan was appointed first administrator of the National Aeronautics and Space Administration (NASA). He also felt the need for help, and he asked Cotter to become a temporary employee until NASA was set up and operating.

It's not surprising that one of Frank's friends once told him he would like to complain to Cotter's boss about him if he could ever figure out who that was.

Although he was not knowledgeable about the technical aspects of space, Cotter knew about government operations and had an intimate knowledge of the Space Act. So when he returned to Bettis, his knowledge and enthusiasm for the space program aroused our interest even more.

This interest was further whetted in 1959 when Al Hoppe from the Westinghouse research laboratories, who had been assigned to work at LASL, visited Bettis and described its fuel development work.

As the design became more definitive, Esselman and Cotter held many evening discussions on the merits of this technology and the growing national interest. In the meantime, Cotter had talks with me on possible Bettis participation.

By early 1959, I was again beginning to worry that Bettis needed a challenge beyond building more naval reactors. Since we failed to obtain a development program for gas-cooled central station plants, and Westinghouse and GE were already in the commercial nuclear plant business, it seemed clear that future electric utility reactors wouldn't be developed at Bettis.

Therefore, what now? The naval program still would require much good technical work to be done, but it was not the same as the early *Nautilus* challenge. And if we weren't to meet the challenge of bringing commercial nuclear power to the world, space seemed to be the big challenge remaining.

One evening, in the spring of 1959, Cotter and I were mulling over this question and Frank said, "Why not take the plunge? Get into the nuclear space business."

Westinghouse Enters the Field

The next day, I called Krasik. "Sid," I said, "I would like to have a conceptual design of a nuclear rocket engine."

"How soon do you want it?" he asked.

"How about a week?"

"You'll have it," he promised, and that's where it began.

Krasik and I took his design to Mark Cresap, Westinghouse CEO. Cresap was a decisive man, a man of action. With such a leader at the helm, even a large corporation can move quickly. He bought the idea, and then helped persuade Rickover to try to secure such a project for Bettis. Cresap and Cotter told the admiral that they would arrange to have several of his supporters appointed to key congressional committee posts.

Somewhat to my surprise, Rickover agreed to try to get a nuclear rocket contract for Bettis. But our pleasure in this was short-lived. Rickover backed out a few weeks later. He said the project was too technically risky. He may well have realized that running the development program for a nuclear rocket engine would be very time consuming and that this might detract from what he always considered his primary mission—creating the nuclear Navy. He had been very successful with the Navy program and any lack of success with nuclear rockets might have hurt his ability to get continued support for the Navy. In retrospect, I think he made the wise decision.

This made no difference to Cresap, however. He had the bit in his teeth and forged ahead with the nuclear rocket program for Westinghouse. Of course, Krasik, Cotter, and I were happy to go along. We believed that there

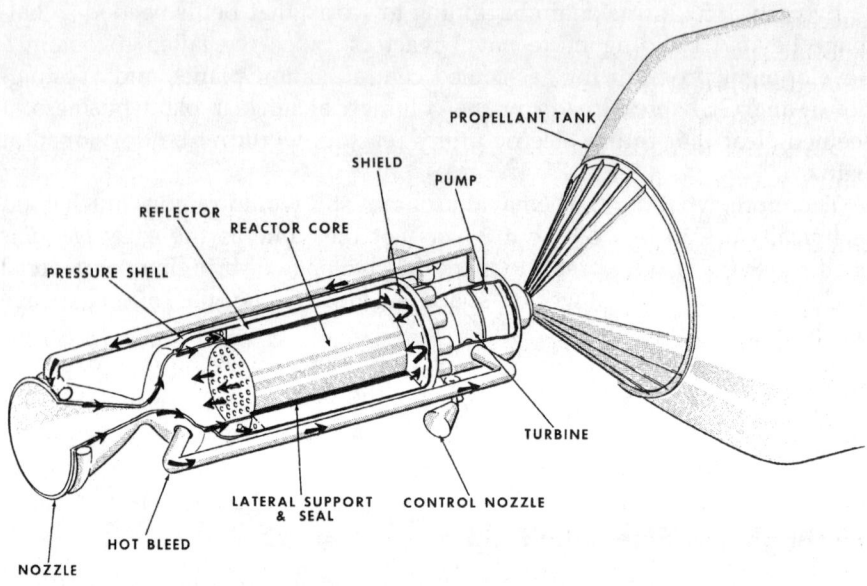

Figure 6.1 Nuclear rocket engine.

was a future for nuclear power in the space program and we were audacious enough to think that we could play a major role in it.

In May 1959, Cresap asked me to leave Bettis and head the project of taking the company into the nuclear space business. I agreed on the condition that he would let me take five men with me and also allow me to run the commercial nuclear activity while still reporting to Weaver.

The five men I asked for were Frank Cotter, Walter Esselman, Lloyd Kramer, Sidney Krasik, and Walter Roman. Without hesitation, all five agreed even though they well realized the gamble they were taking. Richard Cunningham joined the group shortly after it was organized.

Thomas, who headed up the materials department, was not scheduled to be part of the original group but he joined us as soon as the laboratory was formed, having already decided to leave Bettis. He had been offered a job by Bob Gordon, formerly of Bettis, but now at Atomics International. Because he had already made known his intention to leave, there was no further trouble with Rickover over this move

Yes, it was a big gamble for all of us. I was being asked to leave my position as director of the largest government reactor laboratory—at that time it was as large as ORNL and ANL combined, and with a record of success—to take on the extremely difficult job of muscling Westinghouse into the space business.

Actually, right then there wasn't even a nuclear space program to compete for. We would have to try to bring such a program into being.

As to my responsibility for our commercial nuclear power activity, its future was far from clear. We then held the contract for the Yankee Atomic Power Plant at Rowe, Massachusetts, plus a few development contracts, but the economics were worrisome. Shippingport power had cost about 10 times as much as power from conventional plants.

While the cost of power from Yankee Rowe was expected to come down to three or four times that of conventional power, prospects for further reduction depended on some farsighted and optimistic thinking on the part of government and utility executives. This was a commodity traditionally in short supply. Fortunately, for us, it proved to be available this time.

To bring down the cost of nuclear power would require tens of millions of dollars in the development and construction of several generations of plants. Each would cost more than a hundred million dollars and would not be economically competitive. Would anyone be farsighted enough to pay this cost?

The Die Is Cast

The time had come for plotting careful strategy on the Washington front. It hinged around Rickover. The fact that he had vetoed the idea of getting Bettis involved in a nuclear rocket development program made it certain he would oppose what we had in mind, particularly since it would include transferring a half-dozen key people out of Bettis.

We felt it was essential to build some congressional support for our plans before Rickover was advised. Otherwise he would create a furor and torpedo the idea before we could make our plans known. So, in late May, we sent Cotter to Washington to inform key congressional leaders.

Now came the question: Who should advise Rickover about our plans? Mark Cresap believed that Price, the former Westinghouse CEO and then board chairman, would be the best man for the assignment. Of course, there was no contractual prohibition on transferring people out of Bettis and no obligation to notify Rickover beforehand. But we all knew the storm it would cause.

Price undoubtedly was the best choice for this unpleasant task. He had known the admiral for many years and had given the naval program top priority. Because of this relationship, Rickover might go easy with the dignified, highly respected and easy-to-like Westinghouse chairman, right? Wrong.

Price said that never in his life had anyone talked to him the way Rickover did on that occasion. Rickover's reaction can be judged by several statements in the book *Rickover and the Nuclear Navy* for which he undoubtedly was the source.[1] It is claimed there that Westinghouse did not advise Rickover of the move. We did, of course, but not *beforehand*. And it is said that Westinghouse rescinded the move, which we never did. Rickover's initial overt reaction was to take the contract to build the natural circulation reactor and give it to GE's Knolls Atomic Power Laboratory (KAPL).

Rickover Fights Our Decision

Right after the Westinghouse decision was made on this matter and before Price talked to Rickover, I went to the Greenbrier Clinic for my annual physical. When I returned to Pittsburgh Sunday evening, I knew that the exam had shown a growth in my armpit. Happily, surgery later that week found the growth to be benign.

In the interim, however, Rickover was scheduled to come to Bettis that Monday evening for a meeting and we did not know if Price had yet talked to him. It was a tense meeting, to say the least. However, not until after the meeting did he come into my office to express his displeasure about the astronuclear plans and my departure.

After I was transferred from Bettis, he gave orders that I was not to be permitted to enter Bettis under any circumstances. But Rickover was nothing, if not a pragmatist. If he thought he could benefit by doing the opposite of what he had said earlier, he didn't hesitate. Later, when the light water breeder was in trouble and senior government officials were visiting Bettis, he invited me to come to help bolster his position.

Threatened General Dynamics Takeover

Even before I actually left Bettis, Rickover began retaliating for the decision to move my nuclear space team out of Bettis. He informed me that Earl Johnson, president of General Dynamics, the parent company of Electric Boat, would be visiting Bettis to look it over. The intent was to replace Westinghouse as contractor.

Johnson arrived the next morning and toured the laboratory with Lawton Geiger, manager of the AEC's Pittsburgh area office, and me. When the group returned to Geiger's office for a discussion of the Bettis program, I asked to make a few remarks that I addressed to Johnson.

---

[1] Francis Duncan, *Rickover and the Nuclear Navy*, p. 25, Naval Institute Press, Annapolis, MD, 1990.

"There is one thing you should keep in mind," I told him. "The laboratory belongs to the government, but the people are dedicated Westinghouse employees. I am sure most of the key people will not go with General Dynamics. If you want to try a takeover, be my guest."

Johnson later decided not to attempt a takeover, for whatever reason.

Rickover also launched an offensive in Washington. He went to the National Security Council and to many congressional leaders to ask them to put pressure on Westinghouse to back down. The company was faced with the possibility of losing Bettis, and there was a chance that Rickover could gain the power to prevent our transferring people. This would have been a staggering blow to our ability to move forward in the nuclear area.

Cresap Fights Back

To defend against Rickover's Washington offensive, Weaver and Cresap, guided by Cotter, visited Washington almost every day for two weeks to keep the lid on. Cotter had very close relations with Senators Henry Scoop Jackson and Clinton Anderson and with Chet Holifield, Craig Hosmer, Mel Price from the House, and others, including Bobby and Jack Kennedy and Lyndon Johnson. This was very helpful at this time. Fortunately, key people already had been told by us what we intended to do. Among others, they visited Senator Clinton Anderson, who knew and liked Cresap and Weaver. He was sympathetic to our position.

But Rickover scored some points. Cresap received a letter from John McCone, chairman of the AEC, stating that he wanted Westinghouse to know—before starting a new division—that there was no place in the space program for Westinghouse. This, along with Rickover's opposition, was not an auspicious beginning for our new venture.

Astronuclear Laboratory Formed

The Astronuclear Laboratory received its name when it was officially made a Westinghouse division on July 26, 1959. Krasik was appointed technical director and was the lead man in this effort to apply nuclear science and technology to the exploration of space. Cotter became my executive assistant and marketing manager.

There is a big of difference between starting a new organization with a major government contract in your pocket as we did at Bettis and starting with no contract at all as was the case now. When the Bettis Laboratory was established, we had the relative luxury of setting up shop on a former airport complete with a number of buildings that could be readily remodeled. We looked like something important rather quickly. Not so with this new effort. Getting up into space was going to start with operation bootstrap, obviously.

Our first "headquarters" was an office in a shopping center in the Pittsburgh suburb of Whitehall. Then came two moves to larger offices, one above a bar and night club in the same community, then to office space above an A&P store in Castle Shannon, Pennsylvania. This latter location gave us space for expansion—which we sure hoped we would need.

Although Rickover received little support in Washington, in retaliation he canceled the contract for an advanced submarine reactor project and gave it to GE. The *Pittsburgh Press*, on September 3, 1959, reported this in a big headline. The Navy later brought back the operation of the prototype reactor to the Naval Reactors Testing Station in Idaho in the late 1960s. The study of gas-cooled reactors for destroyers was also transferred to GE's KAPL.

The first job, of course, was to win some contracts; even small ones would do in order to develop staff and stay alive until a major contract could be obtained. We competed for every contract in sight, at the same time doing our best to stimulate interest in the application of nuclear energy to the space program.

We had no laboratory. We had no product to sell. Yet we had to be ready to compete for a major NASA development program. We knew that when the time came for that competition we would have to have a sizable and capable organization ready. All of our potential competitors already had large well-staffed laboratories. The ROVER program (AEC/Air Force nuclear rocket) at LASL was well under way, but no program for bringing industry into the picture had been announced. The NERVA project did not become a reality until 1961.

However, on the positive side of the ledger, we did have some important assets. Our group of six people consisted of very talented people—men who were qualified to be division managers in Westinghouse. What we had to sell was, obviously, technical competence and vision. These were people with a track record of major accomplishment. We sold hard on those points. The appeals of our mission also gave us the ability to attract top engineers and scientists when the time came to expand.

We also had strong corporate backing—no small thing in a game that could get very big, very fast.

The Potential for Space Nuclear Power

What was our market for nuclear power in space? Here's how we saw it.

A nuclear rocket engine could make it possible for man to land on the moon and Mars. The Apollo program, which was to get men to the moon and back using chemical rockets, had not yet begun.

A nuclear rocket could be boosted into orbit by a chemical rocket, then the nuclear engine could be started. The vehicle could then go into orbit around the moon. A chemical module could land on the moon and then return to the nuclear-powered vehicle, which would return to an earth orbit. From orbit, the crew could be brought back to earth by a landing module, similar to the lunar-orbiting module.

It was not clear at the time that chemical rockets could get us to Mars, certainly not by a direct flight. With the nuclear rocket engine, however, a direct manned mission to Mars was possible—and with a mission time at least 100 days shorter than with a chemical engine. With a nuclear engine, spacecraft weights of one and a half to two million pounds would be possible. This mission, with such a nuclear engine, is again being contemplated, some 30 years later.

A manned mission to Mars might well be reaching the last frontier. If not impossible, it is certainly highly unlikely that man will ever travel out of our solar system, and the other planets are not very inviting. The nearest star, Alpha Centauri, is light years away. Even with the highest velocity we can conceive of with the current knowledge of physics, it would take more than a lifetime to reach that star and explore its planets, if indeed there are any.

What Makes Rockets Go?[2]

Ever since their invention by Chinese makers of ceremonial fireworks, rockets have relied on fire to make them go. They have been propelled, like the pistons in your car, by the energy liberated when chemicals burn. Progress in Russian and American rocketry since World War II has not altered that fact. Rockets began as fireworks and they are fireworks still!

Now, after a few thousand years of staying the same, this basic truth was about to change. Rockets that do not rely on chemical combustion were proposed, leading to these questions: How will they differ from existing rockets? Why are they needed? The clearest way to answer those questions is for us to review the whole subject of rockets, the old kind and the new.

A rocket propels itself by throwing a stream of particles backward. A gas, whether flaming or not, consists of particles. According to Newton's law, for every action there is an equal and opposite reaction. The rocket is not moved by the stream of particles; it is moved by the reaction to the expelling of the stream of particles.

Stand on a raft in the water and throw rocks at the shore. Your feet will thrust the raft toward the opposite shore. The harder you throw, the greater

---
[2]Based on booklet prepared by Los Alamos Scientific Laboratory and published by the Space Nuclear Propulsion Office, Nevada, June 1964.

the thrust. If you throw all of the rocks at the same velocity, then the thrust will be equally strong whether you throw five 10-pound rocks or fifty 1-pound rocks per minute.

A rocket has to have particles to throw, energy with which to throw them, and some kind of apparatus for the throwing. In other words, it needs a propellant, it needs an energy source to heat and expand the propellant (or a fuel-oxidizer combination), and it needs an engine.

The chemical kind of thrust—the Fourth of July kind—meets these requirements with beautiful simplicity. The engine is a can, releasing energy in the form of heat. The heat expands the gases produced in combustion. These same gases, in expanding and rushing out through the open end, act as the propellant.

The advantages of this system are many. No engine could be simpler, and you get your propellant particles for nothing, so to speak, since they come from the fuel you had to carry anyway. It would be hard to improve on anything so neat—hard, but not impossible.

Theoretically at least, when you have three essentials for rocket propulsion, you also have three areas for improvement. Is there really room for improvement in current fuels? Engines? Propellants?

The best fuel for flying, other things being equal, is the most concentrated one—the one that gives you the largest quantity of available energy per pound. By this standard, uranium-235 is some ten million times better than chemical fuels. Remember, though, that we are being theoretical. The practical picture is less rosy, as we shall see.

What about engines? Engines might be improved by the use of lighter materials or materials that stand high temperatures better, but there is nothing about nuclear propulsion that promises to make such improvements any easier.

How does one improve propellants? For that answer, let's go back to that pond, that raft, and those rocks. A rocket engine ejects millions of propellant particles per second. The lighter those individual particles are, the more speed the engine can give them. Their dimensions are submicroscopic, but particle mass is no less important for being on a miniature scale.

The weight of an object is the pull of gravity on its mass. Weight, therefore, varies with the force of gravity. Mass, on the other hand, is independent of the force of gravity. The mass of an object is the same on the Earth, the Moon, or in outer space. It is equal to its weight divided by the acceleration of gravity. Mass has been called a measure of the quantity of material present. The resistance of a mass to a change in its velocity or direction is called *inertia*.

Since the rocket can throw particles of small mass at higher velocities than particles of great mass, the way to improve propellants is to reduce the individual particle mass. This is not a way of getting something for nothing. The additional thrust will require additional power from the engine, and use up fuel faster. But it turns out that this is a price worth paying. Rocket performance, in terms of what percentage of the take-off weight can be delivered at the destination, is benefitted by the bargain. The best index to the kind of performance just described is something called *specific impulse*, which is a measure of how much good your propellant is doing you—how many pounds of thrust you are getting for each pound-per-second of propellant you spend. Specific impulse increases with exhaust velocity, which goes up as particle mass goes down. Exhaust velocity—and, hence, specific impulse—goes up also with propellant temperature.

So we want particles of small mass. But the exhaust stream of chemical rockets is already so finely divided as to be in the form of individual molecules and atoms, so what more can be done? Chemical rockets cannot use propellants of low molecular mass because they must use the products of fuel combustion. These products are relatively heavy. Even oxygen, without which combustion cannot happen at all, consists of molecules sixteen times heavier than the hydrogen molecule, which itself consists of two hydrogen atoms together.

Thus rockets propelled by the ejection of combustion products have serious limitations, especially for long-range high-payload missions in space. Nuclear rockets do not use combustion products as propellants. They can use hydrogen, the lightest molecule of all, delivered as a working fluid through the coolant channels of a nuclear reactor for expulsion from and propulsion of the flight vehicle and they can perhaps even pry the twin atoms of the hydrogen molecule apart—although nobody is counting heavily on this bonus.

First Years at Astronuclear

Krasik was the director of the Astronuclear Laboratory, and Bob Wells was the manager of the atomic power division, the commercial nuclear power activity. Both of these men reported to me. Bob Wells was soon transferred to Westinghouse headquarters. I then ran the atomic power division directly for a while in order to get to know the people and programs more quickly. However, most of my time and efforts were devoted to Astronuclear. Dr. W.E. Johnson was made manager of the atomic power division in 1962.

The eight of us, with a few clerical employees, then got down to work. Since we were in such unpretentious quarters, we spent a little extra on furniture. This let us put up a front when we had visitors.

The Basic Business Plan

Our principal objectives were simple to understand. We wanted to help persuade NASA to go ahead with a major program for a nuclear rocket engine. We also wanted the contract to develop and build the nuclear reactor, which would be the heart of that engine, if not the complete engine itself. Everything else we would do would be supportive of those two objectives.

Although I had ultimate responsibility for everything, Cotter and I orchestrated the marketing and Krasik concentrated on the technical side. Of course, the dividing line was faint many times. We had to call on the technical people in the marketing effort since what we were really selling was technical competence.

One of Cotter's functions was to help write speeches for various important people. In his job as staff director of the JCAE, he had done a lot of this for various members of Congress and now he continued to give this help when called on, even doing some volunteering. In each case, he succeeded in inserting a strong plug for a bigger nuclear space program, including a rocket engine.

We needed all the exposure we could get, and one good way to get it was testifying before congressional committees. In hearings you get a chance to make statements that are widely circulated and can perhaps influence the committee to move in the desired direction. We also visited every government office having anything to do with nuclear energy in space and any government agency where our expertise fit. We also visited all major space contractors to persuade them that we would be an asset on any team they were putting together, and that we could help them as a subcontractor.

We were pretty well known by congressional leaders and government agencies as a result of our work at Bettis. I had met with and testified before congressional committees many times during my Bettis days, but among the major defense and space contractors we were an unknown quantity. The only thing that even got us in the door was the reputation of Bettis. Visiting contractors was an intimidating experience. They each had several thousand technical employees, major computing and laboratory facilities, and large development contracts. We tried not to think about our size and facilities and just charged ahead.

They also were doing development work that would lead to future contracts. These programs were largely funded by the percentage of contracts set aside for them to spend on independent research and development. The AEC did not have a similar provision in its contracts. In any event, we were a bootstrapping operation and as we had no contracts, we would not have had any independent R&D money anyway.

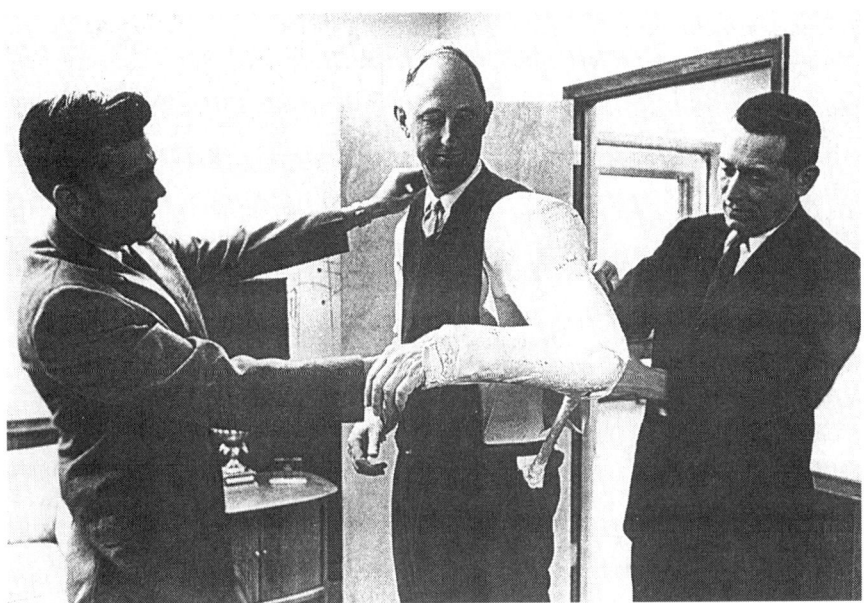

Figure 6.2 Simpson prepares to testify about nuclear power in space before a congressional committee.

In the fall of 1959, Cotter persuaded the JCAE to hold hearings on nuclear power for space and arranged for me to testify. When I entered the room, Ken Davis, head of the AEC's reactor development division asked me, "What are you plumbers doing at this hearing?" That was the AEC attitude concerning Astronuclear.

You had to use devious tactics to get in your two cents worth at those congressional hearings. For example, I was asked to testify, along with engineers from all the principal space companies, before the Space Committee of Congress on March 11, 1961. The engineers were giving their opinions on the space nuclear rocket program. My appearance suffered, and so did I, from a skiing accident in which I had broken a shoulder. I had to wear a full body cast from neck to below the hips. There was a stick from my hip to my elbow to support my left arm. What a mess.

Well, it was a panel hearing, and everyone was grabbing for the microphone to get their pitches in. The president of North American Aviation was sitting to my left and I soon discovered that when I leaned my superstructure to the left a little bit, he couldn't reach the mike. Strangely enough I got to speak more than he did.

Then at the evening cocktail party in the Capitol building, most of the committee members were able to remember my name very readily even though I didn't know most of them. I was the guy in the body cast. Our PR people wanted me to continue wearing the cast indefinitely. It's an ill wind that blows nobody good. But there were drawbacks. The cast was so heavy that I couldn't get out of bed without help. I had to call for a bellman to help me out of bed the next morning.

Building Staff and Getting Contracts

Astronuclear's plan was to develop a top-notch technical organization that could successfully compete and develop large noncommercial nuclear power projects. Our objectives included the capability to be a leader in nuclear space and aerospace propulsion, space power generation, and other compact power systems. Selected projects were sought to build the Astronuclear expertise in the basic disciplines and skills required for a well-rounded team.

The early days of the laboratory were challenging because everyone had to expand their technical expertise. The technical horizon of each staff member had to expand to cover a wide range of engineering and scientific disciplines. Expertise in technologies such as rocket engine design, cryogenics, aerodynamics, nozzles, and materials for both extremely high and extremely low temperatures had to be added to the reactor development background gained at Bettis.

We brought in engineers and scientists whose backgrounds complemented those of existing staff. The additional people gave us the capability to probe more deeply into all aspects of a nuclear rocket development program. We were unsuccessful on a proposal submitted for a SNAP-type space electric generation system.

Then we began to land some small contracts. Esselman and Roman were successful in obtaining an Air Force contract, which included the conceptual design of several nuclear engines. This project gave us about one year's support for the study of potential nuclear rocket engine applications and its scope included engine design, shielding, control, and operational aspects of a nuclear stage. Another small contract with the Point Mugu Naval Station included consideration of the water launching.

Significant help on the design of the engine components was made available from other Westinghouse divisions. Consultation with R.P. Kroon, engineering manager of the aviation gas turbine division, on the design of the turbopump was particularly valuable. He was the designer of the first United States aircraft jet engine.

Westinghouse was the source of many additions to the Astronuclear staff. The aviation gas turbine division supplied engineers with jet engine design experience. They included J. Kenny, mechanical design; F. Retallick, analytical expertise; H. Faught, program management; David Goldberg, materials engineering; Arnold Redding, engine design; Robert Weiler, engine test; and F. Henning, quality assurance. Hank McCreary came from the British submarine *Dreadnought* project, and Nick Pollack and Sam Cerni from Bettis. All these people played key roles in the future NERVA program.

Early transfers from the atomic power division included initial thermoelectric know-how from Joe Danko, Gerry Kilp, and Bill Blankenship, and welding engineering capability from Gerry Lessman.

Early on we also received significant assistance from other company units. J.H. Bechtold of the research and development center had in his division a small group of metallurgists under R.T. Begley. They had several contracts to develop refractory-metal alloys for high-temperature service beginning in 1958. Bechtold transferred the contract and Begley, R.L. Ammon, and L. France to Astronuclear, but allowed the group to function at the center until our facilities were in place. This work resulted in the development of new tantalum- and niobium-based alloys. The program continued until about 1970, and the alloys found uses in the space program.

Westinghouse's Blairsville, Pennsylvania, group, which was being phased out as an alloy and process developer, turned over to us a contract to operate the Wright Paterson Air Force Base (WPAFB) extrusion facility. This proved to be a good window on the base's materials activities.

We were fortunate to find Larry Moberly of the research and development center. Since the war days, he had been involved in collaboration with Stackpole Carbon in the development of graphite motor and generator brushes for high-altitude aircraft. His background helped us understand the processing of graphite for the NERVA fuel element.

The materials department program is a good example of how we were preparing ourselves for the future. D.E. Thomas[3] was manager of the department until 1966, serving both NERVA and non-NERVA efforts. The senior managers who constituted the initial Astronuclear staff were all solidly behind the concept that a strong in-house materials capability would be necessary to support the contemplated range of engineering development projects. They had learned this from their Bettis experience. Thomas, with no staff initially, set out to build the materials department.

A survey was made of the types of materials the laboratory would be using. It pointed to fissile materials and radioactive isotopes capable of withstanding high temperatures, structural materials of the refractory-metal class capable of high temperatures and liquid metal environments, and

---

[3]D.E. Thomas, Private Communication dated July 8, 1991.

graphite-based materials for high-temperature fuel elements as the main areas of concern.

Efforts of the materials department could be classified into three categories: (1) materials support of the NERVA proposal and contract, (2) materials support of non-NERVA engineering work, and (3) contracts to develop materials technology seen to lead to future system and component work. Obviously, the first order of business was to support the NERVA preproposal and proposal effort. Thomas was assisted in this effort by Al Boltax, Dave Goldberg, Al Hoppe, and others.

While the primary goal of Astronuclear was the application of nuclear power to space, other projects were pursued in the interest of keeping the laboratory alive and the people challenged. Cunningham conducted a program on special reactors for Navy applications. Lloyd Kramer was working on the Pluto Ram Jet concept. John Guzek also assisted Westinghouse at East Pittsburgh in the design of a nozzle of a plasma arc installation for the Tullahoma Wind Tunnel.

The first major project we tried to obtain was the Supersonic Low Altitude Missile (SLAM). This work had been going on as Project PLUTO at the Lawrence Livermore National Laboratory (LLNL). It involved a supersonic nuclear ramjet-powered missile that could cruise almost indefinitely, at about a hundred feet above ground, thus escaping radar detection. This was possible because the topography of the entire area of operation was stored in its computer. It would carry multiple nuclear warheads, each assigned a target. It could cruise for long periods around the perimeter of the Soviet Union and could be ordered to attack or recalled to its base.

We made the proposal to Brigadier General Irving Branch, chief of the aircraft nuclear propulsion office who reported both to the AEC and the Air Force. He assured us we would receive the contract for the nuclear engine. However, the Air Force canceled the project as we were about to get the contract.

The National Nuclear Rocket Program

The development of fission reactors for the naval program and civilian electric power applications quickly focused on the less complex technology of the thermal-spectrum, water-cooled reactors. Considerable interest and attention at other laboratories continued to be directed toward the characteristics and application of alternative reactor concepts.

Substantial R&D was performed on the engineering feasibility of gas-cooled and liquid-metal-cooled reactors as well as on the physics and safety characteristics of epithermal- and fast-spectrum cores. Technological progress on these advanced concepts was sufficiently encouraging by the mid-1950s that major R&D programs were established. Their purpose was to apply these technologies to a variety of sophisticated missions, including aircraft engines and space power and propulsion systems.

The nuclear rocket program ROVER was started in 1955 as a joint effort of the AEC and the Air Force with early efforts located at both LASL and LLNL. The original intent was for these rockets to power ICBMs. When chemical rockets proved capable of those missions, the Air Force lost interest and withdrew its support. Development continued at LASL under AEC auspices. When NASA was established in 1958, the former Air Force responsibilities were transferred to it because of the potential for space missions.

In March 1957, a specific R&D approach was selected. The AEC decided to proceed with fabrication and testing of reactors using uranium-loaded graphite fuel elements to heat hydrogen to temperatures useful for rocket propulsion.

The first experimental reactors were developed by LASL and were called KIWIs, named for a flightless bird, because in this program the reactor was not intended to fly. Several KIWIs were built to demonstrate feasibility and prove the nuclear rocket engine principle.

To get flight-type engine test data on a reactor, four study contracts[4] were awarded in the airframe industry in September 1960 after the successful completion of two KIWI-A gas-cooled reactor tests by LASL. Complete design and test of RIFT (Reactor-In Fight Test) were to occur after we had brought NERVA through the engine-testing phase. It was to be launched by a Saturn booster to obtain nuclear-engine operational and design information. Later, the RIFT tests were contracted to Lockheed, but RIFT was canceled before it got off the ground owing to budget restrictions on NASA. All of the study contractors recommended the use of a Saturn-class launch vehicle and a test of the nuclear stage in a ballistic trajectory.

The NERVA Project[5]

Space Nuclear Propulsion Office[6]

America finally got serious about the potential of nuclear power in space in 1961, when NASA issued a request for proposals on the NERVA engine.

---

[4]H.R. Schmidt, "The Nuclear Rocket Engine and Flight Program," *IRE Trans. Nucl. Sci.,* Vol. NS-9, No. 1, January 15, 1962.
[5]Much of the NERVA material was furnished by D.E. Thomas, W.H. Esselman, and W.H. Arnold, Private Communications.
[6]H.B. Finger, Private Communications dated September 16, 1991, and February 28, 1992.

NERVA stands for Nuclear Engine for Rocket Vehicle Applications—a name undoubtedly chosen by a committee.

The ROVER and NERVA programs were to be closely meshed, each reactor experiment building on the knowledge gained in preceding tests in order to evolve a continuing improvement in technology.

The project was to be administered by the Space Nuclear Propulsion Office (SNPO), which was a joint office of the AEC and NASA set up in 1960. If that sounds like the joint AEC-Navy office that Rickover headed to develop the nuclear-powered submarine, it should. Glennan of NASA and McCone of the AEC deliberately patterned it after the Rickover operation.

But they did not put a Rickover in charge. They chose a quite different, but capable manager named Harold Finger from NASA's Lewis Laboratory to head SNPO. His deputy would be Milton Klein from the AEC's Argonne National Laboratory.

The roles and responsibilities of AEC and NASA were assigned to the SNPO. In this arrangement, it was clear that the AEC's responsibility was for reactor and nuclear aspects of the system. NASA's was for the overall engine system, including developing the non-nuclear subsystems such as turbo pumps, jet nozzles, and engine controls and integrating them with the reactor and its controls. NASA also had responsibility for integrating the engine into the flight-vehicle stage and developing mission requirements. SNPO developed agency budgets in accordance with these defined responsibilities.

We felt such a setup was desirable since it had served Rickover so well. Due to our experience at Bettis, several JCAE members asked our opinion, and we encouraged this at every opportunity.

This office had line authority over the Cleveland extension, SNPO-C; Albuquerque extension, SNPO-A; and the Nevada extension, SNPO-N. It exerted technical direction over the program work at LASL and the support work at the Marshall Space Flight Center and the Lewis Research Center in Cleveland.

Nuclear power in space now had a head. Our job was clearly focused—to compete for the NERVA contract with every ounce of energy and every bit of knowledge we could muster.

Preparing the NERVA Proposal

This is what the Astronuclear Laboratory had been preparing for since its inception. Several years of design studies, detailed analysis and materials development and study for NERVA-type nuclear rocket engines were aimed at preparing a proposal that would beat competition. This effort prior to a request for proposal was an important part of our success.

The prospect of nuclear power in space was an exciting one, so we were able to attract exceptionally good technical people. Consequently, the staff expanded rapidly, reaching nearly 100 for the writing of the NERVA proposal, 184 by the end of 1961, and peaked at 1994 in September 1963. The RFP was issued February 1, 1961, and proposals were due by April 1961.

I had picked men who were capable of being division managers and who also could perform the necessary engineering and write proposals themselves. It is not at all unusual for a good technical man, who becomes a manager, to spend so much time on administrative duties that he loses much of his ability to do technical work. By the time the RFP was issued, we had at least two years of solid design studies on essentially all aspects of the engine and graphite-core reactors.

Who were our competitors? Oh, just a few "fly-by-night" companies such as GE, Pratt & Whitney, North American Aviation, Aerojet General, and Thiokol. These major companies in the defense and space field had large engineering organizations and were experienced in the general field. We knew little about rockets or space vehicles, but we had one area of expertise greater than the others—nuclear technology. In that respect, Westinghouse was number one.

Believe me, we pulled out all the stops—not only technical effort but also marketing and political savvy. This was, after all, a government contract. Cresap was great; he was a partner in getting us started and he also wanted to win. He gave us a budget for preparation of the proposal and several times I had to go back for more money. I always got it. But he never forgot the bottom line; with his last approval, he sent me a letter stating, "I have given you everything you asked for, just win."

Once after that, Krasik told me we were overrunning our budget again. I said, "Sid, go ahead and spend the money. If we win, nobody will care. If we lose, we will be down the tube anyway."

The proposal was the most elaborate Westinghouse had ever put together. It cost $250,000 to prepare the proposal including $40,000 just to print it. This was on top of the expense of setting up Astronuclear, which came to more than a million dollars. In those days that was a lot of money, especially for a fledgling organization.

We had a great team on the job. Krasik, in addition to being the laboratory technical director, handled reactor physics. Esselman was engineering manager, and Thomas was the materials department manager. The whole laboratory of about 100 people worked on the proposal. Detailed designs were developed for each part of the engine and there was a carefully crafted development and management plan.

In addition to the staff members I've already named, there were other notable contributors to the proposal: Frankie Frisch from the atomic power division with his extensive mechanical design experience; Dick Cunningham and Al Bethel contributed to the management and development plans; Don Drawbaugh, physics; Sam Cerni, Anthony Bournia, and Joseph Weaver, reactor thermal design; David McCutchan and Dick Doncals, reactor nuclear design; Hank McCreary, systems design; Fred Henning, quality assurance; Frank Spurrier and Jack Fisch, mechanical design; Paul Blake, reactor control; Arnold Redding, engine design; Nick Pollack, fuel processing and Al Rothman, flight safety. An all-star lineup!

Then came the big day. Krasik and I made the presentation to the evaluation board in Washington and Sid did a masterful job. He had already convinced Finger of his technical competence, and his presentation convinced the other evaluation board members.

After this top-level presentation to the evaluation board, there was an in-depth review by NASA and AEC, to make sure that we didn't just have a smooth-talking top team, with a thin veneer of competence. SNPO had some 30 reviewers questioning Krasik and some 10 of our key technical people. Yes, Sid Krasik and Al Rothman went over the details of everybody's presentation until about 3:00 a.m. the day of the presentation. Again, Krasik was great and the rest of the team demonstrated technical knowledge of nuclear rocket reactors in depth.

Winning the NERVA Contract

The results of the evaluation board showed how close the competition was.

Rocketdyne Division of North American Aviation placed first for the entire system. Aerojet placed first for the engine system, and Astronuclear placed first for the nuclear reactor. Finger and the evaluation board felt that there were two rather distinct parts to NERVA—the nuclear reactor and the overall engine, which included non-reactor components such as pumps, nozzle, and controls.

While coordination would be required between these two parts, they would be handled by separate departments, even if one contractor got the whole thing. So the board decided to place the contracts with Aerojet as prime contractor and Astronuclear as a subcontractor for the nuclear reactor.

That decision was fine with us, since the nuclear reactor was the key to the rocket. We received the NERVA contract in June 1961, and there was cause for celebration and a great feeling of relief. A lot of us, from Cresap on down, had put our reputations on the line.

My staff and I took a real deep breath after the cheering stopped. Our careers would have been seriously set back if we had not won the contract, since there were to be no second chances—no other large nuclear space contracts. For me personally there would have been the stigma that I was just Rickover's man and without him could not succeed. But now I gained unusual prominence in the corporation as a repeat winner. There's nothing like pulling off a winner, when the boss's reputation is also on the line.

Of course, the key to success was the staff I picked at the beginning. This may have been the first time Westinghouse started a new activity with five or six senior men—a vice president and five potential division managers and no one else. One senior chief and seven, even excellent, Indians would have failed. Too often corporations assign personnel to programs based on their current size, not on potential. The Astronuclear Laboratory eventually grew to more than 1900 people.

The other key to our success, I think, was that we were the only group in the country with a winning track record in nuclear power. The organic-moderated, the sodium-cooled, and the homogeneous reactor development programs had met with limited success, and the nuclear-powered airplane project had been terminated. The Navy-proven light water reactors in submarines, aircraft carriers, and Shippingport were unqualified successes.

The contract with Aerojet was unusual in several respects. There was to be direct communication between Astronuclear and SNPO and its various offices. SNPO was to evaluate both companies for incentive awards. The SNPO-C project office managed the activities of the two contractors, Aerojet and Westinghouse.

Finger's management style was quite different from Rickover's, but it was equally effective. Finger had agreed to inform Rickover before hiring any former Bettis people, and in the spirit of that agreement, he notified Rickover of the decision by AEC and NASA to award the nuclear portion of NERVA to Astronuclear. Rickover asked Finger if there was a clause in the contract prohibiting Westinghouse from moving anyone from Bettis to Astronuclear. Finger said no and asked Rickover if he had included such a clause in his contract with Bettis.

"No," replied Rickover, "but you should have such a clause in your contract." Finger said he thought that wouldn't be appropriate. However, there was a clause in the contract covering key employees.

Finger did not face the bureaucratic and political problems Rickover did in getting the submarine program authorized, but in the end, his program didn't make it all the way to outer space as Rickover's did to sea. It is most unfortunate that, although NERVA was an outstanding technical success, the program was canceled before its completion with a flight vehicle.

Although Finger's staff at SNPO came from three different agencies—NASA, AEC, and USAF—the people worked well together as an effective management team. Their allegiance was to the program, and the office was fully accepted by NASA and AEC and the congressional committees as being responsible and faithfully representing each agency.

Astronuclear Finally Gets a Home

The Astronuclear employees had a joyous summer of 1961. The day the NERVA contract was announced we came out of the blocks at full speed on all aspects of the project. First priority was to establish a laboratory.

We were able to acquire a site at Large, Pennsylvania, which was formerly the Old Overholt distillery. Parts of it had been used for overflow of Bettis activities. The buildings were old, but with modifications could be made adequate for our needs. In November, we moved to Large with a staff of about 150. Most of the new people were from other Westinghouse divisions. Later additions were mostly from outside the company. By 1963, the staff had grown to about 1100 people, and we were going strong.

One of the major problems during the fast growth period was that NERVA tasks had to be accomplished while we were hiring people. This was handled largely by letting out various contracts to subcontractors and consultants, including Westinghouse Research Laboratories for studies on beryllium combustion by Paul Snyder and Ed Boes and graphite behavior by Mike Manjoine.

Much of the early flight safety work was subcontracted to Arthur D. Little for including beryllium samples in a simulated Saturn spill with a toppling of the nuclear stage; to Cornell Aeronautical Laboratory for the study of the ablation of fragments of graphite during reentry; and to Combustion Explosives Research & Picatinny Arsenal for methods of intentionally destroying the NERVA core before reentry.

Hiring of staff was also aided by people being made available due to the shutdown of the Westinghouse Testing Reactor and the Aircraft Nuclear Propulsion Project.

Extensive modifications to the buildings had equipped us with facilities for materials, mechanical testing, high-temperature fuels testing, reactor safety, and control-drive laboratories, fuel fabrication, and reactor assembly. Significant space was renovated for computer facilities, both digital and analog. Up to that time all simulation and control studies were performed with analog computers.

During the next year, facilities were in operation to perform mechanical property and materials testing of metals and graphite, from cryogenic to high-temperature conditions. A test cell was built to heat fuel elements electrically, simulating in-core temperature and flow conditions. Component test capabilities included those for vibration, reactor safety, and control-drum drives. Facilities to develop and produce fuel elements were constructed at Cheswick, Pennsylvania, and a reactor assembly building was built at Large. A reactor critical capability was set up at Waltz Mill, Pennsylvania. Simultaneous with the establishment of the Large facility, a test organization was established at Jackass Flats, Nevada, and staff was being assigned to LASL.

Old Overholt would never have recognized the place, believe me. Not a keg or bottle to be seen. Much credit for the smooth transition and organizing of these laboratories goes to Daniel Wolf, who was in charge of Astronuclear administrative functions.

Organizing for Action

Finger exercised a lot of ingenuity in organizing his resources for NERVA. He established a joint organization, involving both Aerojet-General and Astronuclear. For us, Krasik was responsible for overall technical effort and Esselman was reactor development manager. Roman was his deputy.

Aiming for a fast start, we moved to increase staff as quickly as possible. Esselman and Roman directed the design effort. Joe Kenny headed mechanical design and Frank Retallick analysis. However, more key people were required.

We started an immediate recruiting drive. Technical society meetings were good sources of recruitment. Astronuclear had a challenging story to tell, so there was great interest in the project. Sid Krasik's philosophy was to get the best people possible. He insisted that we not hire a good and qualified man to fill a real need, but to hold out for a superior individual. Krasik also personally interviewed all of the key people, at least in the early days of the laboratory. As a result, well-qualified people joined Astronuclear from all types of organizations. He reviewed all of the staffs performance evaluations with their section managers yearly.

Our recruitment turned heavily toward Westinghouse divisions. Mike Manjoine was recruited from the Westinghouse Research Laboratories, and used his years of experience to set up a mechanical testing laboratory.

We were growing fast.

Howard Arnold joined the NERVA team as deputy manager for analysis on March 1, 1962, after a short period with Nuclear Utility Services. In this position he directed analytical efforts throughout the project. About the same time, William Jacobi came on board to head an advanced studies group. Flexibility and a team approach were essential ingredients during this growth period. Each person had to fill as wide a range of disciplines as possible.

It had been evident from the start that we would need someone to direct the entire reactor testing program, and we were fortunate to be able to recruit Max Johnson. Max had been general manager of the Westinghouse Test Reactor (WTR), and also had 10 years of operating experience at the National Reactor Testing Station in Idaho. When the WTR was closed down, Johnson became deputy manager for test operations. In this job he was required to build and direct staffs at two widely separated sites, Jackass Flats and Pittsburgh, where design of the test car and core handling equipment was undertaken.

Johnson developed his test organization around a nucleus of people who had been salted into the Los Alamos test site for training. Perry Davison, who had built the Westinghouse reactor evaluation center team from scratch, was particularly impressive. He became our ace in the hole in all future turf wars with Aerojet at the site.

Esselman, as manager of engineering, had three deputy managers: Roman for design, Arnold for analysis, and Max Johnson for test operations. This team directed NERVA engineering and development during the years of growth. Thomas continued to direct materials and fuel element development efforts. Prime requirements during the first NERVA years were a capable staff and flexibility to keep up with many changes.

Rapid growth also creates management problems. As the NERVA staff grew from 50 to more than 1200 in 18 months, some managers found themselves overloaded. So Krasik suggested the deputy manager arrangement. As each deputy identified a need, he would pitch in and see that it was addressed in a high-quality way. Walt Roman was a steadying influence on the younger managers and he taught them how to delegate work assignments.

Esselman recalls, "Personally, I was fortunate to have Roman as a deputy at the beginning of the project. Frank Retallick worked on any problem requiring analysis. When Howard Arnold joined Astronuclear he provided a strength on all of the analyses. Johnson relieved me of the almost-weekly trips I was taking to the Nevada test facility." Even with this arrangement, Esselman had to stretch his technical expertise to cover many of the gaps.

With fast-changing needs—conceptual design, testing facilities, fuel facilities, manufacturing-engineering capability, and core assembly facilities—new skills were required at almost every step. Esselman felt that the three-deputy-manager organization resulted in a team approach that got the job done with a minimum amount of internal rivalries.

Joe Gallagher from ALCO became our mainstay on nuclear and thermal analysis. ALCO was also the source of Jack Ravetz, who played such an important role in our nuclear analysis. On the test side, Dave Bolender joined us from the USS *Savannah* project. A substantial number of staff came from the nuclear-aircraft project. The first flight safety manager was Al Rothman, who wrote the flight safety section of the NERVA proposal and went on to build the staff up to 17 engineers. One of the most talented engineers he hired was Joanne Bridges, a Ph.D. chemist from Gulf Research Laboratories. Howard Kraig, who headed our flight safety program later was from General Atomics. When he left, Joanne Bridges was made manager of flight safety. A female manager was rare in U.S. industry at that time.

Ed DeZubay, who had earlier been at the Westinghouse Research Laboratories, returned from Curtis Wright. He used his background in heat transfer and fluid flow to set up the fuel element testing facility. This facility would regularly put 1 MW of electrical energy in a fuel element to test the integrity of the coatings. A larger high-temperature test facility was built at Waltz Mill for experimental purposes. J. Holmgren, who came from Vitro Laboratories, made important contributions to these testing capabilities. Dick Thomas, who later became head of systems engineering, came from Babcock & Wilcox.

The philosophy of detailed conceptual design followed by a thorough analytical and component test program reflected the systems approach we learned at Bettis. This also matched the SNPO perception of how a project should be conducted.

Everyone agreed that the enthusiasm evident in this project had been felt by the participants only once before—during the first few years at Bettis. It was a family affair and the tone of this creative positive mental attitude was due largely to Sid Krasik.

The Challenge of NERVA

While the extensive Westinghouse background in reactor development provided a solid base for the NERVA project, it was immediately evident that many new technological problems would challenge our skill and experience in building a nuclear propulsion reactor to operate in space. The technical story of how these challenges were met is described in Chapter 14.

Among them, high operating temperatures in a potentially corrosive atmosphere, extreme variations in temperature across the reactor, and rapid temperature changes were foremost, making this reactor design a formidable task. The interplay of fuel, coatings, and reactor operating temperatures taxed the capabilities of a broad range of material specialists and reactor designers.

Development of this reactor would require closely coordinated programs for design, analysis, experiments, and testing. As in the Bettis naval reactor program, our development philosophy called for testing successively the

Figure 6.3 Close-up view of the NRX-A2 thrust nozzle.

components, subsystems, and systems before testing a complete reactor. Considering the cost of a test that includes an entire operating reactor, we did not want to discover problems at that expensive level that could have been found through careful testing of parts and components. Three particular areas appeared to offer the principal challenges. We had to make significant progress toward solutions there. We were gratifyingly successful. These areas were:

1. *Structural, nuclear, and thermal design to enable the reactor to withstand a hostile environment:* We had to produce a reactor that would operate in extremely severe conditions. It had to withstand such shocks and stresses as startup from vacuum conditions, thermal shocks from the initial liquid hydrogen flow, and rapid increases to temperatures of 2500°C. The most extreme conditions for any machine ever built by man.

2. *Development of fuel that could survive in high-temperature hydrogen:* The fuel elements had to maintain their integrity in corrosive and high-temperature hydrogen conditions. We quickly realized the magnitude of this problem, when initial graphite fuel elements survived only a few minutes of hydrogen exposure in a high-temperature, electrically heated test.

3. *Understanding reactor dynamics and control characteristics to achieve startups from source level to full power in about 60 seconds:* The question was what would happen during startup as the liquid hydrogen entered the nozzle and proceeded to the reactor. What would be the condition of hydrogen as it entered the reactor core, and what would be the resultant reactivity effects? These factors all had to do with reactor performance and control.

Howard Arnold remembers his initial briefing from Sid Krasik, when he came on board as Walt Esselman's deputy for analysis. Sid said that the most challenging analysis problem would be to work out the space-time kinetics of the reactor, subject as it was to the large reactivity feedback of the hydrogen whose density would change rapidly as it changed from liquid to gas. After much work on sophisticated models, it later became clear that simple point models adequately represented the reactor's dynamics on an analog-hybrid computer. However, sophisticated nuclear and thermal hydraulic models were necessary to get detailed power distributions and tailor flow orifices.

To judge the progress made in overcoming these and other many difficult problems, consider a base point of early 1964. At that time, no reactor had been operated at the high-power and high-temperature conditions needed to achieve the NERVA performance goal. A number of feasibility questions still existed. The main one concerned structural integrity during high-temperature and high-power operating conditions. A goal had been set that the engine would operate successfully for 20 minutes at full power and temperature.

A measure of the challenge of the reactor's thermal design can be realized by the fact that the reactor core had power densities of more than 1 kW/cm^3 and heat fluxes 10 times those experienced in water reactors.

Coupling these conditions with the large temperature rise across the reactor of 2222 K/4000°F introduced unique problems.

For example, a power density variation of only 5% results in a 200°F change in outlet temperature. A change of this magnitude would have a significant effect on performance of fuel elements and reactor components. These factors, coupled with the closeness of fuel material operating temperature to its physical limits, required much attention to the detailed thermal and nuclear design. Statistical methods were used to calculate hot-spot temperatures, and fuel elements were selected for each position in the core.

In retrospect, these technical problems were more difficult and technically sophisticated than those of the naval program, but at Bettis we were pioneering completely uncharted waters.

Figure 6.4 Sid Krasik (at right) receives a commemorative photograph of the NRX-A2 test from Walt Esselman. The photograph is signed by members of the test team.

The initial few months of the project were devoted to studying alternative design concepts of the ROVER KIWI-B4 technology that were most adaptable to the needs of the flight engine. In November 1961, a proposal was made to SNPO to proceed with the tie rod design for the NRX-A1. Studies continued on the other two alternative support systems, but they were never tested. SNPO approved the proposed approach. The resultant design was released for fabrication November 15, 1962, seventeen months after start of the project.

The tragic death of the brilliant Sidney Krasik from cancer in 1965 was the big negative of the Astronuclear years. He became so seriously ill in 1964 that he could not run the laboratory and was appointed consultant instead to the Atomic Defense & Space Group, reporting to Vice President Weaver, because we needed his advice. He died October 18, 1965. W.E. Johnson, who had been general manager of the atomic power division, was appointed vice president in charge of Astronuclear to succeed him. Woody performed in excellent fashion, as did his later successors, but Krasik was an irreplaceable individual. His reputation for technical competence was a big factor in our winning the NERVA contract.

We had a lot of excellent help and cooperation from Los Alamos about which much could be written, because it contributed much to the success of the NERVA program. Its contributions included improved fuel elements, fuel supports, and reentrant tie tubes to replace tie rods. LASL carried out a number of successful tests and made significant contributions to the training of the Westinghouse people.

Continuing Development at LASL

While the NERVA team was developing the flight-type system. LASL continued to extend technology to higher power density, higher power, and higher fuel temperatures and operating durations. Its efforts on the ROVER program continued with development of advanced concepts and technology to improve reactor performance.

LASL designed a series of reactor tests called Phoebus, which had the goal of operating between 4000 and 5000 MW. Such a reactor could produce a thrust of 200,000 pounds and could accomplish more ambitious missions with shorter run times.

The first of this series was Phoebus 1B operated at about 1500 MW. It served as a test of design features to be used in Phoebus 2A. This reactor, tested in July 1968, achieved a power level of 4200 MW. Total running time at full power was 12.5 minutes, cut short by exhaustion of the liquid hydrogen supply. Considerable time was needed for performance mapping and other experiments. Available liquid hydrogen at the test site became a consideration when testing reactors of this size.

At the beginning, working relations with LASL had to be established as quickly as possible. Shortly after the project started, Astronuclear sent 10 engineers to the LASL to work with their staff for about a year. This step proved to be extremely worthwhile—both from a technical and personal interrelationship viewpoint. Good friendships were established that were of continuing value throughout the project. Later about another 10 engineers were assigned to LASL.

Our key LASL contacts were Raemer Schrieber, who started the ROVER program and later became associate laboratory director; Rod Spence, who was head of the reactor development program; Keith Boyer who was head of the LASL site test program; Frank Durham, who was the main contact on the mechanical design; Mel Bowman, who was our main consultant on coatings and carbides; and Jim Taub, who was very active in fuel element processing.

NERVA Test Program[7]

Countdown began about two years before a NERVA test date. This process began with the post-mortem examination of the previous test. Questions abounded: What design changes should be made? How did the fuel elements perform? Did reactor control perform as predicted? Can the next run time be extended?

During this period, designs were released, components fabricated, and the reactor assembled. The Nevada team assembled the test car and prepared the facilities. As the run date neared, scheduled steps shortened to days, minutes, seconds and finally startup. NRX tests were conducted with the reactor mounted in an up-firing position. NRX test dates and progress are shown in Table 6.1.

NRX-A2 was a proof of principle. NRX-A3 was marked by a disastrous beginning. Startup proceeded smoothly but after three minutes at 100% power a complete loss of hydrogen flow occurred—we had just had our first experience with a full-power scram. Thermal conditions in the core and nozzle were as severe as might be imagined. The only encouraging note was that no parts were seen flying out of the nozzle. It was a sad day at Jackass Flats.

Investigation showed that the scram was caused by a loose connection in a turbine overspeed circuit. During the next week Frank Retallick, Dick Thomas, and Joe Gallagher analyzed the reactor transient and concluded

---

[7]"Technical Summary Report of NERVA Program," TNR-230, Vols. I–VI, Westinghouse Astronuclear Laboratory, Pittsburgh, PA, July 15, 1972.

Table 6.1
NERVA Test Series

Date	Test	Results
	NRX-A1	Bundled core concept did not experience a flow-induced vibration.
9/24/64	NRX-A2	Proof of principle; six minutes at power. Control tests: frequency response, fixed drum flow control 30 to 60 MW.
4/23/65	NRX-A3	Scram after three minutes at 100% power; restarted and ran for about 16 minutes at full power. Restarted for three minutes at full power. Control tests: fixed drum flow control tests from 400 to 700 MW; fixed drum startup from 0.1% power to 30 MW.
3/2/66	NRX-A4/EST	Ten bootstrap startups of bread board engine. Control test: fixed drum flow control startup to 250 MW, used on subsequent startups.
6/23/66	NRX-A5	Startup to low-power hold position in 40 seconds using period control and then to 800 MW with fixed drum flow control. Power level trimmed to full power; ninety seconds from source level to full power. Operated at full power for two periods for a total time of 30 minutes.
12/13/67	NRX-A6	Achieved 60 minutes at full power. Postmortem examination showed engine could be operated for significantly longer periods.
6/11/69	XE'	Conducted by Aerojet in down firing test stand ETS. Engine started 28 times for a total time of 3 hours and 48 minutes. Eleven minutes were at full power. All objectives met or exceeded.

reactor temperatures had been high enough to damage the tie rod support system. Sections of the tie rod assemblies were corroded but sufficient structural integrity existed to restart the reactor. About this time a disturbing bit of news was reported. A hole was seen in the nozzle.

Milt Klein called a meeting in Washington. Esselman and his staff were convinced that the reactor could be restarted but the nozzle had to be replaced remotely in the MAD building. At the last minute Chuck Rice of Aerojet brought in new information from Nevada. A check with a jury-rigged remote examination tool showed that the hole in the nozzle was an optical illusion. The test could be restarted.

NRX-A3 was restarted and ran for about 16 minutes at full power. Fixed drum control tests were conducted in which the reactor power varied with the hydrogen flow. All test objectives had been achieved.

NRX-A4/EST reactor tests included tests of the engine systems. Ten bootstrap startups using hydrogen flow control were accomplished.

Control experiments were completed on NRX-A5. The reactor was brought to a low-power hold position in 40 seconds using the LASL-developed period control and then to 800 MW with hydrogen flow control. The control drums were fixed in position. Power level was then trimmed to full power. Total time from source level to full power was about 90 seconds. This method of startup became standard for NRX-A6 and XE. The control problems had been solved with a simple startup and power control scheme.

NRX-A5 was operated at full power for two periods of 15 minutes. An ambitious goal of a 60-minute run was set. In December 1967, NRX-A6 was operated for 60 minutes at full power. Postmortem examinations showed that the reactor could be operated for significantly longer periods.

NERVA XE' engine tests were conducted by Aerojet in a specially constructed down-firing test stand ETS. The engine was started 28 times for a total operating time of 3 hours and 48 minutes. All technical objectives were met or exceeded.

These tests showed that all the NERVA objectives had been met and the potential existed for further performance improvements. These advanced requirements were introduced into the flight engine design.

The reactor tests were spectacular sights, with a 1000-MW reactor plume augmented by the subsequent combustion of the exhausted hydrogen. The flame and the sound were quite an experience, even at the control center's distance of several miles. Of course, such operations would no longer be allowed after passage of the above-ground test ban treaty at the end of that decade. All future reactor tests would have to be performed in a ducted enclosure, similar to that used for XE'.

The tests were only allowed on days when weather conditions were right, and when the prevailing breeze was generally in the uninhabited direction—north-northwest. When test operations were anticipated, a site committee presided over by LASL and the AEC manager would meet the previous evening to decide whether the conditions were auspicious. On the day of the run, they would have to give the go-ahead for start of the cryogenic cooldown and subsequently for commencement of reactor operation. Once in 1966, we found these people to be temporarily unavailable, and we later learned that the personnel who normally monitored our plume as it drifted downwind had been temporarily assigned to monitor the cloud from the first Chinese bomb test.

The proof is in the pudding—in NERVA the testing and the pudding were good.

Good relations were developed on testing operations with Keith Boyer and many others. In this case, Perry Davison and Max Johnson played key roles in developing a coordinated test team at Jackass Flats. Davison, who was assigned to work with LASL, established excellent relations with their people at the test site. Initially, Davison—and then other Astronuclear personnel assigned to Jackass Flats—were integrated into the test team for the KIWI tests.

For the reactor and overall engine tests, a joint test team composed of Aerojet General and Westinghouse personnel was set up at Jackass Flats, Nevada. We had lead responsibility for the reactor tests (NRX) and Aerojet General for the engine tests (XE). The reactor was assembled at the Large site and shipped to Nevada for test. The test vehicle was a modified railroad car.

From 1964 to 1969, structural integrity and performance were proven by the successful power operation of 12 NERVA and ROVER reactors. The final tests of the project were complete engine tests conducted by Aerojet General in September 1969. These tests took place in a specially constructed engine test stand in which the engine was fired downward into a simulated space environment.

Relations with Aerojet and AEC/NASA

The relationship between Aerojet and Astronuclear was never smooth, each was frequently trying to improve its position. In the early part of the program, the most important thing was to be sure there was a satisfactory nuclear reactor. That was Astronuclear's realm. When that was assured, the engine, the vehicle stage, the launching, Aerojet's scope, etc., would become the dominant areas. Of course, Aerojet was also responsible for developing and providing the turbo pumps, jet nozzle, and other non-nuclear parts critical to the reactor tests. Therefore, the tension with Aerojet was always present.

Aerojet had a nuclear operation at San Ramon, California (now PG&E's solar research center) that they were trying to use to take over our scope. Krasik's tactic was to show the SNPO (Harry Finger in Washington and the field group at NASA's Lewis Laboratory in Cleveland under Schroeder) that we were competent and technically correct. The struggle continued throughout the program, but we generally got our way after demonstrating that our people could do the job. This became particularly acute at the test site, where we had very competent people who held LASL and SNPO's respect, but Aerojet continually tried to take over by pulling strings and reading contract clauses.

We eventually won that battle hands down. Hal Faught, who had come from the Westinghouse aviation gas turbine division, worked for Krasik as program manager in parallel with Esselman. Faught was a key man in our relations with Aerojet, since he handled all the work scope and funding aspects of the program. Aerojet kept trying to use their position as prime contractor to limit the Westinghouse communication with the customer.

However Finger had provided for direct communication between Westinghouse and SNPO, as well as that between Aerojet and SNPO. Hal had to deal with this, while keeping communications with the customer flowing.

Dealing with the customer was also complex because there was an SNPO office in Cleveland, which appeared to us as more of a NASA entity, and the headquarters operation in Washington, which seemed more AEC than NASA, even though both were staffed by NASA and AEC personnel and SNPO was clearly part of both agencies. They had their own interface problems.

Faught was also very good at working with funding from the two agencies that had different budget cycles and sources of funding. Later Arnold became Faught's understudy in 1966, after he had been the manager overseeing the Nevada test operations from NRX-A2 through NRX-A5, and replaced him as manager of program management a year later.

Finger and his key subordinates knew all the issues involved and worked hard to address them and make sure they did not have any damaging influence on the program. Actually, those differences did at times raise issues that led to significant and constructive discussion to resolve them.

The problems in relations between Westinghouse and Aerojet were probably no more than would be expected and those differences had their positive side. They undoubtedly caused both organizations to work harder.

Analysis Techniques Development[8]

Throughout the program, great emphasis was placed on pretest prediction and post-test problem-solving analysis. For this reason, extensive techniques were developed by all analytical disciplines—controls and performance, thermal, nuclear, and structural. Computer programs were written for coupled solutions for neutronics, thermal-hydraulics, and controls. Frank Retallick and Joe Gallagher deserve most of the credit for the conception and direction of NERVA analytical efforts. As discussed in Chapter 14, Don Miller, Bill Brusalis, and John Swanson developed mechanical analysis techniques including a finite-element stress analysis code, which was truly innovative and the precursor of some programs that find wide use in industry today.

LASL, contrary to what you might expect, generally used a much more experimental approach to a new technological issue. We, on the other hand,

---

[8]D.L. Black, G.H. Farbman, and J.F. Wett, "NERVA Program History and Technical Summary," paper presented at Obinsk, USSR, Conference, May 15–19, 1990, Westinghouse Electric Corporation.

Figure 6.5 NRX-A2 test at Jackass Flats.

approached problems more from an analytical viewpoint. We tried to predict what would happen, even when we thought a test would not be completely successful. In the nuclear design area, LASL relied extensively on their KIVA, a flexible mockup critical facility. Our critical facility was not available until after we had performed extensive design analyses. Even then, it was less flexible in operation, so we used it as a confirmatory rather than exploratory tool. Besides, our charter was not just to make a reactor that would work in Nevada but to provide an engineering basis on which flight-engine reactors could be designed. This dictated flexible analysis techniques. The team of Gallagher, Ravetz, and Doncals adapted codes being used in the commercial nuclear industry, or they wrote new ones from scratch where necessary.

Now we were ready to design and build an actual flight engine. Its first test was scheduled for early 1973, but the program was discontinued before the reactor for this flight engine could be tested.

A Potential NERVA Mission[9]

As space missions become more ambitious, the advantages of nuclear power propulsion will become increasingly evident. NERVA's high performance and multiple restart capabilities make it the logical choice for the more demanding orbital operations, lunar missions, and planetary exploration. Advantages of nuclear propulsion can be illustrated by a description of a manned mission to Mars that was studied by NASA in the 1960s. The NERVA flight engine design with the performance demonstrated by the test series was the basis of the mission studies.

Power: Propulsion for a manned mission to Mars would come from the NERVA nuclear rocket engine with its superior specific impulse. Electric power for spacecraft operations would come from a thermoelectric conversion system with no moving parts. Such a system is capable of generating 20 kW of power utilizing a reactor heat source and liquid metal coolant.

Vehicle assembly: Chemical rocket propulsion would be used to boost the modules of the spacecraft into orbit. The studies were based on the use of an upgraded version of the Saturn V booster rocket. A three-stage nuclear-powered Mars vehicle would be assembled in Earth orbit. The first stage propels the vehicle from Earth orbit to Mars; a second stage brakes it into a Mars orbit; a third stage is used to propel the space ship from a Mars orbit back to Earth.

Earth orbit: Assembly of the vehicle in Earth orbit is begun about four months prior to the start of the Mars trip. The quarters for the assembly crew and the first and second stages are put into an Earth orbit by chemical rockets. When the vehicle is assembled, the assembly crew is brought back to Earth and the flight crew is sent up to the vehicle.

Mars trajectory: The trip to Mars will take about seven and a half months. The three nuclear engines have fired for approximately 25 minutes to impart the speed necessary to get the vehicle to Mars. The three-engine booster stage is separated and the long unpowered coast to Mars begins. After a 20-day stay on the Mars surface, the return trip will require another seven months.

Retrofire: Vehicle and mission equipment must be braked into Mars orbit. The nuclear engine performing this task will operate for about 30 minutes. In Mars orbit, the spent braking stage will be separated.

---

[9]William Jacobi and Paul Dickson designed the mission described here.

Descent operations: The Mars "lander" is separated from the main vehicle and the astronauts board it for the descent. Electrical power demands rise at this point with two vehicles in operation, so a second thermoelectric converter may be needed.

On Mars: Once the astronauts have landed, the power requirements reach their peak, as the astronauts will need power to conduct their exploration. Necessary power will be supplied by an additional compact thermoelectric converter carried aboard the lander vehicle.

Return blast-off: After the landing mission is completed, the astronauts return to the spaceship in the lander—leaving behind scientific instruments for monitoring weather, seismic activity, and so forth. Nuclear-powered thermoelectric equipment is ideally suited for operating remote monitoring equipment.

Rendezvous: After rendezvous with the spaceship the astronauts leave the lander. They approach the spaceship from the rear to stay within the shadow of the shield and compact converter at the forward end of the ship. After the lander has been abandoned in Mars orbit, it will provide power to a signal booster transmitting data and commands between Earth and the experiments.

Homeward journey: The final propulsion stage fires for 20 minutes, is separated, and the astronauts coast toward Earth. Work schedules will gainfully employ the crew during duty hours on the long journey home. Navigation, data processing and transmittal, vehicle maintenance, plus regular periods of exercise and adequate recreation will reduce the psychological strain of the long journey. Favorable weight allowances inherent in utilizing nuclear propulsion permit additional equipment to be carried to make life more comfortable for the astronauts. Power from a thermoelectric conversion system helps provide Earth television, music, hot meals and hot water.

Earth reentry: Earth rendezvous and proper reentry angle will be assured by on-board trajectory calculations and midcourse corrections. Finally reentry and landing are achieved—a year and three months after takeoff.

While such an ambitious mission could be performed with chemical engines, the ground takeoff weight would be significantly greater. For example, in years of favorable planetary alignment, the NERVA system could accomplish this mission with a one-million-pound vehicle in orbit. This weight would be increased to a five-million-pound vehicle in the least favorable years. This mission with a chemically propelled vehicle would require three times the orbital weight in favorable years and five times in unfavorable years. Therefore, a NERVA-based system would reduce required Saturn V launches by a factor of three to five depending on the year of launch.

Non-NERVA Contracts and Development

Staffing of the non-NERVA engineering design and analysis side of the laboratory was on a low-key basis until after the NERVA contract had been obtained in 1961. In that year Dr. William Shoupp became technical director for the part of the organization later called systems and technology. His responsibilities were almost completely outside the NERVA project. In the early years, much of the activity involved relatively small and short-range efforts to mount proposals or to prepare for expected request for proposals (RFPs), as well as to meet contract requirements.

"Shouppie," as he was affectionately known, was in 1962 to leave Astro upon his elevation to vice president of the Westinghouse Research Laboratories, in which position he continued until his retirement in 1973. Thus it was no accident that from the beginning, fruitful relations between Astro and the research laboratories prevailed. They continued under George Mechlin, a Bettis graduate, who followed Shoupp.

The systems and technology operation received support from several sources. The divisional budget provided funds for product development and for bid and proposal efforts. In addition, the company apportioned certain funds to its divisions to be used for their benefit at the central research laboratories, the apportionment being based on a three-factor formula. We soon discovered that the Astronuclear share depended more on close relationships than on any formula. By the late 1960s, we were getting more than our share, enough so that other divisions might well have complained had they been sufficiently alert.

Of course, the main objective was to obtain contract support from various government agencies and to obtain payment for work performed for other Westinghouse divisions.

Engineering personnel were largely obtained from related divisions of the corporation. Often our needs were best satisfied by the high level of expertise and breadth of knowledge to be found in advisory engineers, consulting engineers, or technical managers. R.T. Begley, J. Danko, W.M. Jacobi, A. Mei, and L.B Kramer were among the early staff members, most of whom remained to grow with the laboratory, as may be seen in the organization charts. Others who made significant contributions were R.L. Ammon, F.R. Lorenz, D. Roberts, T. Varljen, C. Rose, C. Kim, and J. Sadler.

This part of the organization required marketing activity to keep pace with the customer thinking and schedules as well as to aid in preparing proposals. Technical managers and experienced engineers were also intimately involved in direct contact with prospective contracting agencies on a technical basis.

In September 1966, Woody Johnson assigned Don Thomas as engineering manager of the systems and technology organization, reporting to Bill Budge. By this time, the non-NERVA billings were almost $3 million per year. Billings having reached that level, a more formal organization was necessary. The organization was tightly knit, with close contact between technical and marketing people. Materials assistance was provided by the materials department, then under Dave Goldberg. A large array of product development, bid and proposal, and contracts was carried out. The principal customers were NASA, DOE, and DOD. The largest program was the thermoelectric project. By November 1969, various programs were becoming large enough to call for a more project-oriented organization, and Carroll Sinclair was appointed manager of systems and technology. Thomas was responsible for the product development program and for the corporate-supported research laboratory programs carried out for Astro.

The systems and technology programs were growing in spite of a declining market, reflecting the same downward trend in government spending for unmanned space related projects that was to cause the demise of the NERVA project. By 1971–1972, systems and technology programs were declining and it was necessary to look elsewhere for support, as is described in the next section.

Whatever Happened to Astronuclear?

Finger left the SNPO in 1967. He was followed by Klein, who had this position for several years. A major part of the contact with SNPO was with their office at Lewis Laboratory in Cleveland. R. Schroeder was in charge of this office throughout the entire project. Later Dave Gabriel became head of SNPO.

The NERVA project came to a sudden end when it was canceled by the government in 1972, since there was no Mars mission in the NASA program. With the loss of NERVA, Astronuclear's legacy of technical and management expertise remained in place or was transferred to other corporate units. The laboratory became advanced energy systems.

In the 1970s, George Mechlin, vice president and director of the Westinghouse R&D Center, and Max Johnson began to consider using advanced energy systems as a place to translate research results into products, building upon a practice that was already the rule. Examples were the process for producing dendritic-web silicon ribbon for solar cells and the process for phosphoric-acid fuel cells.

The silicon ribbon project fell short of achieving an economically competitive product and was terminated. But it served as a useful background in the acquisition of Integrated Power Systems, a company that sells solar-cell power systems. The phosphoric fuel-cell project led to production of several 400-kVA systems, but marketing potential was limited.

The most successful project, from the point of view of spinoff produced, was the nuclear waste disposal study in the early 1970s. This escalated into the West Valley contract, then the Idaho Chemical Processing Plant, then the WIPPS (Carlsbad, New Mexico) project, Hanford, and Savannah River. This essentially became the whole Westinghouse environmental resources group.

By 1981, Howard Arnold, then general manager of the advanced reactors division, began a diversification program as it became clear that cancellation of the Clinch River Breeder Reactor was likely. Max Johnson, then general manager of advanced energy systems (the new name for Astronuclear), had laid the groundwork in the area of spent-fuel disposal through the use of the former engine-maintenance, assembly and disassembly (E-MAD) facility in Nevada, built for the NERVA engine program and used only for XE-1. Arnold headed a task force to look for any and all opportunities to use the capabilities of both divisions. This led to the successes mentioned earlier.

After receiving the contract for the Idaho Chemical Processing Plant, Max Johnson moved to Idaho as general manager. The advanced reactors division was merged into the advanced energy systems division, with Arnold

Table 6.2
Astronuclear Statistics

Year	Number of Employees	Sales Billed ($1,000) NERVA	S&T	Total
1959	8			
1960	73	0	5	75
1961	184	808	55	863
1962	881	3038	192	13230
1963	1994	37627	554	38181
1964	1739	38252	1215	39467
1965	1683	39161	1545	40706
1966	1885	39795	2864	42659
1967	1910	39039	4148	43187
1968	1420	30405	5476	35881

as general manager. This division was part of the advanced power systems business unit under John Yasinsky.

But stay tuned. NERVA, or a version of it, may yet be recalled to active duty for space exploration. If there is to be a manned mission to Mars the only practicable propulsion engine is a nuclear one. The technology is sound, but its time has not yet come.

Table 6.2 summarizes some of the information about Astronuclear from 1959 to 1968. Employment reached a peak of 1994 employees in September 1963.

CHAPTER 7

Early Commercial Nuclear Power

Westinghouse Enters the Field

Gwilym Price didn't give atomic power top priority back in 1948 just to help Rickover and the Navy build a submarine, and he didn't give Charlie Weaver the authority to cherry-pick Westinghouse for prime engineering talent just to strengthen the company's postwar business.

No, it's my guess that he had his eye on the future of one of the firm's core businesses—commercial electric power equipment. That, after all, was the business that led George Westinghouse to found the company back in 1886. It was Westinghouse, with Tesla's patents that he bought, who led the move to alternating current that revolutionized the industry and put new vitality into the industrial revolution.

The move to distributing electricity in New York was not without its difficulties and some high drama. A small but extremely vociferous minority charged that the plan could kill people in a few seconds and destroy large sections of the city by fire. The press, which was pretty irresponsible in those days, ran sensational stories on what it called "the murderer." The feeling of panic was encouraged by interested parties and groups.

New York's Mayor Hewitt was urged to take the law into his own hands if necessary to stop this thing, and when he failed to do so, one journal called for his arrest as accessory to "a carnival of avoidable suicide." An ex-governor of the state wrote to the mayor urging him not to let the plan go into action anywhere within the city limits of New York.

What George Westinghouse was proposing, of course, was alternating current. His proposal would spread cheap power throughout New York and keep the city from strangling in a network of low-voltage direct current lines. He might have expected applause, but what he got was an emotion-charged crusade.

Westinghouse explained patiently over and over again that alternating current was safe. "The transformers," he said, "are so constructed that the primary high voltage street current can never enter the home."

Despite this, the attacks increased. Even Thomas A. Edison attacked alternating current (ac) as unsafe. He wrote an article against it in the *North American Review*, a leading magazine of that day. He was freely quoted in newspaper stories as being positive that no known method of insulation could render ac wires safe if they carried more than 200 volts. Their use underground in subways, he said, would not lessen the danger, because the high-tension current would burn out tubes and enter the dwellings through the manholes.

Given this background, then, we should not have been surprised that there was opposition to something as new and mysterious as nuclear energy.

In the late 1940s, Westinghouse was one of the world's leading suppliers of apparatus for generation, transmission, and distribution of electricity. Price intended to keep it in that leadership position. He saw nuclear energy as a key power source of the future.

The question of how to move the military application of nuclear energy to a peacetime business for Westinghouse was something to which both Price and Weaver were giving much thought. In the first place, how much could be done legally? The Atomic Energy Act of 1946 had given the newly formed Atomic Energy Commission (AEC) monopoly of production and ownership of fissionable materials. But, in 1951, the AEC opened the door a crack by starting an industrial participation program. It established four teams to study the feasibility of commercial atomic power: (1) Detroit Edison and Dow Chemical, (2) Pacific Gas and Electric and Bechtel, (3) Union Electric and Monsanto Chemical, and (4) Commonwealth Edison and Public Service of Northern Illinois. Weaver assigned Ray Witzke of the Westinghouse East Pittsburgh divisions and D.W.R. Morgan, Jr., of the steam division in South Philadelphia, to monitor the work of those teams.

The door to commercial power opened a little wider with the passage of the Atomic Energy Act of 1954, which permitted private development of nuclear power with certain restrictions. Among these were the prohibition against owning fissionable or fusible material. But private industry now could enter the nuclear field.

By late 1953, Mark I was operating successfully, construction of *Nautilus* was under way, and we had started on development and construction of the first central station nuclear power plant at Shippingport, Pennsylvania.

Thus, it seemed opportune to begin preparation for entering the commercial nuclear field.

For Westinghouse to move into this legal opening, Weaver faced the possible, actually, the probable, opposition of its only nuclear customer, so far—Hyman Rickover. The admiral had been dead set against letting any of the key Bettis people have responsibilities outside of his programs. Weaver discussed with Price the desirability of Westinghouse making a push, and with the CEO's strong backing, Weaver persuaded Rickover to let him form a small commercial group, independent of Bettis. Price provided the necessary corporate financing for this initiative.

Actually, Ray Witzke and Dave Morgan's monitoring of the utility team studies, which began in 1951, provided a starting point. The next year Weaver gave Witzke the assignment of organizing an industrial atomic power study group for Westinghouse. He set up shop in McKeesport, Pennsylvania, in 1952, where the group was joined by E.U. Powell and A.R. Jones. Later they moved to larger quarters at the former Old Overholt distillery in Large, Pennsylvania, which the Astronuclear Laboratory eventually took over.

To augment these forces, Weaver's group had a windfall. He was able to tap the growing body of "Rickover rejects." That is not a term of dishonor, by any means. There were some very good people at Bettis who, for one reason or another, fell into disfavor with the admiral—a very easy thing to do. Those people he actually was glad to see go. No matter how Rickover had evaluated someone a few weeks before, if they left on their own or were moved by Westinghouse, he immediately considered them completely worthless. To be a Rickover reject became something of a badge of honor in the Westinghouse commercial nuclear power organization as time went on. This was how Budge, Creagan, Frisch, Johnson, Rengel, Shoupp, and Voysey were freed to join.

On October 1, 1954, Bill Shoupp joined the group as manager, Dr. W.E. Johnson, who said he couldn't work with Rickover any longer on the Mark I prototype, was assistant manager, Dr. R.J. Creagan was manager of nuclear analysis, and R.L. Witzke became manager of plant analysis. Creagan had played an important role at Bettis. Among other things, he had performed the critical experiments for Mark I at the Argonne National Laboratory (ANL).

So it was largely due to Rickover's reverse influence that Westinghouse was able to put together a cadre of talented people to pursue the atom commercially.

Although Westinghouse completely financed this commercial activity at the beginning, the group soon began to obtain small study contracts from a few electric utility companies including Consolidated Edison, Commonwealth Edison, and Pennsylvania Power & Light (PP&L).

One of the first customers, Jack Busby, of PP&L, saw the potential of nuclear power for his business. In July 1955, he gave Westinghouse a contract for study of a homogeneous reactor, which was organized as the Pennsylvania Advanced Reactor (PAR) project. Remember this was just shortly after the start of the Shippingport project.

Oak Ridge National Laboratory (ORNL) also was attempting to develop a homogeneous reactor, but after many experiments they ran into lots of problems. And after more development effort over a number of years they abandoned the homogeneous reactor. Following considerable development, we also decided to abandon this type of reactor in 1962. Its principal disadvantage was the contamination problem associated with possible leaks of the highly radioactive fluid, which made maintenance difficult, and the slurry that had been selected caused corrosion problems.

Even so, the commercial effort gained strength. In August 1955, the growing Westinghouse study group was designated the Commercial Atomic Power Activity (CAPA) with division status. It reported to Weaver who, as vice president of the atomic power divisions, still reported to Price. In consultation with A.C. Monteith, the utility group vice president, Weaver decided that a person with marketing experience was needed to generate more utility industry support. As a result, Carroll Roseberry, a marketing manager, was put in charge of CAPA.

Bill Shoupp was named head of the technical activity. This was a bitter disappointment to him, as he had expected to become the CAPA manager. This activity was located at the old Westinghouse Research Laboratories in Forest Hills, another Pittsburgh suburb.

Encouraged by Congress having opened the door to commercial development, the AEC started a five-year program in the early 1950s to foster development of five reactor types—pressurized water, boiling water, sodium graphite, fast breeder, and homogeneous. According to a study for the Joint Committee on Atomic Energy (JCAE), light water reactors were considered to have the least promise of the five, because of the low-temperature steam produced. The boiling water reactor (BWR) was thought to have advantages over the pressurized water reactor (PWR) because, by not having a steam generator, it appeared to be simpler and could have somewhat higher efficiency.

Westinghouse chose to bet on the long shot of PWRs mainly because of our successful experience so far in the naval program, and partially because of interest shown by some of the utility groups such as those of Commonwealth Edison and Consolidated Edison. We were not in a good position to compare various alternative reactor types, but our experience with Shippingport made us comfortable with our decision.

Many other reactor types, i.e., combinations of coolants and moderators, were tried under the AEC program. Most were unsuccessful and were dropped.

Atomics International's sodium reactor experiment (SRE) began operation in 1957, but it was "no sale" in the market place. Their 75-MW(e) sodium-cooled and graphite-moderated Hallam plant was completed in 1962 but was shut down in 1964, after failure of several system components. Development of that reactor type was stopped. They also developed a small sodium-cooled reactor for space applications. It was launched, but after a few weeks a control problem shut it down also.

Perceived advantages of safety and simplicity led to attempted development of organic-moderated reactors. One such reactor experiment (OMRE) was operated at the Idaho testing station. Then, as a part of the AEC's second-round demonstration program, an organic-cooled and -moderated 11-MW(e) reactor was built at Piqua, Ohio, by Aerojet General, and was funded by the AEC. Neither of these was successful. A number of other combinations were also attempted with little success.

Light water clearly seemed to be winning the day. When we reevaluated our choice of pressurized water in the early 1960s, our original decision was confirmed. The BWR had the advantage of eliminating the steam generator and pressurizer, offering higher steam pressures and thermal efficiencies. However, it had counterbalancing disadvantages such as water chemistry problems and the necessity to shield the turbine and condenser. If the choice had been close, we would have had a tough decision, realizing the problem we would have selling BWRs against General Electric (GE). Later, even more exhaustive analyses confirmed our decision.

GE and the BWR

The race to license and build the first American commercial nuclear plant probably began as both Westinghouse and GE were pondering the choice of which reactor type to bet on. As we were selecting the PWR, GE[1] was also struggling with the problem of what type of reactor to select for commercial exploitation. The Schenectady group that had been working on an intermediate-energy neutron spectrum reactor presumably was an advocate of such a reactor in the earliest days of GE's decision-making process. However, this reactor had been studied enough that GE knew it would not breed. The problem was that there are more capture resonances than fission

---

[1] R.J. Creagan, Private Communication dated July 7, 1991.

cross-section resonances in uranium-235 from about 1 to 100 electron volts and up. However, even if it did not breed, this type was still a contender.

They had a lot of experience in the use of sodium, both for the intermediate-spectrum breeder reactor and in the naval program (the SIR was a sodium intermediate energy spectrum reactor), and they had facilities for any further sodium development needed. The sodium-cooled reactor in the *Sea Wolf* had not yet been removed.

When GE's vice president in charge of Hanford was transferred to Schenectady, however, he was instrumental in the decision to exploit water-cooled reactors. By then, Walter Zinn of ANL had built and operated the Experimental Boiling Water Reactor (EBWR) at ANL near Chicago. Sam Untermyer,[2] who had been in charge of the design of EBWR at ANL, was now at GE. It was he and Bruce Prentice who designed the Dresden-1 reactor, which GE sold to Commonwealth Edison.

No doubt GE regarded our PWR as its obvious competitor for the utility market. While the BWR eliminated the steam generator and pressurizer and offered higher steam pressures, and consequently higher thermal efficiencies, its lower power density and various instabilities may have been downplayed. And much of the apparent simplicity in the concept disappeared with complete engineering design. GE no doubt was reluctant to follow Westinghouse with a PWR.

The marketing battle to win Commonwealth Edison's first order was decided by—what else—price. That's with a small "p," unfortunately. At the CE offices, Roseberry, Shoupp, Witzke, and Creagan met with several of the utility company VPs and agreed on a joint study program to be performed in Pittsburgh by a task force including both Westinghouse and CE engineers.

Things looked pretty good for Westinghouse. However, at that time, on the top floor of the same building, the presidents of GE and CE were talking money. They concluded that no one really knew what reactor plants would cost, but GE said it would build a nuclear power plant (Dresden-1) for the utility for $250 per kilowatt and would swallow any costs beyond that. This offer was accepted.[3] So it was that Dresden became the first commercial nuclear plant in the United States, beating out the Westinghouse-built Yankee Rowe by a few months.

Because Untermyer and GE were not sure of themselves, they designed a reactor that could go either BWR or PWR. Thus Dresden-1 was designed with a steam generator that could be used if necessary, and a direct cycle

---

[2]B.R. Prentice and S. Untermyer, "Dual Cycle Boiling Reactor for Electric Power," presented at ASME Spring Meeting, Boston, Massachusetts, June 21, 1955.

[3]R.J. Creagan, Private Communication dated July 7, 1991.

that could provide steam directly to the turbine. The pumps were such that the reactor could be nonboiling or boiling. Such a compromise reactor cycle was called the dual-cycle boiling reactor.

The dual-cycle boiling reactor provides a method for removal of additional heat from the reactor without exceeding maximum allowable steam volume in the core. This is accomplished by removing heat from the saturated recirculating water. Part of the recirculating water is taken from the reactor and passed through the primary side of a steam generator. Lower pressure steam, produced on the secondary side of the steam generator, is admitted to the turbine at a point of reduced pressure. Cooling water passes through the primary side of the steam generator and then is pumped back into the reactor.

GE conducted development programs in its Vallecitos, California, BWR in parallel with Dresden-1 design and construction. The original concern over BWR instability never materialized so GE went to high-void, single-cycle system BWRs.

Water chemistry problems were underestimated at Dresden, and as time went on more shielding blocks had to be erected around various steamlines and the turbine. For many years, there was a contest between Dresden and Yankee to see which would generate the most kilowatt hours. Yankee won, even though Dresden started earlier with license No. 1 and a higher initial megawatt-electric rating.

Belgian Thermal Reactor (BR-3)

At this point, I need to add a word about our work on Europe's early reactors.

Commercial nuclear power excitement was stirring outside of the United States. Edison Volta in Italy announced plans for a nuclear power plant on December 20, 1956, and on August 1, 1957, Westinghouse signed a license agreement with Fiat in Italy.

Our first contract abroad was signed with Societe Cooperative Electro Nucleare of Belgium on April 23, 1957. It was for a small [11.5-MW(e) gross][4] nuclear power plant. Originally called the Belgian Thermal Reactor (later named BR-3), it was to have been the central attraction for the Brussels World's Fair of 1959. But then somebody decided the King's palace was too close to the fair, so it was relocated. Also, due to construction delays, it was not completed in time for the opening of the fair.

On July 25, 1957, Westinghouse signed a license agreement with the Belgian firm Ateliers de Constructions Electriques de Charleroi (ACEC). The

---

[4]*Nuclear Engineering International*, July 1980.

hardware to be used was the first CAPA designed for actual nuclear power plant construction. The fuel chosen resembled the Shippingport blanket, having cylindrical UO^2 pellets in cylindrical cladding. Zirconium was difficult to obtain outside the naval program. We had about 10,000 pounds of zirconium, but unfortunately it had been contaminated with tungsten during processing and this contamination made it impracticable to convert it into tubes. So stainless steel was used for cladding material. Also, the highly enriched uranium used in the Shippingport seed was unavailable and inappropriate for an export reactor anyway due to nuclear weapons proliferation considerations, so a uniformly low-enriched core was chosen.

These factors dictated an experimental program in support of the physics design, so Westinghouse purchased the Waltz Mill site and constructed the Westinghouse Reactor Evaluation Center (WREC) to perform light water reactor critical experiments in support of BR-3 and eventually Yankee and other PWRs. Perry Davison, later to be the head of our on-site NERVA nuclear rocket reactor testing in Nevada, was put in charge of the facility's design and operation. Howard Arnold, working at that time under Harvey Graves and Bob Creagan, was assigned to specify the parameters to be studied and to do the design of the experimental program. This facility was later used to perform heavy water reactor criticals for the CVTR, and also graphite reactor experiments in support of NERVA. The Waltz Mill facility was also used for the Westinghouse Testing Reactor, for NERVA reactor fuel endurance testing, for a coal gasification test program and the liquid metal breeder reactor program, and it is still in use as a field service training and support facility for Westinghouse PWRs.

The critical experiments were done at the Waltz Mill site and served as a model for Yankee, since stainless-steel-clad slightly enriched uranium pellets were used. The Yankee criticals were the size of the BR-3 core and were run in the same building.

This Belgian reactor operated routinely and satisfactorily for many years. Union Meunier du Haut Katanga, which became the operator of the reactor, appeared to want to use this plant to enhance its capability for building a submarine propulsion plant. It was made leak tight by welding all sealed outlets, such as those around valves, so heavy water could be used as the coolant-moderator.

The BR-3 core was redesigned so it could start out as a full heavy water core and then, as burnup occurred, some of the heavy water could be replaced by light water. Reactivity would be controlled by addition of light water as well as with control rods. BR-3 operated successfully for a full year without a shutdown. Everything worked well, and thus this was the first actual operation of a "spectral shift reactor." This type had been advocated for many years by Milt Edlund, who moved from ORNL to Babcock & Wilcox in Lynchburg, Virginia.

Yankee—Westinghouse's First U.S. Commercial Order

Despite our plans, expertise, and early experience in naval reactors and Shippingport, Rickover was convinced that Westinghouse was making a big mistake in entering the commercial field. He thought that only he could successfully build a nuclear plant; that only his style of uncompromising, hard-driving management could conquer such a difficult technical challenge. He sincerely believed, I'm sure, that our success thus far was really his doing.

If the truth were known, there probably were more than a few Westinghouse people who had doubts. But Weaver wasn't one of them, nor was I. We didn't think you had to be a self-styled S.O.B. with military authority and government money to get the job done in the commercial market. And we knew that the technical talent and know-how that solved Rickover's problems came from Westinghouse, not the other way around.

The fact that our first big commercial job—the Yankee plant at Rowe, Massachusetts—was to be called the "nuclear success story of the year" in 1961 by the AEC validated our confidence in the Westinghouse capability.

It was the second privately owned nuclear plant in the country, being completed in July 1961, just shortly after Commonwealth Edison's Dresden-1 boiling water plant.

The challenge now was different—not just technical as was the case with Shippingport and the Nautilus, but also economic. Now we were entering the world of competition for the industrial and consumer dollar. Although Yankee had a very high power cost, much higher than fossil alternatives at that time, it has since proved to be economically competitive.

The seed for Yankee was planted in 1952, when many utilities and industrial companies formed study groups to explore the use of nuclear reactors for the generation of electricity. One of these was formed by Dow Chemical and Detroit Edison. After a few months they asked the New England Electric System (NES) and several other utilities to join the group.

Early the next year, 1953, NES also began working with Monsanto and the Fluor Corporation on a nuclear plant. This work would form the basis for their submission two years later of a proposal under the AEC's new power demonstration reactor development program. What they proposed was a three-loop, 100-MW(e) reactor with aluminum-clad fuel enriched to about 4%. This design favored increased plutonium production at the expense of power production. Plutonium at that time brought a high price of $30 per gram and upward depending on the content of the isotope plutonium-240. A high plutonium-240 content made the plutonium unsuitable for nuclear weapons.

Actually, representatives of the New England utilities had incorporated Yankee Atomic Electric Company in Massachusetts on September 17, 1954, shortly after President Eisenhower signed the amended Atomic Energy Act into law. The CEO for both companies—NES and Yankee—was Bill Webster. It was he who brought Westinghouse into the picture.

The reason he did so was because he was concerned about the economics of the proposed plant. How stable would plutonium prices be? If the proposed plant favored increased plutonium production at the expense of electric power generation, it would depend for its success on a continued high price for plutonium. Webster asked Westinghouse to review power production economics. He also had the plant design reviewed by Al Weinberg of ORNL and ANL director Walter Zinn. Both men were also on an AEC reactor development division advisory committee.

Westinghouse regional vice president in Boston, Les Lynde, arranged for Webster to meet with Weaver on June 8, 1955, and more meetings followed on July 3 and August 8. We also approached Webster from a different direction, when Shoupp and Creagan talked with Roger Coe, also of NES and Yankee, at a Bettis information meeting on Shippingport. It was announced there that Shippingport was to be a seed-and-blanket reactor—the blanket consisting of natural uranium dioxide pellets in zirconium tubing. Shoupp and Creagan talked with Coe about design of a slightly enriched reactor using uranium-dioxide fuel. They got another crack at Coe at the American Nuclear Society (ANS) charter meeting at Penn State, in July. At that time the AEC was still evaluating the Yankee-Monsanto-Fluor proposal.

Following up talk with action, we proposed to NES a reactor designed for electric power production that would significantly change the NES approach. To get higher thermal efficiency and longer fuel lifetime, aluminum-alloy-clad fuel elements were replaced by uranium-oxide-pellet fuel in stainless steel tubes. The design rating was increased to 134 MW(e) (four loops).

The Westinghouse core was first designed for minimum enrichment, 2.8%, but after the critical experiments this was raised to about 4% to provide longer lifetime. For a power reactor, a long-life core is desired, so the processing costs of the core can be written off over more kilowatt hours. To produce weapons-grade plutonium, i.e., to keep the 240 isotope of plutonium low, the core must be designed for a shorter life. Westinghouse reactor fuel was designed for a burnup of about 8200 megawatt-days of heat output per metric ton of uranium, rather than the much lower burnup of the Monsanto design.

The Contract

Yankee liked what they saw and submitted the Westinghouse recommended four-loop plant as a revised proposal on August 22, 1955. This was timely because early that month the AEC's Chicago operations had advised Yankee that its original proposal had been rejected. Since Yankee already held the Westinghouse proposed design, it was able to submit a revised proposal promptly.

Ken Davis, head of the AEC reactor development division, was critical of using a stainless steel cladding that required 4% enrichment. Therefore, he had his assistant, Lou Roddis, meet with Yankee and Westinghouse. At this meeting, NES people were treated almost as criminals for designing a core to maximize production of power and to utilize the economics of gaseous diffusion plants to show favorable economics for nuclear power—all at the expense of the limited supply of enriched uranium. However, by this time Westinghouse had proposed the revised design to Yankee.

At first it was not well received. The AEC would have preferred a reactor type "with more promise," but there were few serious alternate proposals for the AEC's demonstration reactor program. After much negotiation and after the JCAE sent a letter stating that it favored construction of the Yankee plant, the AEC signed a contract in February 1956.

Yankee now had the green light to build a nuclear power plant and Westinghouse lost no time in teaming up with Stone & Webster (S&W) to submit a construction proposal to them. The R&D program that led to building of Shippingport had been under way for a little more than two years and much of the technology being developed for the Shippingport blanket was the basis for design of the nuclear plant for Yankee

Selection of Yankee by the AEC was in large part due to skillful and aggressive pushing of the proposal by Webster of NES. The government provided a minor amount of development funding and required payment only for uranium fissioned, with no charge for just having the enriched uranium in our possession. This was important, because fuel cladding of the first core was stainless steel, requiring higher uranium enrichment but fissioning no more uranium. Stone & Webster was the architect/engineer/constructor.

Several months of negotiation with the AEC, carried on by Krause, Allen, and Coe of NES, resulted in the first contract under the power reactor demonstration program on June 6. It specified $5 million for research and development, which Yankee immediately subcontracted to Westinghouse. It then negotiated a three-party contract for design and construction of the plant. There never were separate contracts with Westinghouse or S&W.

Entering into such contracts back then was scary business. We were dealing with a lot of uncertainties on costs and a lot of money. It was Weaver who had to make the final decision to enter into such a contract. Fortunately, he had the full confidence of his top management and a lot of faith in the competence of his technical people. The situation was very similar at Yankee, where it was Webster who made the courageous decision to proceed with construction of a commercial nuclear power plant.

The details for our role in Yankee were hammered out at the Copley Plaza in Boston where Weaver, Creagan, Witzke, and our regional vice president, Leslie Lynde, met with the CEO and other senior officials of Stone & Webster. They refused a fixed-price contract but offered to perform plant engineering design and construction for cost plus one dollar. The next day, the same people met with Webster to hear him outline his desires for the Yankee plant. Weaver was in general agreement and a tentative agreement was reached on how to proceed, subject to working out many details. Westinghouse met later with Stone & Webster to discuss scope and contract terms, which were agreed on by Webster.

Yankee had chosen Monsanto for the 100-MW(e) power plutonium plant. However, for the 134-MW(e) power plant on July 1, 1956, Yankee signed the three-party contract with Westinghouse and with Stone & Webster for design, manufacture, and construction. We were authorized to proceed with the R&D program and to order long-lead time components. Westinghouse was to furnish all equipment for the nuclear steam supply system and all other components it usually supplied or subcontracted, as well as all electrical equipment, at negotiated prices.

Weaver had persuaded Webster to buy the Westinghouse nuclear plant, and he negotiated the contract for a specific price, plus shared savings and shared costs for overruns.

Laying the Groundwork

Bill Webster was a man ahead of his time. As though able to foresee the public acceptance problems that lay ahead of nuclear power, he moved to build public confidence among people of his industry and his community.

He gathered presidents of the various New England utilities together to make decisions and to keep them informed. He also provided Washington and New England political figures with persuasive justifications for his actions. And he kept the citizens of Rowe, Massachusetts, informed and on his side.

They loved the plant. Why? Because Yankee Atomic Power paid all taxes locally. The only problem facing the citizens was how best to spend the money on schools, roads, and other community needs. Tax benefits were so impressive that *The New York Times* devoted several full pages of pictures in a Sunday edition to Rowe's low millage.

Monroe Bridge, the small town on the other side of the Deerfield River, received no tax or other benefits from the Yankee plant. So Bill Webster bought them something they wanted most—a big red fire engine. They were happy.

Instead of resenting the plant and the influx of new people, the reserved New Englanders welcomed them warmly. They elected one of the Yankee Atomic wives to be president of the PTA. There were many other activities where energy and new ideas were provided.

When fishermen on the Deerfield River expressed concern over the hot water from the plant that they feared would kill fish, Glenn Reed, the Yankee plant manager took action. He conducted public tests by inserting thermocouples to show that river water was hotter when it was discharged from the adjacent hydro plant than when Yankee sucked up cold water from deep in Sherman Pond, heated it through condenser tubes, and discharged it down river.

Yankee Atomic Electric Company acted as a line organization. Although the board was kept informed, Webster and Coe, the technical head for Yankee, made all final contractual and technical decisions for Yankee. These were later ratified by the board of directors. Webster let Coe chair most of the technical meetings. Technical aspects of the Yankee project were handled by Coe and Glenn Reed, who was the first Yankee plant manager. Glenn went on to become the operations manager at Point Beach in Wisconsin where there were two Westinghouse PWRs.

Reed had been loaned to Westinghouse to work in Idaho on construction of the land-based prototype for the *Nautilus* and later was sent to ANL's BORAX reactor in connection with the NES-Dow Chemical-Fluor project. He played an important role in the design, construction, testing, and operation of Yankee.

Their counterparts in Westinghouse were Roseberry and later Robert Wells, division managers, and Shoupp, the technical director with overall responsibilities; he delegated these to plant manager Witzke and reactor engineering manager Creagan. Shoupp, Creagan, and Witzke handled the technical side for Westinghouse, referring some contractual items to Roseberry and later Wells. Witzke interfaced with Stone & Webster to guide plant design, while Creagan was project manager during the conceptual design phase of the nuclear reactor. Later, Al Voysey was the Westinghouse project manager.

Bill Abbott coordinated much of the reactor administrative effort for Creagan and Harvey Graves did the reactivity analysis. Working for Graves were Howard Arnold, in charge of design methods development and George Minton, who was performing core layout studies that led to the current fuel cycling schemes used in PWRs. Erhling "Frankie" Frisch designed the magnetic-jack control-rod-drive mechanisms, because earlier designs could not lift the much heavier Yankee control rods plus followers. Frisch provided mechanical design of fuel and control rods, as well as the entire reactor mechanical design.

In several respects the reactor differed significantly from Shippingport and the naval reactors and much original engineering was required. Naturally this led to some surprises, but as mentioned elsewhere in this chapter, the design team was able to handle them. Three of these areas were control rod worth, resonance capture by fission products, and the Doppler coefficient of reactivity. These are discussed in Chapter 15, which covers the commercial reactor technical story.

Chemical Control

Aware of the N.I.H. (not invented here) syndrome, Webster had the plant design reviewed by MIT and other experts. Creagan successfully defended the design against their criticisms. The Yankee board then approved the Westinghouse design in principle. Largely as a result of these interactions, we subsequently enjoyed close relations with key members of the MIT faculty. These included Manson Benedict, a member of the AEC's Advisory Committee, and T.J. "Tommie" Thompson, a consultant to Yankee and later an AEC commissioner.

Nuclear fuel for the Yankee reactor had been established conceptually, but details were to be worked out during the R&D program. It was at this time that Monteith called Creagan and asked about nuclear fuel. He knew the nuclear fuel was going to be very expensive and that it would be expected to perform in a manner that would produce the desired results of power and margin. Otherwise, customers would be disappointed, unhappy, and downright hard to get along with.

Monteith told Creagan he would be comforted to hear that someone had a conceptual design of such fuel someplace and might even have put it down on paper. More specifically, he asked "Were there manufacturing drawings of such fuel and, if so, where were they and who was establishing the reality, feasibility and cost of manufacture?" Creagan's answer: "Mr. Frisch has made a sketch of the fuel and we are reviewing it for practicability and design improvement, which will be implemented during the R&D program just awarded." After that, all Creagan heard was—CLICK as the phone slammed down.

Westinghouse Management Wants a Review[5]

It was clear to Westinghouse management that this nuclear "first" would either pave the way for future success or block the path, depending on performance. A lot was riding on the few people designing this plant.

---

[5]W.H. Arnold, Private Communication dated June 5, 1992.

We were going to load the fuel and expect it to run as predicted based on the work of just a few physicists. We were planning to perform several small-scale critical experiments (about 10% to 20% of reactor size), but we had no digital computers.

A series of evening meetings were held with Sid Krasik and a few of his people—Al Henry and Bill Jacobi, among others. They seemed to think CAPA was indeed taking a major technical risk by not having larger and more extensive experiments and by not using digital computers, but they believed that at least the CAPA people were competent. Shortly thereafter, we were able to obtain some software in use at Bettis, and we were also able to log time on the East Pittsburgh IBM 650. At one time, Arnold ran an extensive calculation in the lobby of the Westinghouse headquarters building, where the computer was set up temporarily during a strike.

"Whatever it takes," was a slogan of the Pittsburgh Steelers' coach. It also became the watchword of the Yankee designers and planners.

Licensing

Safety was a major concern for the people who would create this first commercial nuclear power plant for Westinghouse. They were fully aware of safety's importance for the future of commercial nuclear power.

In April 1957, the preliminary hazards report was submitted to the AEC. It did not mention "maximum credible accident," because Yankee felt the term meaningless. Instead the words "hypothetical accident" were used. It was analyzed as a partial meltdown, which was handled by the safety injection system (the term "emergency core-cooling system" was not yet used) with no release of fission products. Cliff Beck of the AEC asked whether Yankee considered their hypothetical accident credible and, if not, what was the maximum credible accident? There was much discussion but no agreement. The AEC later changed the term to "design-basis accident."

Public Hearings

Public hearings? Sure, this wasn't a government-sponsored Shippingport. This was the real world. About August 1, 1957, the *Federal Register* stated that the operating license had been reviewed and the intention was to issue a license, but there was a 30-day period during which people could intervene and ask for a public hearing.

The contract with the AEC specifically stated that Yankee would not go forward with the project without third-party liability insurance, which was to be provided in the Price-Anderson Act. This act, which passed during the 30-day period, required the AEC to review each application with the Advisory Committee on Reactor Safety (ACRS); it also stipulated a public hearing before issuing a license. After Congress adjourned, President Eisenhower signed the bill within the ten-day period following adjournment. So early in September a public hearing was scheduled for October. The AEC did not have its own examiner so it used one from the Department of the Interior. The hearing was scheduled for 10:00 a.m., but was convened at 9:00 a.m. without incident or intervenor. It was over in 45 minutes.

On the way out, however, the AEC people ran into Leo Goodman, a United Auto Workers executive who had been active in opposing Detroit Edison's fast breeder. He had just jumped out of a cab and was headed for the hearing. Because the hearing was held early, he claimed that no notice had been given. He spent the day in various congressional offices successfully protesting and succeeded in having another hearing a week later. Again no qualified intervenor showed up, but a "limited appearance" was made in opposition by a woman representing the AFL-CIO Council of New England. She argued that, because of its limited financial resources, Yankee would cut corners in order to save money and an unsafe plant would result.

The construction permit was issued November 4, 1957. Each plant change had to be documented and filed with the AEC as an amendment to the license, which required staff review, an ACRS report, and public hearings. Twelve such hearings were held in 1958 and 1959.

What was Rickover saying about all this? At an important status report meeting for Yankee with the AEC reactor development division and the Schenectady operations office, Chave, of S&W, had to be held down by Coe and Creagan when Rickover said, "Stone & Webster lies about every cost estimate it makes." Stone & Webster had performed the plant engineering for Shippingport, and there had been much controversy over the S&W cost estimates.

That, however, was the extent of Rickover's opposition, because of Bill Webster. He had visited Rickover that morning and reminded him, "You owe me one."

Rickover knew what Webster meant. It was Webster who had cast the deciding vote when a committee of three (Conant of Harvard, J. Robert Oppenheimer, and Webster) had recommended that there be a nuclear submarine program for the Navy. Rickover had graduated from the Naval Academy in 1921, one year after Webster.

The vital hearing for the Yankee operating license was held on March 3, 1960. On that day there were ten inches of snow in Maryland. Sam Jensch, the hearing examiner (who had arrived on snow shoes), recessed the hearings until the following day. Then in three days, all vital issues were resolved, except that the final design was given only conditional approval.

During the next three months, several design changes were made, and each required an amendment to the licensing application, an ACRS report, and a public hearing. These changes were handled in hearings through May and June. Plant completion was certified at a public hearing on July 8, 1960. Fuel loading started the next day. Three weeks later (July 27, 1960), a provisional full-power license was issued [392 MW(th), 115 MW(e)]. Yankee tried to incorporate an automatic increase to 500 MW(th), but the AEC would not agree.

In addition to the hazards report, Yankee filed a request for AEC approval of plant design, operating procedures, technical specifications, and financial qualifications—to clear away as many procedural matters as possible in advance of plant completion. This request was granted only in part.

Westinghouse's ability to license and build plants with a minimum of extra expense and time was repeated on Conn-Yankee and San Onofre 1 and led the company to be willing later to offer turnkey plants, a subject that is covered in Chapter 8.

Operation of the Plant

Criticality was achieved at 8:19 p.m. on August 19, 1960 and the first electric power flowed from Yankee to the New England grid at 2:37 a.m. on November 10, 1960. Full licensed power was achieved on January 17, 1961.

A 500-hour test run was completed successfully in February. With the help of the in-core instrumentation system, the AEC approved on March 31 a power increase to 485 MW(th) [145 MW(e)], which was reached in late June. The plant was declared "commercial" on July 1, 1961.

Following the first refueling in May 1962, additional power increases were approved up to 540 MW(th) [185 MW(e) gross, 175 MW(e) net].

During this refueling it was decided to install a new set of follower rods, because unacceptable wear had taken place between the zirconium followers and the stainless steel end pieces of the absorber section. New followers were therefore urgently needed, but zirconium sponge, all of which was under Rickover's control, was not available. Due to the efforts of Webster and Westinghouse, Rickover was persuaded to release enough zirconium sponge to manufacture needed control rod followers for the second core. For the third core, Rickover let Yankee have enough hafnium for the cruciform control rods.

The first Yankee core was chemically processed at the Nuclear Fuel Services plant, West Valley, New York. The first commercial plutonium was sold to Germany by Jim Tribble, later president of Yankee, at $23.50 per gram.

Yankee Costs

What about costs? Webster received a mid-$40 million estimate for Yankee's predicted cost from Stone & Webster but he wanted to avoid having to go back to the financiers for more money. So he used $57 million for his financing. Actual costs turned out to be around $39 million, not including $4 million for the fuel, plus an overrun on the R&D contract.

Webster included in plant capital a first core cost of between $1 and 2 million of working capital; the result was the preplanned $42 million commercial cost. To achieve this objective, he ran the plant for about five and a half months and credited to capital the value of the power produced, at 10 mills per kilowatt hour, thus reducing expenditures to the $42 million figure.

Success Story

How long does it take to build a nuclear power plant? The newest plants have taken 10 to 12 years. Yankee was built in less than 5.

From November 30, 1955, when Yankee purchased 2000 acres in Rowe, Massachusetts, until first criticality was achieved on August 19, 1960, less than 57 months had passed. Twenty-one months later, on May 18, 1962, when core I was removed, it had generated 1,229,105,000 kWh, with a plant factor of 75%. This was in spite of some five months of operation at gradually decreasing power levels during the core stretch-out period. This was achieved by decreasing reactor temperatures to obtain more reactivity lifetime.

Yankee's first core was evaluated for chemical and isotopic content after burnup. The results were used to establish the amount of plutonium produced by Yankee. This work was done at the Westinghouse Waltz Mill site near Pittsburgh and was financed by the AEC.

During the first 18 months, the reactor was at temperature and pressure, ready to produce steam 95% of the time, with only two spurious scram signals being generated. The plant had a cumulative capacity factor of 77%, better than many current plants. Such high availability is even more impressive when one considers Yankee's many technical firsts, which are now standards in the nuclear industry.

Among them are slightly enriched (3.5% uranium-235) uranium-dioxide pellets, nucleate-boiling heat transfer, boric acid dissolved in main-coolant water for reactivity control (which resulted in better control rod positions for reduction of hot-channel factors), the low-pressure coolant-purification system, and a reactor vessel with no inlets below core level to reduce leakage possibilities and permit reactor-vessel constant temperature.

It was indeed a nuclear success story.

Yankee operated until October 1991, when the plant was voluntarily shut down to negotiate with the Nuclear Regulatory Commission (NRC) about the potential effect of pressurized thermal shock. There is a possibility that if a reactor vessel is cooled below the nil ductility transition temperature

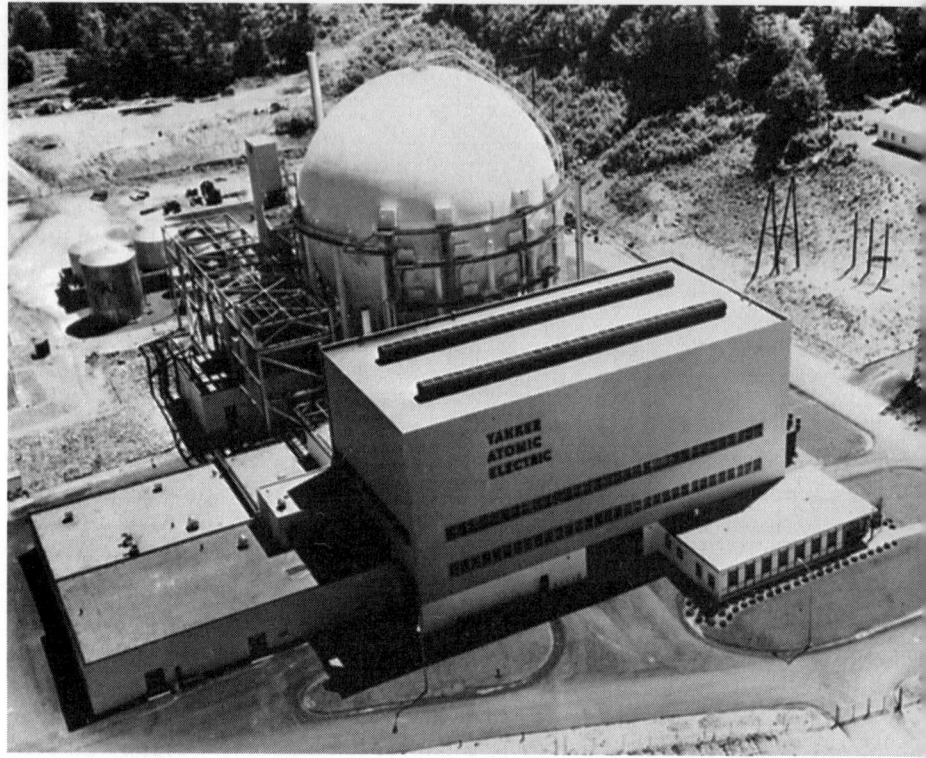

Figure 7.1 The Yankee-Rowe nuclear power plant. (Courtesy of Yankee Atomic Electric Company)

(NDTT) by localized emergency cooling water during a small-break loss-of-coolant accident while the vessel is still under pressure, stress at low temperature might fracture the vessel.

Yankee proposed to keep two reactor-coolant pumps running so cold emergency-injected coolant would be mixed with hot-reactor coolant water, but NRC questioned whether the 99% reliability requirement of the coolant pumps has been conclusively proven. The discussion was important because Yankee expected to set a pattern by being the first utility to seek reactor license renewal, even though many years remained on the original license.

However, Yankee has decided to shut the plant down permanently.

Westinghouse Test Reactor (WTR)[6]

The new commercial nuclear power industry lacked some of the essential tools necessary for the job. That's why the AEC urged them to build—and Westinghouse did decide to build—a nuclear testing reactor. This 60-MW(th) reactor was the first commercial testing reactor. Monty Schultz was the technical director for the project under Edmund T. Morris, general manager.

The construction permit for the Westinghouse Test Reactor (WTR) was issued on July 3, 1957, and it began operation at Waltz Mill, Pennsylvania, on July 1, 1959.

It was a low-pressure low-temperature water-cooled reactor. The fuel elements for this reactor were totally different from those used in today's power reactors. Instead of an assembly of 1200 or more tubes, 12 to 13 feet long and containing pellets of low-enriched uranium dioxide, the fuel elements for WTR were 4-feet long and consisted of three concentric aluminum metal cylinders alloyed with fully enriched uranium metal and clad with aluminum.

The WTR operated on commercial contracts in which various materials were inserted into the core and removed at the end of a 21-day cycle. Shielded piping (loops) and associated control and monitoring consoles were located on the floor of the vapor container to allow the material being irradiated to be exposed to the various environments.

The expended fuel from the WTR was removed from the reactor through a 16-foot-deep canal located under the reactor. The canal penetrated the reactor containment structure and was used to transport the irradiated fuel to the adjacent building for storage, or to transport customer experiments, etc., to the Hot Cell Facility located at the southern end of the building.

After a cooling period, the expended fuel was removed from the canal in a four-element cask and transported to the Transfer Building for storage in the 26-foot-deep pool prior to shipment to Idaho Falls in a 16-element capacity cask for reprocessing.

We kept WTR busy for five years, but not without problems. At 8:34 on Sunday, April 3, 1960, there was a partial meltdown of some of the fuel rods. This caused radioactive krypton and xenon gases to be released to the atmosphere. The site was immediately evacuated and plant personnel surveyed the surrounding area in Westmoreland County to determine the extent of the radioactive release. While radiation levels were high within the site, they were not above allowed levels at monitoring stations outside

---

[6]"Westinghouse People Make It Happen at Large and Waltz Mill," Westinghouse Electric Corporation, 1986.

the site. Westinghouse submitted a report to the AEC on July 7, 1960. However, the accident went virtually unnoticed by the press and the public at the time. There were no injuries and no adverse health effects reported at the time, nor have any been reported since.

The AEC required improvements to be made in the venting system before allowing the reactor to be restarted in October 1960. The reactor operated without further untoward incident until it was shut down.

Unfortunately, shortly after it was completed, the government, despite having urged us to build the WTR, built two other testing reactors. The AEC built the Engineering Test Reactor (ETR) and NASA built one at Plumbrook in Ohio. This took away most of our customers causing the WTR to be uneconomic, and it was retired in March 1962 just weeks after Max Johnson had become general manager. Johnson had been moved from a senior management position in the Naval Reactor Program in Idaho to his first general management job.

The AEC-approved plans for the retirement of the WTR included removal of all irradiated fuel and shipment for reprocessing, the sealing of all penetrations into nonradioactive areas, and administrative control of future access to the retired facilities. The WTR containment is still a visible landmark at the Waltz Mill site.

The former WTR head tank, a large tank supported on tall legs like a city water supply, was subsequently disposed of, and other reactor support facilities were decontaminated and decommissioned to make productive space available for active projects.

The Nuclear Regulatory Commission has begun a nationwide program to clean up and decommission closed licensed sites. Although there were no adverse health effects, during the cleanup process following the accident, thousands of gallons of water were used that became contaminated with radioactive material. Some of the water has contaminated the soil and groundwater. This cleanup must be completed before final decommissioning can be completed and will be quite costly. However, we maintained the site as a licensed facility. This enabled us to use the facility for the work of the nuclear services inspection division in handling radioactive material.

Carolina-Virginia Tube Reactor (CVTR)

On the theory that it is never good to stop exploring for better ideas, Westinghouse did not let its success with the pressurized water reactor block other possibilities. One of these other concepts that we explored was the Carolina-Virginia Tube Reactor.

This experimental reactor used zirconium tubes to contain pressurized heavy water, rather than employing the conventional pressure vessel. It required development of zirconium tubing that would withstand the stress. It also demanded some method of connecting high-pressure zirconium tubing to stainless steel pipes. A welded joint wouldn't work because of the different rates of expansion. So Art Thorpe designed a mechanical juncture that did not leak and could be disconnected to refuel the reactor.

Frisch thought up a unique fuel design that worked perfectly. The fuel was contained in assemblies that slid inside the zirconium tubes. To separate fuel elements, zirconium wire was spiral wrapped around the individual fuel rods and secured at the ends of the fuel assembly. Control rods were inserted between and parallel to the zirconium pressure tubes. The zirconium pressure tubes were contained in a vessel containing heavy water at low pressure.

The project was managed by Phil DeHuff, but used the same functional organization that was working on the PWRs. Perry Davison managed the critical experiments at Waltz Mill in the same facilities used for BR-3 and Yankee. Howard Arnold did the physics design, Long Sun Tong did the thermal analysis, and Frankie Frisch and Al Thorpe the mechanical design.

The reactor was completed in 1962 and operated successfully for the contracted five years. Design studies were performed to see whether a scaled-up version would be commercially successful. Although several variations were judged to be technically feasible and nearly competitive, it was decided not to proceed. One negative factor was the cost of heavy water inventory and makeup. Another was the lack of a ladder of experience and momentum that even then existed in PWRs. Certainly the Canadians have made a success of this reactor type. Ours would have differed in the use of slightly enriched uranium, which the Canadians have so far avoided although they continue to study it.

The Saxton Reactor

The 5000-kW(e) reactor at Saxton, Pennsylvania, was built by Westinghouse and used as a test bed for some of the chemical shim operation requirements. It served to confirm Yankee operation and help with licensing applications.

The Saxton plant was bought in January 1959 by Lou Roddis, a former Rickover aide, who became president of Pennsylvania Electric Company, a subsidiary of General Public Utilities and later president of Consolidated Edison. He appointed Bill Layman, formerly second officer on the *Nautilus*, as general manager. The Saxton plant used the turbine-generator of the coal-fired plant, located next to the nuclear plant. This dictated the steam pressure for the nuclear plant.

The supercritical once-through tube reactor (SCOTTR) prototype fuel elements and pressure tube were also installed in the Saxton reactor. The mixed-oxide (mixed plutonium oxide and uranium oxide pellets) prototype fuel elements were operated in Saxton to demonstrate that plutonium could be recycled in a PWR. Jim Wright was a key man in this project.

Nuclear Power Overseas

On December 20, 1956, Edison Volta announced plans for a nuclear power station in Italy and we received the order for it. On this plant we had trouble with the reactor vessel internals not being able to withstand the force produced by the coolant flow. They had been made in several sections due to the small neck of the pressure vessel, and then bolted together after insertion. For later designs the reactor vessels had larger necks.

On September 4, 1958, the Westinghouse International Atomic Power Company was formed, and news of a license agreement with Societe Franco-Americaine d'Energie Atomique was released. This was followed shortly by a letter of intent on July 28, 1960, and then a contract to build a 240-MW(e) plant at Chooz, near the Belgian border. This unit was the forerunner of the entire series of PWRs now supplying more than 75% of France's electricity.

The Edison Volta plant, called SELNI, and the Chooz plant, called SENA, were just in time to absorb the energies of the team coming off the design work for Yankee. They also did studies of larger PWRs, and soon we received a contract from the AEC to do generic design work on 330-MW(e) PWRs. On January 23, 1959, Westinghouse said it was ready to build such nuclear plants and on September 30 announced a program to build turnkey plants.

San Onofre-1

We were convinced that Southern California Edison (SCE) was the right utility for the next step after SELNI and SENA in the logical development of the PWR. Working with Walter Maytham, regional vice president in San Francisco, Weaver contacted the SCE senior officers and first met with them in Los Angeles September 8, 1959.

Present at that meeting were Harold Quinton, chairman; Jack Horton, president; and Jim Davenport, vice president. Later on the same day, Weaver met with Perry Yates and John Keilly of Bechtel who worked with us from that time on in selling SCE on building the plant. Seemingly endless and often most discouraging meetings took place at least once a month, often without result. Bill Budge showed his tenacity by working tirelessly to get SCE to go ahead with an order.

One rumor had it that SCE was negotiating for a gas pipeline contract and was simply using nuclear power as a bluff. The bluff was called. Frustrated, Westinghouse and Bechtel offered to build the plant with their money. Finally, Jack Horton broke the logjam and decided to go ahead with the plant. It wasn't until his check for the initial payment was picked up, photographed, and deposited, that we breathed a sigh of relief.

This San Onofre deal started the ball rolling toward what we really needed—contracts for large plants. We got a letter of intent from SCE for San Onofre-1 on April 25, 1960. Later the same year the letter of intent came from Connecticut Yankee Atomic Power Company. These plants were rated at 436 and 565 MW(e), respectively. Well, that was more like it. While these letters of intent certainly were welcomed, we realized that there would be a lot of tough negotiations before definitive contracts were signed. And there were.

Getting the San Onofre-1 Contract and the Site

Getting the San Onofre-1 plant under contract and built was as complex a problem as we had ever encountered up to that time. Everybody got into the act—the utility, Westinghouse, Bechtel, the U.S. Congress, and the United States Marine Corps, to mention a few.

The AEC provided $12 million under the Large Reactor Development (LRD) program for the plant design. This program was managed by Chuck Roderick, with the AEC looking over his shoulder very closely. They had to approve his quarterly reports. Then the LRD program was canceled after about $2 million had been spent, because the San Onofre-1 plant contract was not getting signed. Later the LRD contract was reinstated by the AEC and provided the basis for the design of both San Onofre and Connecticut Yankee.

Just when we thought we were getting close to signing the contract, Jim Davenport, SCE's executive vice president, who had been dragging his feet, seemed about to stop altogether. So Cotter and I visited Chet Holifield to try to get his help. After we explained the situation, and since the JCAE wanted this plant built as much as we did, Holifield agreed to help.

While we were sitting in his office, he called Davenport and told him that if he did not go ahead with the contracts promptly, SCE would never again get any government development funding in the nuclear area. Coming as this did from a California congressman and the co-author of the Gore-Holifield bill,[7] it got Davenport's attention. He could see the prospect of the government building a large nuclear plant at TVA or some other government facility, if they thought private industry was going to cop out. The contract got signed in a hurry.

---

[7] A bill to have the federal government build demonstration nuclear plants on government sites. The bill was sponsored by Al Gore, Sr., and Chet Holifield.

While the development contract was very helpful for Westinghouse at the time, it created a problem later. The AEC also got their money's worth, as other reactor manufacturers, such as Combustion Engineering and GE were able to get access to these data.

Frisch concentrated on design changes that would improve commercial PWRs over Shippingport technology. One of these was the "St. Peter in Chains" (SPIC) fuel assembly. Instead of brazing the fuel rods together in a rigid bundle, with ferrules as spacers, he came up with the concept, now universally used, of the rods being held by friction between spring fingers so they could expand differentially without warping the bundle. The fingers mixed the water flow, thus reducing the hot-channel factor. This permitted greater power output before departure from nucleate boiling (DNB) would be limiting.

This important bit of design work was done by Frisch, who also during a two-week illness, while he was at home in bed, designed the Yankee fuel. He used cross-section paper on his drawing board under his drawing paper to keep lines straight. We were thinking of putting a bed in his office after that.

Incidentally, that spring clip design was the subject of a lawsuit for years. Combustion Engineering had formally requested many Westinghouse-developed designs. Bill Abbott, who had been reactor engineering manager at Westinghouse moved to CE and was very familiar with what Westinghouse had developed. This included fuel element innovations such as the spring clip design. The issue was finally resolved when an AEC review revealed that Frisch had charged time to the AEC contract while doing the development. So the government won the U.S. patent rights. The solace to Westinghouse was that it was given the foreign rights so they could be included in what we sold to licensees.

Other design work involved earthquakes. Because of the earthquakes California experiences, we came up with some seismic designs. This provided us with a head start during our involvement in the Kansai plant at Tsuruga, Japan. Of course, earthquake tremor frequencies and directions are different in Japan than in California.

Getting the site for San Onofre was a problem of its own. The AEC was trying to get some land from the Marines at Camp Pendelton, but the Marine Corps was adamantly refusing to give up any land. And they had Senators Anderson and Jackson on their side. The problem it seemed went back to the building of the N Reactor at Hanford and GE's opposition to including a generator to produce electricity. Production of electricity by a government-owned nuclear plant seemed to many as a possible step toward more government production of electricity, such as at TVA. Just what this had to do with Camp Pendelton is far from clear, but there it was.

Cotter and I had an appointment with Chet Holifield late one evening to try to get his help in resolving the site problem. We finally got him to agree that things had gone on long enough, and that he would get the two senators to call off the hold on getting the site. Chet talked to them, and they agreed. Another good evening's work. It pays to have friends.

But not too fast. The senators were no longer objecting, but the Marine Corps had not yet agreed. John Conway and Holifield had a meeting with the Marine Corps commandant. The meeting got nowhere, but afterward Conway talked to the commandant in the hall and learned what his problem really was. He objected to the AEC insisting on a site in the middle of Camp Pendelton's shoreline. It would split their area for landing maneuvers in two. He told Conway that he had offered another site. Conway had never heard of this nor had Holifield. With this clue Weaver then met with the commandant and finally settled the matter. We got the site. But what a hassle!

We also had technical problems, some of which involved such things as fuel element corrosion and fatigue failure cracks. These appeared to be extremely serious at the time, because they meant shorter fuel life or perhaps switching to stainless steel cladding and higher enrichment, which would have added staggering costs.

In one meeting, the manager of our fuel department asked how much he would be allowed to spend on extra development. I told him there was almost no conceivable amount that was too much, if it would solve the problem because our exposure was in the billions. He would only be limited by his own ingenuity in coming up with a worthwhile development—which he did and the problem was solved.

This plant was completed in 1968 for about $175 per kilowatt, within budget and essentially on time. San Onofre-1 has operated successfully for 25 years and produced 52 billion kilowatt-hours of electricity. On August 11, 1992, the plant was permanently retired in accordance with an agreement with the California Public Utility Commission (PUC).

The plant was neither worn out nor unsafe. SCE believed the plant could be cost effective, based on near completion of NRC-mandated retrofits and the performance of similar plants. However, the PUC disagreed, but will permit SCE to recover the remaining $460 million net investment in the plant.

In 1992, the plant completed a site record of 377 days of uninterrupted operation. Overall, San Onofre-1 posted a record to be proud of.

Connecticut Yankee

In 1962 a group of New England electric utilities formed the Nutmeg Electric Companies Atomic Project under the leadership of Sherman R. Knapp, president of Connecticut Light and Power Company. Nutmeg began negotiating with Westinghouse for the construction of a nuclear power plant. These negotiations were carried on for Westinghouse by A.C. Monteith, utility group vice president, W.E. Johnson, Atomic Power Division manager, and W.L. Budge, his marketing manager. They were assisted by Paul Shiring and Bruce Morrison of the Boston office.

As negotiations progressed Nutmeg was supplanted by the Connecticut Yankee Atomic Electric Company with Knapp as president and Roger Coe playing the leading technical role and heading up the project for Conn-Yankee. On October 1, 1963, Knapp and I signed the contract for the plant. The AEC provided $6,050,000 for development with the proviso that the AEC would own the design. The AEC also waived the use charges, which were estimated to have a value of $7,145,000.

The total cost of the plant was estimated to be between $70 and $80 million. The Westinghouse contract was for $43,701,000.

This nuclear plant was very similar in technology to San Onofre-1; however, it had an additional reactor coolant loop, which permitted an increase in rating. The rating was initially to be 490 MW(e) with a provision for later upgrading to 590 MW(e). To accomplish this, the secondary plant was sized for 590 MW(e). The turbine was a newer design with 44-inch last row blades for increased power. The reactor coolant pumps were the shaft seal type rather than the canned motor type as on San Onofre-1.

Mel Yadon was the original Westinghouse project manager, but he moved on to another assignment and was replaced by Don Hunter with Walter B. Thee as site manager.

Knapp continued the policy set by Bill Webster of Yankee Rowe of paying careful attention to environmental problems. He gave a $400,000 contract to the University of Connecticut to explore the effect of hot water effluent on the Salmon and Connecticut Rivers. This was written up in *Scientific American* some years later. There were beautiful infrared pictures showing the flow of heated water, even upstream. An outflow cooling canal was added to provide some cooling before discharge into the river.

Stone & Webster was the architect/engineer/constructor for the plant. The construction permit was issued on May 26, 1964, with latest completion date to be before June 30, 1967. Today, that would be an unreasonably short time for the construction of a nuclear plant, but the core for Conn-Yankee was loaded on July 7, 1967, the plant went critical on July 24, and full power was reached on December 11.

Conn-Yankee has continued to operate quite successfully. It had a cumulative availability from the start until 1985 of 83.10% and up until 1992 of 77%.

This plant also was completed within budget and on time. Incidentally, both San Onofre-1 and Conn-Yankee were profitable. No other company built a plant that was profitable for them for many, many years.

My Responsibilities Change

In 1962, my career took a new turn. Mark Cresap made me vice president, engineering and research. That meant staff responsibility for all engineering in the corporation and line responsibility for the Central Research Laboratories. I put Sid Krasik in charge of Astronuclear and chose Bill Shoupp to be the director of the Central Research Laboratories.

Since July 1959, I had had direct line responsibility for Westinghouse commercial atomic power. Now I had staff responsibility only and Astronuclear reported directly to C.H. Weaver, who was vice president of the Atomic, Defense and Space Group. Although I no longer had any line responsibility for nuclear power, I kept up to date with it as part of my new responsibilities.

Westinghouse had a cross license with Siemens of Germany for all products, but with no payment by either. The only exception was that Siemens made one payment to us at the start for the nuclear portion of the deal. I was one of the Westinghouse representatives that had to meet at least once a year in each country. The meetings in Germany were to be in German, so I set about learning the language.

A lady at the research labs who taught German came to my office once a week for a two-hour lesson. Since the Germans I dealt with spoke excellent English, I never really became proficient in their language. I did learn enough, however, to understand some of it and could at least engage in the social amenities

My year as vice president of Engineering and Research was uneventful. I was on the corporate administration and capital expenditures committees, but mostly I was just learning about the rest of the corporation and getting to know what I thought needed to be done. There was a lot of contact with Don Burnham, who had been vice president of manufacturing, but was now industry group vice president. Burnham had come from General Motors, where he was engineering manager of the Oldsmobile division.

During this period, Mark Cresap contracted hepatitis and was out sick much of the time. John Hodnette was really running the company. In July 1963, Mark Cresap died, and Don Burnham succeeded him. I was promoted to group vice president in charge of the electric utility group, succeeding Monteith.

My new world—the electric utility group—was a good news, bad news situation. On the good news side, it was a business of tremendous scope and importance to the corporation. It included more than 20 plants in cities all over the United States and manufacturing facilities in Spain, Belgium, Argentina, Brazil, the Philippines, Greece, Australia, and France. I was in charge of the design, manufacture, and marketing of everything sold to electric utilities throughout the world. This ranged from a home watt-hour meter to a turnkey nuclear plant. There were about 70,000 employees.

On the bad news side, the utility business was in the doldrums. There were very few orders, and they were at depressed prices due to heavy competition. Also, we were living in the aftermath of the famous electric antitrust cases.

Strangely enough, during the period of the alleged price fixing, Westinghouse was selling our biggest product—turbine-generators—at a loss. Shortly after I started my new job, GE announced an increase in steam turbine prices. They also said that they would not discount from book price, and, if they lowered prices, they would give the same lower price to anyone who had bought within the prior six months. After years of disastrous price cutting, did they really mean it? I knew that their new division manager had come from the meter division, and the meter manufacturers were noted for not cutting prices.

So I decided to believe GE and raised our prices to match and gave orders that under no circumstances were we to cut a price. It didn't take long for the test to come.

We were at the annual corporate Management Council meeting at Sea Island when it happened. We hadn't had a turbine order for a long time and needed business desperately. Consumers Power offered to give us about a hundred million dollar order, if we would just reduce our price $40,000. Some of my people were pressuring me to cut prices—old habits die hard—but we stood firm. We got the order anyway and there was no more price cutting.

Not having been connected with the utility business during the price fixing days was a big help in my relations with utility customers. They were all thinking about building nuclear plants, and I had met many of them while at Bettis. Fortunately for me, orders picked up sharply. Of course, that's the utility business. Feast or famine.

CHAPTER 8

Commercial Nuclear Power 1963–1993

With its spectacular successes—*Nautilus, Enterprise,* Shippingport, BR-3, SELNI, SENA, Yankee Rowe, Connecticut Yankee, San Onofre, and the prospect of major international sales in France and Japan—Westinghouse by the early 1960s had bolted to a position of industry leadership in nuclear power beyond our most optimistic hopes.

It seemed that Gwilym Price's gamble in giving atomic power top priority back in 1948 was one of the shrewdest moves in the company's history. Now the firm was ready to capitalize on this success by dominating the market for nuclear central station power plants all over the world. Add the prospects for supplying nuclear fuel and the potential business was staggering.

Sure, the electric utility industry had always been known as a "feast or famine" proposition for the equipment manufacturers. It had been that way, I guess, ever since the days of George Westinghouse, but it looked now like there was nowhere to go but up.

For the next 14 or 15 years that's exactly what happened. In that period from 1963 to 1977, Westinghouse sold 53 large nuclear plants in the United States and 42 overseas. In addition, 83 nuclear plants were sold by our foreign licensees—all sales on which we collected license royalties.

That's far more utility business than we had done in the previous 77 years of Westinghouse history, and it appeared to be only the beginning. The best forecasts estimated that growth in the demand for electricity would be from 5% to 7% and even in 1977 the forecast was for about a 4% growth rate.

But then some totally unexpected things happened. Orders for new power plants virtually stopped. In the 15 years since 1978, not one order for a nuclear plant has been placed in the United States. The feast ended abruptly and famine set in.

I will go into the famine story shortly, but first let me tell you what happened at the feast.

The Nuclear Power Market Challenge

We had entered the period when nuclear power became a marketing challenge. The major technical hurdles had been overcome. We knew how to build the plants. Now the problem was how best to market them in our country and around the world. It seemed clear that if there were pitfalls, they would be economic, not technical. But we didn't let down on technical progress; the business was too good not to have strong competitors. We knew that General Electric (GE), Combustion Engineering, and Babcock & Wilcox (B&W) were going to provide stiff competition, so we redoubled our efforts.

Mark Cresap, the Westinghouse CEO, died in 1963 and Don Burnham, the industry group vice president had succeeded him. I was made electric utility group vice president. The electric utility group consisted of more than 20 plants in cities all over the United States and manufacturing facilities in Spain, Belgium, Argentina, Brazil, the Philippines, Greece, Australia, and France. The group executive was responsible for design, manufacture, and marketing of everything sold to electric utilities throughout the world. This ranged from a home watt-hour meter to a turnkey nuclear plant. There were about 70,000 employees in the group. Don was three months younger than I, so there seemed little possibility of my ever becoming CEO, but in my new job I felt I could have a profound influence on the future of the company.

My new position once again had given me line responsibility for all commercial nuclear power and, with nuclear power beginning to show real growth potential, I was prepared to take advantage of it. Of course, the utility business was then in the doldrums with very few orders being placed and those at depressed prices. But with the utilities forecasting strong growth and their reserve margins getting low, new plant construction was inevitable.

The opportunity for a creative marketing effort was certainly there, and we set out to make the strongest possible effort. One of the spearheads of that effort was the Westinghouse Future Power Market Forum, a two-day event held every four years at which we hosted the top executives of every major utility in the country. At this meeting in Pittsburgh in 1964, no expense was spared. The heart of the meeting was a thorough and detailed look into the future to predict technological and economic trends important to the electric power industry. We were the ones doing the developments, so we were in a better position to determine the trends in such things as transmission voltages, plant sizes, and the like.

Our largest operating plant was the 175-MW Yankee, although larger units—San Onofre, Connecticut-Yankee, SENA and SELNI—were under construction. Our technological studies convinced us that nuclear power would be economical only with units of 500 to 1000 MW(e), owing to economy of scale. Fortunately, many utility systems were now large enough and growing sufficiently fast to accommodate these larger units.

For quite some time, we had been looking into what would be required for these larger plants, and our development was well along. Other parts of utility systems also came under scrutiny. We decided that with larger generating units, higher capacity transmission lines would be needed. The 345-kV lines would no longer do the job. Voltages of 500 kV and possibly even higher would play a major role. This exercise focused our attention on what would be important for the future and the forum message got it across to the utilities.

Our planning department worked for many months to prepare for these meetings. They created eye-catching visuals and technical scripts written in words utility presidents could understand. Professionals were hired to write catchy songs and stage skits that would get across various points we wanted to make. The show cost nearly half a million dollars, but we had a captive audience that would purchase four billion or more dollars of equipment a year. Broadway never had an audience like that.

Industry meetings also offered important marketing opportunities. Each year the three principal manufacturers—Allis Chalmers, GE, and Westinghouse—would make presentations on the state of the industry at the Association of Edison Illuminating Companies (AEIC) conference, at either the Greenbrier in White Sulphur Springs, West Virginia, or the Boca Raton Hotel in Florida. Because we spoke in alphabetical order, Westinghouse was always last, just before the meeting ended and everyone headed for the golf course. To hold the audience at all, I had to give a "stemwinder." I also tried to make my talk less of a sales pitch and more of an analysis of where the industry was making mistakes and what public relations steps they should take to promote nuclear power.

I spoke every year from 1964 until 1974 and apparently acquired a reputation for the tenor of my remarks. One Saturday after my speech, I was in the back of the elevator and heard a GE man say to a friend, "There's that John Simpson again telling them what they ought to do and the S.O.B.'s are believing him." Well, not always, I'm afraid.

I have to admit that selling nuclear power plants or other big apparatus to electric utilities had its pleasant side. There are only a few hundred men who are responsible for letting contracts of several billion dollars a year. With so few customers and with each order so large, the executives of my group got to know them quite well and we didn't hold back on the entertainment.

Naturally these utility executives liked to hunt, fish, and play golf. Several times we took groups for deep sea fishing off the coast of Mexico at Mazatlan. On one three-day trip, we caught 29 sailfish and marlin and we also included a dove hunt. There were more doves flying that day than any of us had ever seen.

Several times a year, we held hunting outings at Rolling Rock, a 10,000-acre club in the Allegheny Mountains about 50 miles east of Pittsburgh. The club provided controlled hunting of quail, pheasants, and ducks. We started with dinner Thursday evening, hunted Friday and Saturday, and were home by Saturday evening. Our year's customer entertainment schedule also included many golf outings at Laurel Valley near Latrobe, Pennsylvania, or at a regional executive's club, such as Winged Foot in Westchester County, New York. This was tough, dirty work, but somebody had to do it. In fairness, we shared this "burden" by dividing these outings among a large group of our senior people.

Selling in the international market was a little different, but we also gave it our best shot. In 1964, the third Atoms for Peace Conference was held in Geneva. Remember the boat ride? This conference provided a unique opportunity for us to contact potential customers from all over the world. Of course, U.S. customers were there also, so we had a hectic time touching all bases.

At that time our record of nuclear accomplishments looked good. Shippingport and Yankee were performing well, as were BR-3, Saxton, the Carolina-Virginia Tube Reactor and our materials-testing reactor. In addition, we were well along in design of San Onofre, Connecticut Yankee, SENA and SELNI. All of this was buttressed by our nuclear Navy experience. Nobody else had that powerful a nuclear story to tell. Opportunities like this conference gave us more chances to tell our story. You don't sell many nuclear plants in only one or two meetings; it takes a long campaign.

With future prospects bright for the utility market, we were in a pretty good position from a competitive standpoint with many of our product lines, but not all. Westinghouse was first in large power transformers, for example, but lagged badly in turbine-generator sales. And we were not nearly as profitable there as GE, our chief competitor. Westinghouse was behind them in most lines and seemed to have a mind set that they couldn't beat GE. We had beaten them in naval reactors and my principal aim was to be number one in the world in nuclear power.

Getting to that position would take leadership, planning, additional development and manufacturing facilities, and above all good people. Good people are essential for success and we were doing everything possible to bring the best talent into our nuclear organization. We had no deadwood and we were determined not to accumulate any.

The soundest basis for any sale is a superior product. But even with that, the biggest part of the selling job is done by the people down the line in countless meetings and discussions with the customer at all levels. Seldom will top executives buy, if the purchase is not supported by those directly involved with new power generation.

It was at this time in 1964 that Sid Krasik became seriously ill and we needed a highly competent technical person to take his place at Astronuclear. Weaver and I, together with George Wilcox, vice chairman to whom we both reported, chose Woody Johnson. He was performing well as manager of the atomic power division, but his technical competence was badly needed at Astronuclear. He was replaced by Joe Rengel, who was an excellent manager with a good marketing background. Joe was well suited by experience to run the division. Rengel continued to head the division until his retirement in 1978, at which time he was superseded by Ted Stern. By then, the division had grown to be the nuclear energy systems group.

Stern, who came from Foster Wheeler in the early days of CAPA, was one of Rengel's principal technical managers. He later became senior executive vice president of the corporation. Stern must be given much credit for Westinghouse's success in nuclear power, both from a technical and a business point of view.

In the mid-1960s, Don Povejsil, John Stiefel, and John Taylor came to the division in key positions. Taylor left Westinghouse in May 1981 to become vice president for nuclear energy at the Electric Power Research Institute.

New Orders Needed

Things were looking bright, but where were the new orders? San Onofre and Conn-Yankee were well along, but we had not received a major order since those two. A number of utilities were showing interest, but none was stepping up to the plate with an order. In such a situation, if you can persuade just one of them to take the step, it can break the logjam. We decided to concentrate on McGregor Smith of Florida Power & Light and to help with this we called on Tom Fort, utility group marketing vice president.

People in the atomic power division were not experienced in dealing with electric utilities; Carroll Roseberry's experience was limited, Fort brought needed detailed knowledge of people in utilities. He identified the decision-makers, who they depended on for advice, and he provided other information that was so helpful in negotiations.

In the spring of 1964, therefore, Fort arranged a meeting in Miami with Smith to discuss our latest thoughts on nuclear power and to see if nuclear fit into the future of his system.

Smith was a well-known industry leader and a "character" known for his very positive opinions. He started the meeting by stating flatly that oil was the only way to go. He pointed out that oil was about $2 a barrel, adding that he never paid the market price since he always could persuade one of his suppliers to give him a better deal. What's more, he said, the price of oil had been coming down, so why should he bother with anything else?

So we started our presentation from our one-yard line.

We began by telling him there were limits on our natural resources and we were offering him an alternative to fossil fuels. As the supplies of coal, gas, and oil dwindled, prices would rise. Here was a new source with a bright future. Wouldn't it be in his best interest at least to take a look? The discussion went on at length and we could see that he was getting interested. He probably hadn't been as completely sold on oil as he pretended at the outset.

Finally he agreed to let us submit a preliminary proposal for a nuclear plant rated at about 660 MW(e). It would be located at Turkey Point, an ocean site where there were two oil-fired units already. We put together a rough proposal, and after many discussions with his staff took it to him for a review. He was impressed enough to advise us to go further. Then he got in touch with our competitors to see if he could play us against each other.

Smith loved to negotiate. He played the Will Rogers role, telling us that he was just a country boy up against the city slickers. He prided himself on being able to make deals that would give him the best of the bargain. Then he would play his trump card by asking to talk to somebody at the top of the company whom he knew. A.C. Monteith, our former utility group vice president, knew everybody in the industry's old-boy network and could negotiate with the best of them. So we asked Monty to talk with Smith. He did and Smith gave us the go-ahead. We were to supply the nuclear steam supply system and the turbine-generator, and Bechtel was to do the engineering and construction of the plant. A second unit was added a year later.

The logjam was broken and we lost no time in letting the world know it.

The annual Edison Electric Institute meeting in Miami Beach was being held right at this time and we arranged for Don Burnham to announce the Turkey Point order at the luncheon I usually gave for utility presidents.

Burnham made the announcement from the podium at the Fontainebleau Hotel and thanked Smith, who was the guest of honor. Smith was dressed like a cowboy—the only cowboy among that group of dignified utility presidents, I need not add. After he made a few remarks, he pulled a harmonica out of his pocket and played a rousing chorus of "You Are My Sunshine." It brought down the house.

McGregor Smith was one of a kind. He loved to hold 6:30 a.m. meetings, because he thought he would be more alert than others at that hour and would have an advantage. He was over 80 but still would let nothing interfere with his Thursday afternoon tennis game. Walt Dollard, one of our young engineers, frequently played with him. Smith took strong likes and dislikes to people. He liked Ray Witzke, but said most of the rest of our group were thieves and bums. He liked Monteith, probably because he thought he had gotten the best of Monty in their many negotiations. But I doubt if he did. You hadn't lived if you hadn't had a ride with Smith in his swamp buggy, a shallow-draft boat with an airplane engine and propeller in which he cruised through the swamps around Turkey Point, on both land and sea.

Ravenswood

Consolidated Edison of New York had built a 275-MW(e) B&W reactor at Indian Point on the Hudson River and in 1962 decided to build a one-million-kilowatt nuclear plant at their Ravenswood plant on the East River north of the Queensboro Bridge and across from Welfare Island. The proposed plant was expected to cost $175 million and was to be designed by Westinghouse and Stone & Webster. This plant was to also have an oil-fired superheater, although it was expected that such a superheater would not be required on later plants. The nuclear plant was to be enclosed in a shell whose walls were 7.5 feet thick, 167 feet high, and with a diameter of 150 feet.

On December 10, 1962, Consolidated Edison applied to the Atomic Energy Commission (AEC) for a construction permit, and expected to have the plant completed by 1970 and producing power for 7 mils per kilowatt hour.

Although a number of people spoke out against the construction of this plant, most notable among them being David E. Lilienthal, former chairman of the AEC, Con Ed believed that the majority of the people of the city would not be strongly opposed.

In January 1964 the company announced that it had abandoned any plans to build the nuclear plant at Ravenswood. Instead the utility would buy some 2000 electrical megawatts of power from a 4.5-MW(e) hydroelectric plant 1000 miles away in Labrador. The Canadian power was to be delivered at the Quebec New York border for about 2 mils per kilowatt-hour due to heavy government subsidies. On the reasons for its decision to abandon the Ravenswood plant, Con Edison said that "Present prospects for securing large amounts of hydroelectric power from Canada on an economically favorable basis resulted in this deferral of plans to build an atomic power plant." The company mentioned no other factors.

Turnkey Plants

Yes, the logjam was broken and the nuclear orders began to flow in, but there was a catch. The nuclear feast included an ingredient that made some of us a little sick before it was all over. It wasn't turkey, it was "turnkey."

A "turnkey plant" is one in which the supplier takes the order for a specified price, designs and builds the entire plant, then turns it over to the utility customer to operate. GE started this when it sold the Oyster Creek plant at a fixed price, no doubt expecting a loss on that first sale. However, that sale created a great interest in nuclear plants, because it appeared to promise economic competitiveness. The utilities saw it as a major breakthrough. Utilities were inexperienced in nuclear power and preferred that the reactor manufacturers take a turnkey responsibility. GE eventually sold 10 more turnkey plants.

Believing we could repeat our successes with San Onofre and Conn-Yankee, we also began selling turnkey plants and sold six large nuclear plants on that basis in about a year. This was almost four times the total capacity we had sold up to that time. To protect ourselves against cost increases, we included an escalation clause in these contracts, but the escalation proved to be woefully inadequate. These were our turnkey plants:

Rochester Gas & Electric's Ginna plant, 470 MW(e)

Consolidated Edison's Indian Point 2 and 3, 1000 MW(e)

Wisconsin Electric Power's Point Beach 1 and 2, 485 MW(e)

Carolina Power & Light's Robinson 2, 665 MW(e).

We were well into design and construction of Indian Point 2 when we started negotiations for Indian Point 3. We knew by then that our price on the first plant was much too low, so we increased the price for the second—

enough, we thought, to cover cost increases. As it turned out, it still was not enough.

When the bidding after that really got tough—prices with escalation were less than $100 per kW—we simply dropped out of the auction for some of the last orders. We knew we could not build plants at those prices and make a profit. We had won enough orders to acquire the experience we needed, however. We had lost some market position, but that was soon to change.

Why couldn't we adequately estimate construction costs at this time? The whole political and public acceptance environment changed abruptly for the worse just about the time we took the turnkey orders. The age of nuclear activism and environmentalism had come. Before then, the word "ecology" was hardly known. There had never been a real adversarial public hearing on nuclear, much less one with many well-financed intervenors. The rules kept changing, frequently after designs were completed and sometimes after equipment was installed as approved by the regulators.

At Indian Point 2, for example, the piping, including the large thick-walled main coolant piping, was already installed when we were forced to add seismic restraints. This was to prevent the pipe from whipping about in the extremely unlikely event of a pipe break due to an earthquake. The extra cost was huge. Was it necessary? Maybe yes, maybe no. There is some evidence that making the piping stiffer, if anything, made things worse instead of better.

This sort of thing caused big delays in construction; very expensive delays as interest payments on financing continued. For many years it had been possible to predict construction costs with reasonable accuracy. Now these costs rose abruptly and unpredictably. In addition to the regulatory problems, when the industry ordered 18 plants in about a year, it overloaded the nuclear construction industry, which in itself caused costs to rise.

Westinghouse had completed Connecticut Yankee and San Onofre essentially on time and within budget and had made a profit on them. The changes required to obtain licenses for Conn-Yankee and San Onofre cost only a few hundred thousand dollars. Such was not to be the case with the six turnkey projects, where such changes amounted to several hundred million.

The first few, while exceeding construction cost estimates, were started and finished soon enough to avoid big cost overruns. Indian Point 3, however, required a great many major design changes, even after equipment had been installed. This added tens of millions to its cost.

Westinghouse PWR turnkey plants cost from about $120 to $440/kW(e). Within a few years some nuclear plants were costing as much as $4000/

kW(e). While Westinghouse construction overruns were considerable, they have been greatly overestimated by outsiders, and by today's standards would be considered minor. Further, the fuel costs were even lower than predicted and continued on a favorable trend.

Turnkey Summary

In truth, the effect of the turnkey plant sales on Westinghouse has been greatly overstated.

The total contract price for the six turnkey nuclear plants was more than $700 million. Even after taking into consideration the profit on equipment, the total loss was almost $200 million. However, this was spread over several years and was offset by profit made on reload fuel for these plants. Part of the loss was overhead that would have been spread to other orders. The net effect was an after-tax loss of about $10 million. This was a small price to pay for becoming the leading supplier in nuclear power. So despite this overrun, these contracts were extremely valuable. We would otherwise have been out of the nuclear business, because GE would have swept the order board. The cost to them would have been high, but they would have survived, and we would never have caught up. As it turned out, our plants were built at less cost and with less loss and fewer troubles than GE experienced.

Man-Hours of Labor for Various Plants

Due to differences in utilities, architect/engineers, constructors, labor unions, and site characteristics the man-hours for plants varied considerably as evidenced by Table 8.1.

Indian Point 2 and 3 were turnkey plants on the Consolidated Edison system and were built in Westchester County, New York, a high labor-cost area. The second plant did not benefit from being second, primarily because there were so many changes between the two.

By far the best performance was on the second Surry plant; there, man-hours per kW(e) were about half of most of the others. By far the worst were the Salem plants on the Public Service Electric & Gas system in New Jersey. This was a difficult labor area. Zion 1 & 2, on the Commonwealth Edison system, were among the best.

The nuclear steam supply systems were similar for all plants. Many factors are responsible for the labor differences, including the time they were built, the labor market, and the competence of the architect/engineers, the constructors, and the utility.

Table 8.1
Turnkey and Non-Turnkey Plants

Plant	Actual or Estimated	Man-hours	MW(e)	Man-hours/kW(e)	Year of Operation
Indian Point 2	Actual	7,869,744	873	9.0	1974
Indian Point 3	Estimated	7,790,800	788	9.9	1976
Surry 2	Actual	3,846,301	788	4.9	1973
D.C. Cook 1 & 2	Estimated	19,500,000	1020 & 1090	9.2	1975–78
Zion 1 & 2	Estimated	12,100,000	2–1050	5.8	1973–74
Salem 1 & 2	Estimated	32,300,000	2–1106	14.6	1977–81
Beaver Valley 1	Estimated	7,850,000	852	9.2	1976

Source: Nuclear News, Vol. 37, No. 3, p. 62, March 1994.

The Flood Gates Open

These projects gave the industry the experience and confidence to go forward. There was, however, a shift away from turnkey contracts. Westinghouse was willing to take them, if all suppliers were bidding on a turnkey basis with cost escalation. But others were not, and we would not bid a firm price with escalation in competition with a cost-plus-fee proposal.

With turnkey plants not available, some utilities started doing their own engineering and construction, while others employed architect/engineers and constructors. Many utilities, however, lacked the experience to handle nuclear plant construction. This proved very costly to them.

Despite the turnkey episode, the period from the early 1960s through the mid-1970s was truly the marketing manager's dream come true. We received orders for 20 nuclear plants in 1966 and 30 in 1967. We received orders for three plants on a single day—a record. When you sell 95 large plants and profit from 83 more sold by our licensees in such a time span, you are on a roll.

Building Manufacturing Capacity

To take that volume of orders, we had to have the plant capacity to deliver the product, and that became an issue in sales negotiations.

It first arose in our efforts to sell a plant to Commonwealth Edison in Chicago. As described earlier GE had sold them their first plant, Dresden-1, and had sold them four more large nuclear plants. We were having a hard time persuading them to accept another design to contend with. In discussions with their president, Tom Ayers, and chairman, Harris Ward, we were about close to an agreement when Ayers brought up the subject of turbine

Figure 8.1 The control room at Diablo Canyon nuclear power plant. (Courtesy of Pacific Gas and Electric Company)

availability. The utilities weren't only buying nuclear plants, they had ordered large numbers of fossil plants. He feared our turbine factory was overloaded and he wanted a commitment that we would build another turbine plant.

Knowing we were going to have to do so soon anyway, I made the commitment. That same afternoon I flew in the company Jetstar to Philadelphia, where we manufactured steam turbines, to attend the retirement party of their engineering manager, John Carlson. We made it in 80 minutes—a record, I believe. All the time I was worrying about the commitment for a new turbine plant.

My worry was that the appropriation of more than a $100 million to build new turbine plant capacity would have to be approved by the Westinghouse board. What if they didn't approve? I was way out on a limb with the customer. Sure enough, it did get rather sticky at the board meeting. After our presentation, one director asked whether they really had any alternative to approving, because we already had made the commitment to

Commonwealth Edison. I had to admit that they didn't. But I pointed out that, with the orders rolling in as they were, expansion obviously was essential. Our commitment in Chicago, I said, was just a bit premature. The board allowed as how it sure was, but approved the appropriation.

Nuclear turbine manufacturing plants at Charlotte and Winston-Salem in North Carolina were just a part of our commitment to being first in nuclear power, both in engineering and manufacturing. We built plants in Tampa to build steam generators and in Pensacola to build reactor internals. The nuclear fuel manufacturing plant in Cheswick was expanded and a new plant was built in Columbia, South Carolina. Extensive test and development facilities were constructed at Waltz Mill, Pennsylvania.

Nuclear Fuel

As film is to the camera manufacturer, nuclear fuel is to the reactor builder. It's a business that has the potential of continuing sales for the life of the plant. The customer utility was free, of course, to buy future fuel from our competitors, but the utilities thought that they might experience difficulties buying future fuel from anyone but the plant vendor for technical reasons. So they wanted options to buy many years of nuclear fuel at a specified price on purchase of the plant, because they were afraid that we would sell the plants at a low price in order to lock in future fuel sales. We also wanted to sell as many reloads with the initial plant order as possible.

With this big backlog of fuel orders, we were able to increase our development effort, not only in fuel design, but also in its manufacture. With this much effort we could stay on a steep learning curve and make it tough on our competitors. We kept the cost of fuel decreasing even more rapidly than we had predicted. This fuel business has continued to be a mainstay of our nuclear business, along with reactor service, long after plant orders were a distant memory.

Getting on a learning curve on the entire nuclear plant, however, was much harder, because it seemed no two plants were ever really duplicates. Even two plants on the same site had differences due to regulatory changes. We advocated standardization, continually. At one point we proposed a Standardized Nuclear Unit Power Plant System (SNUPPS). We actually sold six of these duplicate plants, which we called a six pack, but the decrease in load growth caused several of these to be canceled. Two of these SNUPPS plant were built—Calloway for Union Electric of St. Louis and Wolf Creek for Kansas Gas & Electric. They have been excellent performers and the plant design was used as the starting point for the British PWR at Sizewell. In the interest of standardization, we also developed a floating nuclear plant. More about that later.

Antinuclear Activists Enter the Picture

In the early 1970s, although nuclear plant sales were still strong, a few dark clouds began appearing on the horizon. Intervenors were entering the picture and delaying the licensing process. In 1974 environmental impact statements were required for the first time. Also in 1977, the Joint Committee for Atomic Energy (JCAE) was terminated. It had been very knowledgeable and was the sole congressional committee responsible for atomic power. We watched the entrance of several new committees, each of which could hold up progress while none alone could get anything done. One cloud concerned emergency core cooling. This was more than a cloud. It was more like a twister whirling across the landscape.

It started with some scale tests on the Special Power Excursion Reactor Test (SPERT) reactor in Idaho that cast doubt on what were then standard industry analysis techniques. A classic case of regulatory ratcheting and panic ensued. We mounted an all-out program of testing in our own labs, testing in cooperation with other vendors under government-approved circumstances, analyses, and studying of possible design changes.

The storm lasted several years, culminating in generic regulatory rulemaking hearings by the AEC. Many of the changes did improve safety, although by how much is open to question. Some of the changes were unnecessary or perhaps even harmful. In any event, the result was many long delays and greatly increased costs.

The costs began to mount as the community of nuclear critics learned the art of delay. They learned in real cases; then with the taste of blood, became vengeful. They learned how to demand and get hearings and keep them going for long periods—due process at its worst. They seldom had any hard evidence of problems. They didn't have to prove anything—just raise questions that were difficult or impossible to answer. In any event, costly engineering analyses were required. It is difficult to prove absolutely that something can't happen. For example, how do you prove that the worst hurricane in 100 or more years can't, under any circumstances (maybe an incipient crack), cause a pipe to break? Perhaps the hurricane occurred during an earthquake.

Many of the antinuclear people no doubt were sincere in their objections, though almost none had any sound technical basis for their stand. Many, however, had been anti-Vietnam war activists and were looking for a new cause—the Jane Fondas of the world. They found it in nuclear power, where they could raise fears of the unknown. They didn't trust the government at all and so had no confidence in anything government officials said. Still others were professional activists making their living by objecting to and blocking nuclear power.

No Westinghouse nuclear plant was stopped and canceled by these activists, although the delays greatly increased costs. With the OPEC oil embargo and sky-rocketing oil prices, the economy went into a tailspin. Utility loads no longer increased at 7% a year, and in some cases they actually declined. This lack of need was responsible for a number of cancellations.

There was the need for very costly increased testing and documentation of everything. As costs kept going up, the economic advantage of nuclear power decreased. For the utilities the earlier plants were clearly very cost effective, but this margin continued to decline.

There was irony in the fact that, as the cost rose—frequently more the result of delays than anything else—the activists causing those delays were in the forefront of those charging that nuclear power could not be economically justified.

For us there was at least one small silver lining to the cloud. During this period, we were able to work out a strategy for selling any required upgrades and improvements on a customer-by-customer basis and emerge unscathed. This was a strategy worked out by Howard Arnold, then PWR general manager, and Frank Bakos, then commercial policy manager in marketing. One example of this approach was in meeting the need to improve core cooling in the event of a pipe break accident. Frank Retallick was put in charge of the technical efforts to come up with alternative fixes that could be offered to customers

Howard Arnold recalls joining as engineering manager in 1970, with about 600 engineers in his group. The financial people wanted him to cut the number of engineers on the basis that the many plants being sold were really duplicates and that all that was needed was good project management. Soon thereafter, it became clear that changes were inherent in the business and that they could be turned into a profitable sideline. Instead of being cut the group doubled in two years. Fortunately, many competent engineers were becoming available, including Retallick from our NERVA program.

Frank brought together a number of ways of getting more coolant into the core in the event of a pipe break or other event that caused loss of coolant. We settled on a twofold approach. For those plants that had time to make a change-over, the fuel was changed from a lattice of 15 by 15 rods in an element to a 17 by 17 array. This resulted in lower temperatures during accident transients. Where this couldn't be done or wasn't enough, we proposed an upper head injection system.

There also was a good market for our services in making improvements and in assisting in maintenance and refueling. That part of the business remained profitable even when new orders dried up.

PWRs Dominate the Market Worldwide

The PWR concept has proved to be the best, and it helped put Westinghouse in the position of world leader in nuclear plants. The PWR is the dominant type throughout the world. Almost all French plants are PWRs. The British are building a PWR at Sizewell that is intended to be the lead plant of a series. Siemens, which builds both, sells more PWRs than BWRs. Only in Japan has the BWR stayed ahead in a close race and this is due to the GE relationship with Tokyo Electric, the largest electric utility in Japan and Toshiba.

The MW(e) of reactors in operation and under construction or on order worldwide are shown in Table 8.2.

The Uranium Suits

You thought the turnkey experience was bad, right? It was nothing compared to the uranium contracts episode. Here again we were victims of escalation that went sour. How were we to know that the tremendous surge in nuclear plant orders would enable the uranium cartel to push prices through the roof?

In the period from 1970 to 1973, when we sold nuclear fuel, our customers sometimes asked us to supply the necessary uranium. We agreed to supply the uranium at a fixed price plus escalation according to a widely used index. So what happened? The price of uranium quickly jumped from around $8 per pound to more than $40 per pound.

Even more importantly, it became nearly impossible to buy uranium at any price, especially at a fixed price. Thus it became commercially impracticable to supply uranium as we had contracted to do. We posed the problem to our law department and they to their outside law firms, and came

Table 8.2

Reactor Type	Net MW(e) in Operation	Net MW(e) in Operation or Under Construction
Pressurized light water (PWR)	210,226	263,601
Boiling light water (BWR)	74,885	83,187
Gas cooled (All types)	12,139	12,139
Heavy water (All types)	18,793	26,713
Graphite-moderated light water	14,785	15,710
Liquid metal fast breeder	1,178	4,908

up with a possible solution. We would avail ourselves of a provision in the Uniform Commercial Code that we thought would excuse us from supplying the uranium.

Al Bethel, who was then vice president of the water reactors division, was assigned to work solely on the uranium problem and was replaced by John Taylor. John had come from Bettis in the late 1960s and led a task force studying nuclear fuel. He then became PWR engineering manager and then manager of our breeder reactor programs. In that position he managed the Westinghouse effort in obtaining both the FFTF and the Clinch River Breeder Reactor.

Of course, our utility customers had lawyers, too. They sued us for specific performance, that is, they tried to force us to deliver the uranium at the agreed price. Not to be outdone, our lawyers counseled us to sue the uranium companies (many of whom were members of an international uranium cartel). We sued for treble damages, charging violation of the antitrust laws. Estimates of the potential liability were more than $2 billion.

This issue was the subject of much publicity and discussion far and wide, and a lot of law firms got pretty rich. I gave a deposition once in New York. There were lawyers representing the utilities and the uranium companies and Westinghouse lawyers, including outside counsel. And I had my own lawyer there. In all there were about two dozen lawyers every day for a solid week. All of the suits were ultimately settled out of court in a series of very complex agreements.

How much did this whole experience cost Westinghouse? We set up a reserve fund for the settlement of something over $900 million to cover the anticipated costs. However, things are not always what they seem. There were factors that kept the blow from being nearly that bad. In the first place there were significant tax advantages because the reserve amount was written off at the time and many of the costs came much later. Part of the settlement involved furnishing equipment and services at a discount but not necessarily below cost. Some of this we would not otherwise have furnished. The fabrication of fuel was very profitable and we received many additions to our fuel reload contracts.

All in all, the cost to Westinghouse was much lower than many feared it would be. Incidentally, today the price of uranium is less than our original $8 per pound price plus escalation.

However, this aged some of us more than a few years.

Despite the two above setbacks discussed, nuclear power has continued to be a consistent major profit generator.

An additional chapter in the uranium story was our involvement in the mining business. We realized, of course, that we were going to need a lot of uranium, and we didn't want to be at the mercy of mining companies.

(Unfortunately, we were later anyway.) We didn't have the expertise ourselves, so a meeting was set up with Gulf Oil so I could explain the nuclear business to them, and get Gulf to join us as an exploration partner.

I guess I made the nuclear business look too attractive. The CEO of Gulf, Del Brockett, was on our board, and the next thing I knew, Gulf had bought General Atomics and was entering the nuclear plant business itself. Don Burnham and I were discussing whether now was the time to call Brockett and ask him to resign, when the phone rang and Brockett resigned. However, buying General Atomics was not one of Gulf's smartest moves—it proved to be quite a burden.

We did get into the business of mining uranium in a minor way. We set up a small mining subsidiary and also entered into some joint ventures with several mining companies. We also had a project to extract uranium from phosphate rock. We were working with a company in Florida that had large phosphate deposits and with the government of Morocco. Morocco had large deposits and was interested in building a nuclear plant. These efforts came to naught as it did not appear to be economically attractive.

Steam Generators

As one would expect with a technology as new as nuclear power there have been some unanticipated problems. One of the most troublesome has been the leaking of steam generator tubes. This has caused many tubes to be plugged and a number of steam generators to have to be replaced, despite having a considerable number of tubes more than required for the design rating. Some of these steam generators had less than 10 years of service.

A major development program was mounted to try to understand all the facets of this problem. In many instances improper water chemistry has been responsible for the problems. There have also been design improvements that help. Because of the many corrective actions taken, the lifetime of the remaining steam generators has been significantly extended.

Westinghouse steam generator-related litigation involves 10 utility customers or groups of customers (Carolina Power & Light, South Carolina Electric & Gas, Duke Power, Eugene Water and Electric Board, South Texas Partners, Commonwealth Edison, Duquesne Light and co-owners, Furnas, Southern California Edison, and Portland General Electric) with similar claims. To date, there has not been a single judgment entered against Westinghouse for steam generator claims and no court has required Westinghouse to pay damages on those claims.

In October 1992, Westinghouse decisively won its long-standing dispute with the Brazilian electric utility, Furnas Centrais Electricas, over steam generators in operation at the Angra Nuclear Power Plant. The ICC arbitration panel awarded Furnas no damages on its $115 million steam generator claim, and concluded that there was no fraud by Westinghouse and no breach of warranty related to the steam generators.

While commercial disputes are inevitable over the course of complex and long-term nuclear projects, most can be resolved amicably without litigation. The arbitrators' decision in the Furnas case should demonstrate that commercial disputes can be more productively worked out through a resolution process than through a lengthy litigation process.

Litigation is an unfortunate but not wholly unexpected consequence of the maturing nuclear industry and an increasingly litigious society. Utilities, faced with soaring electricity rates and increasing pressure for rate relief, seek to contain operation, maintenance, and capital addition costs (a particular concern for aging units) by any means possible. These pressures, coupled with those of a growing and competitive legal community, have given rise to an elevation of legal remedies over other avenues of dispute resolution.

The primary issue raised in the steam generator litigation cases are as follows: (1) the use by Westinghouse of Inconel 600 in its steam generator designs and (2) the claim of an implied 40-year steam generator warranty. It is alleged that Westinghouse used Inconel 600 for its steam generator tubing despite evidence that the alloy was highly susceptible to stress corrosion cracking. The 40-year steam generator guaranties are sought to demand compensation for any steam generator replacement required in advance of what utilities assert to be a 40-year design life.

Despite these allegations, the facts are quite clear that Inconel 600 was the material of choice for steam generator tubing in the industry at the time the steam generators were supplied based on an extensive, industry-wide testing program. Westinghouse and other nuclear steam supply system (NSSS) vendors (including Babcock & Wilcox and Combustion Engineering) designed, produced, and delivered steam generators utilizing Inconel 600 as the standard steam generator tubing material, based on the engineering data, materials, and technology then available. In addition, the French government and manufacturers of both pressurized and boiling water reactors also selected Inconel 600 for nuclear power applications.

As for the issue of 40-year guaranties, neither Westinghouse nor any other NSSS vendor guaranteed, nor do the contracting documents require, that the equipment will last for 40 years. Each component has a design basis that is predicated on contemplated plant operation and the duty the component is to perform, as envisaged at the time of design. Although standard warranties do not exist, most steam generator warranties for the plants at issue expire 12 to 24 months after the plant has passed agreed-on performance tests.

Westinghouse designs have evolved from the U-bend vertical feed-ring type with stainless steel tube material, to the feed-ring type with Inconel 600, preheat designs with Inconel 600, and a new F model design with Inconel 690, thermally treated tube material, stainless steel tube support plates, and full-depth hydraulically rolled tubes in the tube sheet. The unit lifetime has varied dramatically from 12 to 30 years depending on a number of operating conditions, particularly secondary water chemistry. It would be a long story to review the history of steam generator technology advancement, because there has been enormous worldwide attention and continued improvement. The model F units, which have served as replacement units for some of the early units, have operated with excellent results for more than 13 years. These continuing improvements along with diligent maintenance have the potential to make steam generators reliable for decades.

Ice Condensers[1]

The Westinghouse Ice Condenser Reactor Containment System was designed to absorb the thermal energy released during blowdown to a nuclear plant containment building as the result of a loss-of-coolant accident. The purpose of ice condenser operation is to limit containment building pressure to low values. This effect is obtained in containment buildings having interior volumes that approach only one-half of the volume of the so-called "dry-type" containment systems. Furthermore, the sodium tetraborate ice is an excellent fission product absorber and has a large potential capacity for retention of fission products.

The inherent supply of borated alkaline water from the melted ice can be used for core cooling and flooding the reactor cavity or as a source of water for the containment spray system.

The ice condenser is a completely enclosed, refrigerated annular compartment between the crane wall and the containment shell. In the event of a loss-of-coolant accident, door panels at the bottom of the condenser open due to the pressure rise in the lower compartment. The steam contacts the ice, condenses, and limits the pressure.

This system was first used on the two D.C. Cook plants and then on a number of other Westinghouse plants, but some difficulties were encountered with the mechanical design that required modification. While the containment building is considerably smaller, some utilities have found the smaller space a disadvantage during maintenance operations. This design was not used on our later plants.

---

[1]George Masche, "Systems Summary of a Westinghouse Pressurized Water Reactor Nuclear Power Plant," Westinghouse Electric Corporation, 1971.

International Dominance

The international nuclear business has been important to Westinghouse since the early days. Sale of nuclear licenses to Mitsubishi Atomic Power Industries (MAPI) and Siemens for $1 million was a major source of funds at the start. These were sold by Bill Chapman of our international company and Bill Kelly of the atomic power division.

Westinghouse moved to world leadership in nuclear power in relatively short order. We had strong and effective leadership in the International Company under Jose DeCubas. Although our international marketing effort was excellent, a major factor was that the pressurized water reactor proved to be the best type, and it has been so recognized by most utilities everywhere. Not only does the PWR have inherent advantages, its dominance is in no small part the result of the development effort at Westinghouse.

Westinghouse and the French Connection[2]

My involvement with our international sales started in 1960 after I left Bettis. That year, I went to Paris for a luncheon with Electricité de France (EdF), the Commissariat a l'Energie Atomique (CEA), and our French partners. My only reason for going was the luncheon, at which we hoped to assure we would get a contract for the first French nuclear plant. Flying all the way to Paris just for lunch seemed ridiculous, so I went on to Brussels to visit our group there. We did receive the order, which was the first of many. This meeting was the first of a long series, as I negotiated with the CEA, EdF, and Creusot Loire-Framatome for our license and equity participation in Framatome.

The French were interested in working with us on nuclear power, but there is reason to believe they had another motive. At a meeting at the French nuclear research center at Saclay in 1959, when they were considering the purchase of SENA, Arnold and Long Sun Tong were questioned closely by Horowitz, their chief reactor physicist, and were led to believe that the French technical people were firmly convinced that their indigenous gas-cooled reactors were far superior to the PWR for generation of power. The heavy water plant they were also working on was their second choice. They always wanted to learn PWR technology for their submarine program. Rickover had been willing to assist the British in their *Dreadnought* submarine program, but he refused to aid the French. It was almost a decade later before they became serious about light water reactors for power production.

---

[2]W.H. Arnold, Private Communication dated June 5, 1992.

Later, the CEA wanted to cancel the agreement and embark on its own full-scale development of a French PWR. A high-level decision in France, however, led to their being appeased with a full blown liquid metal program, billed as the successor to the PWR. The part of the CEA left to work with Westinghouse soon came around to the view that the cooperative program was an opportunity for their national labs at Grenoble, Cadarache, and elsewhere to build extensive test programs.

The French government refused to permit Westinghouse to hold a majority stake in Framatome. We could have become a very profitable majority owner, but we only were permitted 45%. Being a minority partner, with equipment supplied by companies owned by our partner, was not the best situation. Later the French government forced Westinghouse to sell its part of Framatome.

However, we had a very profitable license agreement and received more than a billion dollars of royalties from Framatome, but we set them up as probably our major competitor. Would they have eventually succeeded in that role in any case? I suspect that they would have, but not as quickly. France needed nuclear power and had the political infrastructure to get it.

We gave Bob Schasseur, our representative in Paris, a commission contract for selling nuclear plants. The percentage was small and reasonable for one or two plants, which is all we expected to get. When we achieved a near monopoly on all French nuclear plants that small percentage made millions for Schasseur. Of course, much the same applied to our license fees, as there was no volume discount there either.

The French were quick learners, but we did an honest job of providing the support they paid for—about a billion dollars, but only a small percentage of their overall cost. Our license lasted for more than 10 years, but now has expired.

Arnold, who managed the technical interface with the French, worked closely with Michel Hug, the EdF manager in charge of procuring all power plants. Hug deserves much of the credit for their overall success. He was personally austere but quick and decisive; one never doubted who was boss. He was a major factor in the 1972 decision to concentrate on water reactors and abandon the French CEA gas-cooled design, which had led to five plants in France and one in Spain. They proceeded with parallel programs of PWRs and BWRs. This pitted Framatome, then a small company under the wing of Creusot/Schneider, against the mighty Compagnie General d'Electricité (CGE), which had the GE BWR license. However, Framatome was the clear winner in this competition.

Westinghouse and Japan

In Japan, although our patent position gave us a virtual monopoly on the technology, both Japanese electric power companies and the Japanese government made it necessary to do business through our licensees. In Japan or any other country, we got very little help from the U.S. government.

Westinghouse, therefore, operates in Japan with or through its licensee Mitsubishi, which had been a licensee of Westinghouse for many years before the war and resumed that license after the war. As a licensee in the nuclear field, Mitsubishi joined with Westinghouse in bidding on the second Japan Atomic Power Company (JAPCO) reactor in Japan. This contract was eventually awarded to GE and its Japanese licensees, Hitachi and Toshiba. The competitive bidding process and negotiations had much the flavor of the Keystone Kops, as Westinghouse was ushered out of the negotiating room just as GE was entering.

Japan was playing both U.S. companies against each other by getting questions from one and demanding answers from the other. One of the evaluating team wrote a book during the negotiation in which the story of the reactor bidding competition was told. He predicted GE as the winner—long before any results were made public.

There were two crucial items to which Westinghouse would not agree. We could not assure that the reactor would be licensable (before licensing laws were written in Japan), and we would not take responsibility for underground foundations (although hundreds of boring samples were available, the area was subject to multiple seismic faults).

The BWR plant was built by GE-Hitachi-Toshiba on the Tsuruga peninsula site by excavating a large hole and filling it with reinforced concrete, so the entire plant could be mounted on the monolith.

Kansai Electric Company purchased the next reactor in Japan from Westinghouse-Mitsubishi, after competitive bidding with GE-Hitachi-Toshiba. There was much negotiation, and loans were received from the U.S. Export Import Bank. One Ex-Im Bank letter stated that Westinghouse warranties and equipment should pass to Kansai and Kansai should pay Westinghouse, rather than MAPI in between. This PWR plant, Mihama 1, was built on the other side of the Tsuruga peninsula by Westinghouse-Mitsubishi and Obayashi-Gumi (Japan's largest construction company); it has done a good job supplying electric power to the Kansai system. Mihama 1 began operation in the same year as the GE Tsuruga plant.

For this first plant, Westinghouse supplied the design and manufactured the components, including the turbine-generator and the nuclear fuel. Mitsubishi supplied increasing percentages of the equipment as more plants were sold. If there was a major change in size or design, Westinghouse furnished components and design, for the first of the new series.

To date there have been 23 nuclear plants of Westinghouse design built in Japan, with a total of over 18,000 MW(e); four of these were supplied by Westinghouse and the others by Mitsubishi Heavy Industries under license from Westinghouse.

Personal relationships were lubricated by bringing the legal limit of Johnny Walker Black Label to Tokyo. Going out on the Ginza upon arrival in Tokyo was mandatory, even if you were dog-tired and with a headache from 10 hours of transpacific flight. It was your Mitsubishi host's opportunity to use his company expense card and visit his favorite night spots.

On one of my visits to Japan, there was in a meeting with Makita, the CEO of Mitsubishi Heavy Industries and the designer of the Zero fighter plane. He excused himself from the meeting with no explanation, so I expressed my dissatisfaction by refusing to meet with his subordinates. That evening at dinner he was not present, but about halfway through the meal the senior Mitsubishi man, who was sitting across from me, muttered a lame excuse and left. A few moments later, Makita arrived and took his seat. The number-two man was forced to leave, as he would lose face by having to move his seat down the table.

Makita had abruptly—and uncharacteristically for a Japanese—left the earlier meeting with me to sign the automobile license agreement with the Chrysler CEO. In order to make amends he joined our dinner after leaving the Chrysler dinner.

The biggest electric utility in Japan and the fastest growing one in the world is Tokyo Electric Power Company (TEPCO), which supplies 50-cycle electrical power to Tokyo. Traditionally, it has purchased electrical equipment from Hitachi and Toshiba, with or without licensee participation by U.S. General Electric. After purchasing many BWRs from GE-Hitachi-Toshiba, TEPCO requested that Westinghouse-Mitsubishi submit a proposal on the next nuclear plant. Because preparing such proposals is very expensive, Westinghouse declined. TEPCO sent a letter to Westinghouse in Pittsburgh and demanded that a proposal be submitted by the parent corporation—not from the Tokyo branch of the Hong Kong office of Westinghouse's nuclear subsidiary in Geneva, Switzerland. After much correspondence, the vice chairman of the Westinghouse board presented the proposal to TEPCO in Tokyo. Shortly thereafter the contract was awarded to GE-Hitachi-Toshiba.

Douglas MacArthur instituted U.S. antitrust laws in Japan when he governed there. He broke up the ruling zaibatsus (Mitsubishi, Mitsui, Hitachi, Toshiba, etc.). The Japanese smiled and complied, but held dinners each month, attended by all corporate officers, at which informal guidelines were set. In 1958, the occupational governorship was terminated.

Evidently, there remained some concern about antitrust legislation, especially since the Export-Import Bank of the U.S. was involved. Comments were made that competitive bids tended to keep the price lower, even though TEPCO never bought equipment from Westinghouse or Mitsubishi. Since then, separate corporate entities have been allowed by the Japanese government to reunite—for example, the New Amalgamated Mitsubishi Metallurgy Mining and Shipbuilding Company.

Japan bought its first nuclear reactor from England. This reactor (JAPCO No. 1) was graphite moderated and fueled with natural uranium. Because of seismic conditions, which were explained to the English later, the entire graphite structure had to be redesigned and strengthened at great expense. Later, Japan obtained permission to build a uranium-reprocessing plant at the same site (Tokai Mura) as the natural uranium reactor. Also at this same site, a plutonium critical facility has been constructed; experiments there determine static and dynamic properties of plutonium assemblies.

The plutonium facility is an exact copy of the Zero Power Plutonium Reactor (ZPPR-3) built by the AEC at the Nuclear Reactor Test Site in Idaho, and plutonium inventory has been loaned to Japan by the United States. It is technically feasible that natural uranium used in the graphite reactor could produce plutonium. This could, in turn, be reprocessed at the same site and then serve in fast breeder reactors or weapons.

Selling in the Pacific Area

During August and September of 1968, J.J. Kreuthmeier of our international sales company and Creagan attempted to sell reactors in New Zealand, Australia, Philippines, Taiwan, Korea, and Japan.[3] The dates are important, because AEC representatives had just visited these countries to persuade them to sign the Nonproliferation Treaty (NPT). This was a treaty on the nonproliferation of nuclear weapons signed in Washington, London, and Moscow on July 1, 1968; ratified by the U.S. Senate on March 13, 1969, and signed by the president of the United States on November 24, 1969. It went into effect on March 5, 1970.

Arms were twisted and promises made by the AEC. The countries were told they would be given much technical information about peaceful uses of nuclear energy, plus a sum of money that could be used to buy a swimming-pool reactor. American Machine and Foundry offered such a reactor and sold it to many countries.

Direct Quotes from the NPT

> Preamble: ... convinced that, in furtherance of this principle, all Parties to the Treaty are entitled to participate in the fullest possible exchange of scientific information for, and to contribute alone or in cooperation with other States to, the further development of the applications of atomic energy for peaceful purposes.

---

[3]R.J. Creagan, Private Communication dated August 1991.

Article IV, Section 2: All the Parties to the Treaty undertake to facilitate, and have the right to participate in, the fullest possible exchange of equipment, materials and scientific and technological information for the peaceful uses of nuclear energy. . . .

Some countries signed the treaty, took the money, bought a swimming-pool reactor, accepted a shelf of books, and expected a flow of technical information on nuclear power. Then these countries discovered they had to buy commercial licenses to acquire the technology from Westinghouse, GE, or some other company. Usually, the signed NPT was meaningless until ratified, and the countries did not ratify because they thought they had been misinformed.

In this environment, the two traveling salesmen from Westinghouse were met and treated in an ambivalent manner. Rather than purchase a reactor from Westinghouse, the countries said they thought the United States would educate them so they could manufacture and build their own reactors. The fact that their manufacturing facilities were inadequate did not seem to be a subject for polite conversation.

As one example, the Australians said they were a member of the Commonwealth and hence would buy their reactor from England or Canada, unless Westinghouse could persuade the two countries to sign papers saying they could not furnish a comparable product. There were Commonwealth-published, 6-foot shelves of tariff regulations with tax requirements ranging from 20% to 40% that would make any commercial competitive negotiations impossible. Another factor was that a Commonwealth citizen who managed to purchase a reactor from Great Britain gained considerable prestige.

While all countries were in favor of minimizing the spread of nuclear weapons, they felt the NPT was a self-serving device that protected countries with the bomb, (US—1945; USSR—1949; United Kingdom—1952; France—1960; and China—1964) and held other countries to a promise not to compete in the international weapons game.

An ironic footnote is that when the Chinese were seeking to purchase a reactor from the West in the late 1978, Secretary James Schlesinger gave Department of Energy permission for Framatome to bid, as required by the terms of the license agreement. However, we were precluded from bidding ourselves. The French won the contract and the reactors at Daya Bay are now in operation. Essentially the same thing happened in South Africa, where two Framatome reactors are in operation.

Westinghouse in the United Kingdom and Spain

Our first cooperation with the United Kingdom was on the submarine *Dreadnought*, as mentioned earlier. The British have now abandoned their gas-cooled reactors and are building a reactor to Westinghouse design. At this point, the British have placed a hold on building any additional plants after the first PWR until at least 1994, at which time they will evaluate their position on nuclear power.

One of my operations was Westinghouse Spain (WESA); Mike Meircord was my representative with this company. WESA's president was Andres Martinez Bordieu, whose brother had married Franco's daughter. The head of our sales force was Carlos Alvarez de Toledo, a cousin of King Juan Carlos.

In early 1966, a trip took me to all our factories in Spain: Cordova, Bilbao, Santander, Reinosa, and Madrid. Most of the trip was by car, which gave me a chance to learn a lot from Meircord about operating a manufacturing company in Spain.

We were able to sell eight nuclear plants in Spain. Six have been completed and two are indefinitely delayed. The first of these plant orders was closed by Bill Chapman of our International Company and Woody Johnson, when Woody, in a long evening telephone call with Chapman, agreed to the final terms and conditions.

Westinghouse in Korea

Korea has for centuries suffered from a lack of natural resources, fuel in particular. With the advent of nuclear power they saw a way out of this dilemma. Korea had been importing 90% of its primary energy, and with an expanding need Korea was alarmed by the oil crisis of the 1970s. The plan is to reduce primary energy imports to 70% by the end of this decade. Korea expects to produce 40% of its electricity from nuclear power.

In 1970, Korea Electric Power Corporation (KEPCO), the national electric utility, purchased its first nuclear plant from Westinghouse, Ko-Ri 1 rated at 556 MW(e). The plant was completed in 1978. KEPCO has since bought another Westinghouse nuclear plant, Ko-Ri 2 rated at 605 MW(e). This nation has also bought a CANDU heavy water plant from Atomic Energy of Canada. Other plants were sold by Framatome and by Combustion Engineering/Korea Heavy Industries & Construction.

KEPCO also bought two 895-MW(e) plants and two more of 900 MW(e) from Westinghouse.

Westinghouse in Yugoslavia

The 620-MW(e) Krsko was sold to Savske Elextrarne Ljubljana of Slovenia and Elektroprivreda Zagreb of Croatia. This plant began commercial operation in 1983. One example of the many problems encountered on the Krsko project was that Yugoslavia wished to supply at least a $1 million of equipment. However, because of poor quality they were unable to qualify as a supplier. They then wanted Westinghouse to assure them that the equipment would be sold abroad, which we were unable to do.

Other Countries

Westinghouse also sold five nuclear plants in Belgium with ACEC, three in Sweden, two in Switzerland, two in Taiwan, one in Brazil, and one in the Philippines.

The principal market today and for the next few years lies overseas.

CHAPTER 9
=========

Offshore Plants, Fast Breeders, and Fusion

If ever there has been a field of endeavor in which the practical engineer and the visionary thinker have found common ground and pursued common objectives, it has been in nuclear power.

We deal in this chapter with three stories of that nature—nuclear power plants that float offshore, the reactor that produces more fuel than it consumes, and the harnessing of nuclear fusion wherein ordinary seawater could provide the fuel. To the layman these might sound like visionary dreams. But no, these developments are facts, not fiction and although none of the three has yet taken center stage, they stand in the wings.

Offshore Power Systems

The floating power plant concept emerged from the convergence of two needs—the need for standardization to beat the problem of high cost in building nuclear plants and the need for sites on which plants could be built, particularly in the populous eastern states.

A principal reason for the high cost of nuclear plants in the United States has been, and still is, the lack of standardization. Each electric utility wants special features for its plants. The many architect/engineers and constructors were building plants differently. And the various sites proposed for these plants required differences in plant construction.

We had tried for many years to achieve standardization with little or no success, but we continued to encourage our people to come up with new ideas. Al Collier and others in the Westinghouse nuclear energy systems group had decided that building standardized modules offered a possible answer. Thus, in 1970, Westinghouse began to test utility reaction to this method of construction.

One of the companies that showed interest was Public Service Electric & Gas in New Jersey. At that time, they were concerned with the difficulty of finding acceptable sites for future plants. Richard Eckert, a Public Service executive whose job it was to find new power plant sites, was reaching the point of desperation. In that densely populated state, nearly all appropriate sites had already been used. Any remaining possibilities were being closely scrutinized by the growing body of environmentalists and intervenors.

It was Eckert, therefore, who first thought about taking advantage of the great open area that adjoins New Jersey—the Atlantic Ocean. Why not float a plant offshore? His wife was openly skeptical. So, no doubt, were his associates at first. But there were some earlier precedents for generating power offshore. In 1929, the aircraft carrier Lexington went to the aid of Tacoma, Washington, when that community temporarily ran out of power. The ship anchored offshore and fed electricity to the city. During World War II, ships also had been used as power plants in several instances. So Eckert was intrigued when Westinghouse began talking about building standardized modules for nuclear plants. Rather than modules, he said, why not consider larger plants that could be put on a platform that would be floated offshore?

The advantage was clear. Construction could take place in a factory situation with the same supervisors and workers for all plants, instead of starting with a new crew at each site. This allowed crews not only to master their own jobs, but to learn how to work together effectively. Construction techniques would be far superior to what was possible at the normal onsite job, with better tools and handling equipment. Complete duplication of plants would be possible, unlike the each-site-different situation and thus we could benefit from the learning curve effect.

The small group from Westinghouse and Public Service talked to Ted Stern and division manager Joe Rengel. Then they were told to go see me about their idea. It was intriguing to say the least. A new concept that would solve the standardization problem and the site problem at the same time. Now, what to do with it?

I realized we would have to develop the concept sufficiently to get a customer for at least two such plants before we could go ahead. The idea would have to be developed in considerable detail and exceptionally well because utility executives were not noted for taking chances on new schemes; neither were Westinghouse executives.

We would have to convince Don Burnham and the Westinghouse board that we could meet our cost objectives and make a profit. After the construction cost overruns on the turnkey contracts, that would not be easy. This would involve a lot of money. Nonetheless, I decided this was an idea worth pursuing and decided to go after it vigorously.

Organizing Offshore Power Systems

Before going for corporate approval, we had to be prepared with an organization, a detailed facilities plan and, ideally, an order or letter of intent from the first customer.

Finding someone to head up this new activity—initially the special projects division, later Offshore Power Systems (OPS)—was easy. Zeke Zechella was a natural choice. He had been in Navy construction, had built the A1W in Idaho and installed the eight reactors for the carrier *Enterprise*. Zeke thrived on challenge. When we were developing the nuclear rocket engine, he told me that if we ever needed a project manager for a station on the moon, he was my man. Just the kind of guy we needed to direct OPS. Collier would become the engineering vice president.

In early 1971, Zechella was named general manager of the special projects division which was set up in Monroeville, Pennsylvania, near the nuclear energy systems headquarters. Zechella, Collier, and Buck Lee recruited an all-Westinghouse team of experts to analyze the offshore project. Public Service was showing continued interest, but they weren't ready to give us a letter of intent.

One thing we lacked was marine know-how. We didn't have any experience in building ships or platforms that float, so we wanted a shipyard to join us in this project. Zechella and Lee went to Newport News Shipbuilding and Dry Dock Company, a Tenneco subsidiary. Sure enough, they were interested and that summer, Zechella and Bud Ackerman of the shipyard shook hands on a deal. Of course, Tenneco approval was still needed so Zeke went to Houston and made a presentation to the Tenneco board of directors. They voted to join us and form a 50/50 joint venture with Zechella heading it up.

Offshore Power Systems Was in Business

Newport News sent some 15 or 20 people, more later on, to team up with us in designing the plant and manufacturing facility, the site search, and marketing plans. We relied heavily on their cost estimators to interject a shipyard background into our numbers. This was a very cooperative effort and we quickly overcame the "we" and "they" syndrome. This team stayed together and formed the nucleus of the OPS startup organization.

Finding a place to build these floating plants was high on the agenda. After checking up and down the East and Gulf coasts, we narrowed the decision to either Norfolk, Virginia, or Jacksonville, Florida. Jacksonville had more going for it. It provided a shorter run to sea and there was deeper water. Also there was a lot of good land available and we received a warm welcome from the city fathers. So we decided to build our facility on Blount Island in the St. Johns River near Jacksonville.

After looking at man-made islands where we could float the plants in and then bring in sand for foundation material, we finally decided the best option was to locate them either in river estuaries or about three miles offshore—just inside the three-mile limit—and let them continue to float. The design was imaginative to say the least.

The offshore site would require an artificial breakwater surrounded by porous armor that would break the force of the biggest wave without damage to the breakwater itself. The design called for the breakwater to be covered with big concrete dolosse—the singular is dolos—made in the shape of children's jacks. They are shaped like the letter "H" with one leg rotated 180 degrees. Weighing from 40 to 60 tons each, about 18,000 would be required for each off shore site. They would be formed ashore and put in place to build a U-shaped breakwater big enough to hold two 1200-MW(e) floating nuclear power plants.

Behind this would be a back breakwater to prevent waves from bouncing off the shore and back to the nuclear plants. Public Service contracted with the University of Florida at Gainsville to construct and test a model of the breakwater with two nuclear plants. This was done in their Coastal Engineering Laboratory, a vast model basin, with the capability of generating waves that simulated the worst storm expected in a million years. It also was used to determine the result of collisions of 1100-foot tankers with the protective breakwater. This provided the necessary data to design the U-shaped breakwater and the mooring system. It was estimated that it would take four million man-hours over four years and five million tons of material to build the breakwater.

The mooring system would consist of large concrete blocks anchored to the bottom, from which there were ties to each corner of the plant platform. A universal joint at each end tying the plant to the concrete blocks would make it possible for the plant to move up and down with the tide but would restrict its lateral motion. Power was brought to the land by underwater cables.

Planning the manufacturing facility at Blount Island was an exercise in creative thinking. We learned that the widest slipway in any shipyard was only 140 feet. What we needed was a graving dock for building the floating power plant platform about 400 by 400 feet and 40 feet deep. The platform would be moved from the graving dock to the plant assembly positions in

Figure 9.1 Artist's sketch of a floating nuclear power plant.

the slipway and then to the testing dock—a giant marine assembly line, so to speak.

Initially, the manufacturing facility was to have a capacity to produce four plants a year. In the 1971 to 1977 period we were scheduled to complete an average of seven land-based plants per year and the market seemed to be growing, so this was not unrealistic. However, this objective was downsized to one plant a year as the oil embargo impact reflected lower and lower demand for power—but that's getting ahead of the story.

The manufacturing facility had a 565-foot graving dock, where the platform and portions of the power plant below the platform upper deck were assembled. After assembly of these components, the dock was to be flooded and the platform floated out to one of the two positions along the 1000-foot slipway. Components and subassemblies would then be moved from the production shops for installation. Finally, the completed plant would be floated to a testing basin, where all plant hot functional tests (short of fuel loading) would be conducted.

James Turner, who joined OPS from Tenneco, was the vice president for operations for OPS' Blount Island facility.

Public Service Orders Two Plants

With these plans on paper, all we needed in our efforts to obtain Westinghouse corporate approval was an assured customer. Joe Stadelman, our division marketing manager, took care of that by getting a letter of intent from Public Service. The key players for Public Service were CEO Bob Smith, President John Betz, and Engineering Vice President Dick Eckert.

It was now time to try to get the necessary corporate approvals for this venture. Realizing the financial risk involved, although I had confidence in the OPS people, it seemed like a good idea to get a second opinion on the feasibility of the whole plan, so we assembled a group of senior people to review all of our plans and cost estimates. Thus armed we got approval of the corporate capital expenditure committee. Our presentation to the Westinghouse board won approval in late 1971. When the Tenneco board OK'd the joint venture shortly thereafter, we were off and running.

Public Service ordered two 1200-MW(e) plants to get us started. The contracts for Atlantic 1 and 2, as they were called, were signed on September 18, 1972, aboard a yacht off the coast of New Jersey. Delivery of the first plant was scheduled for four years later. Each plant would cost about a billion dollars.

We had a tough time getting marine insurance, much to my amazement. We wanted to buy insurance for a sea voyage of only several weeks, but all the marine insurance companies in the world had never insured a vessel worth about a billion dollars.

Wisely, the utility took great care in selecting the site for these first two plants. They considered nine possible locations. Farthest north was off Long Branch, less than 20 miles from New York City. Farthest south was off Cape May peninsula. The latter, about 140 miles from Newark, seemed too far away, but waters deepened the closer you got to the New York Bight. Deeper water would add to the cost of the breakwater. Some other sites were considered too close to resort or vacation areas. Public Service finally selected a location 11 miles northeast of Atlantic City with seven fathoms, 42 feet for the land lubbers, of ocean depth. Just seaward of the site was an extensive underground hill of sand whose summit was less than 30 feet below the surface of the ocean. It would help serve as a partial breakwater, giving the man-made breakwater a helping hand.

OPS spared no necessary expense or effort preparing for this first-of-a-kind effort in nuclear plant design and construction. It was one of the first in Westinghouse to use computer-aided design (CAD) and one of the first organizations anywhere to apply CAD to design of a power plant. Three-dimensional coordinates of all components and connecting systems, including piping, duct work, and wiring, were entered into the computer. This enabled us to determine in advance whether there were interferences, and eliminated costly changes that usually occur when interferences are found during construction.

We estimated that about 115,000 manufacturing drawings would be required to satisfy the level of detail needed for each plant. The installed cost of the CAD system, including everything, was about $500,000. Savings in production of the manufacturing drawings were at least four to five times that much.

After the CAD system was completed and put into operation, any design engineer could access it and have it produce any drawing. Also, for example, if he wanted to change a check valve to a gate valve, he could type in the symbol for a gate valve and the CAD system would make the change.

You've seen big cranes, but you probably have never seen one the size of the giant we acquired for Blount Island. We bought it from Krupp of Germany for $24 million erected. It was so big that if one rail was positioned on each side of the Capitol building in Washington, the crane could pass over the top of the dome. It was sized to lift the 850-metric-ton containment dome of the floating nuclear plants. That crane could lift a thousand metric tons with a span of 675 feet, and had a crossbeam that was 40 feet in depth.

While we expected the first nuclear plant to require four years for manufacture, we believed this could be reduced to two years through creative

Figure 9.2 The Offshore Power Systems crane.

productivity. This saved a lot on interest cost during construction and the utility also was in a much better position to judge need for capacity if the manufacturing period was that short. To make the concrete containment dome we planned to pour the dome in a prefabricated form in the shop and place it into position in one move, using the large crane. In the normal construction program, a frame is constructed for pouring the concrete for the dome of the containment building. Not only is this expensive, but at this point all work inside must be curtailed.

The cylindrical portion of the containment structure was slip-formed with a continuous pouring process. Forms would be raised as the concrete set, with the previously poured concrete acting as a support.

Putting piping in place and welding it is a major expense. At OPS, large sections of piping would be welded and x-rayed in the pipe shop and then positioned in place by the large crane. The number of welds to be made inside the plant would be dramatically reduced, as would the number of defects.

In the electrical shop, major sections of wiring would be put together using unique fixtures and tested by special equipment to reduce testing time. The control room, four decks high, was to be fabricated in the shop as a single module with all equipment in place and then positioned in the plant by the large crane.

By building plants in a factory on an assembly line in this fashion, we believed we could eliminate at least eight million man-hours of construction work—at about $40 an hour, including overhead. Our estimate of man-hours for a conventionally constructed nuclear plant of this size was about 19 million.

A good feature of the contract with Public Service was that it provided for progressive payments as time elapsed, independent of actual progress. Thus, payments always were ahead of costs. Also, if the contract was canceled, Public Service was obligated to pay all our expenses for the factory as well as for the plant construction.

Now we were really rolling. We moved many Westinghouse people to Jacksonville from Monroeville, Bettis, Idaho, and Newport News. We also hired a number of new people. At the peak, we had about 1000 people working at OPS; about 800 of them were technical personnel. Then the bombshell—the OPEC oil embargo.

OPEC Oil Embargo Brings OPS to a Halt

When the embargo hit in 1973, it threw the economy into a tailspin. In addition, the need for conservation of energy became apparent. Energy conservation became a clear necessity and quickly reduced the nation's use of electricity.

Public Service's main industrial load was petrochemical facilities around Newark and, of course, that load was greatly reduced. Commercial and housing loads also dropped. The utility's load decreased, instead of increasing at about 7% a year, as it had been doing for decades. At that same time we were starting on the Atlantic 1 and 2 projects, Public Service was building Hope Creek, a GE nuclear generating plant with 1031-MW(e) capacity. Much equipment already had been completed for that plant and some delivered to the site, so its cancellation would have been very costly. Public Service decided to finish it.

This meant that Public Service didn't need our plants at that time. They asked us for a two-year delay on the first plant to which we agreed. However, they thought their load would resume growing at about 1200 MW(e) per year, but that didn't happen. Eventually, both orders were canceled. We had received letters of intent from both the Jacksonville Electric Authority and the Southern Company, but these were also canceled because their load growth also did not materialize as projected.

OPS had to lay off about 600 people and went into a holding mode, although we did continue construction of the manufacturing facility. It was about 90% complete when the final cancellation came through. Strangely enough, in that two-year delay period, we sold Public Service two more plants, increasing the order to four before the entire order went down the drain. We continued work on the design and completed most of it.

We were able to close the Blount Island plant on very favorable terms since our contract with Public Service was quite specific. It was a "take or pay" contract, which meant whoever canceled would make the other party whole.

At the time of termination, there was very little Westinghouse money involved, as we had been receiving progressive payments that were sufficient to cover all our costs. There were, however, about $400 million in outstanding equipment contracts and about $125 million in construction contracts for the facility.

Public Service owed us an estimated $90 million in termination costs. They thought that figure was too high, so we offered to settle the contracts with suppliers for cost plus a fixed fee. The utility did not believe the New Jersey Public Service Commission would permit them to enter into such a contract.

Zechella finally reached an impasse in negotiations with Bob Smith, CEO of Public Service. At that point, Gordon Hurlbert, president of the Westinghouse Power Systems Company, was brought into the negotiations. He suggested we agree to a $60 million termination fee, with Westinghouse getting Blount Island. They accepted this offer. We were able to settle all terminated contracts for about $30 million, giving us a profit on that transaction and then we sold the island for about $17 million after closing down OPS.

We made every effort to get our workforce of about 800 placed in other jobs. We advertised that the people were available and many companies came to interview them. Most technical employees either were transferred to other Westinghouse locations or hired by other companies. Many of them stayed right in Jacksonville.

It's Not Over 'Til It's Over

The story could still have a happy ending. Just as we were closing down OPS, we received approval from the Nuclear Regulatory Commission (NRC) for eight plants. We didn't have a customer, but we were the first and only applicant to receive approval to build eight identical plants.

As part of the application, OPS had to satisfy regulatory review by a number of federal agencies, including the NRC, the U.S. Coast Guard, and the Environmental Protection Agency (EPA). Environmental justification was presented in a half dozen or so volumes, each several inches thick, collectively known as the floating nuclear plant (FNP) Environmental Impact Report.

Figure 9.3 Artist's sketch of a floating nuclear power plant as seen from the shore.

EPA in turn coordinated with many other agencies, such as Fish and Wildlife and environmental groups, such as the National Resources Defense Council and the Sierra Club. These reviews included not only thermal effects, land use, construction impact, biologic and aquatic changes, etc., but also radiological evaluations of normal and abnormal releases to the food chain. In the final analysis, the coordinating agency's conclusion was that the floating nuclear plant could be built, sited, and operated with acceptable environmental impact.

Two floating nuclear plants, moored in their breakwater three miles offshore although well above the horizon, would pose minimal visual impact. Since they are more compact than similar units constructed on land, floating nuclear plants utilizing cooling towers at a river or estuary site would be either comparable to or less aesthetically offensive than the land-based plants.

If we could build prelicensed nuclear plants in two or three years at a fixed price as low we believed possible, the nuclear option could be revived. This concept still has merit.

To paraphrase that eminent philosopher Yogi Berra, "It's not over 'til it's over."

Breeder and Fusion Reactors

To a lot of people the idea of perpetual motion and the breeder reactor have one thing in common—incredibility. The difference is, we know how to build a breeder reactor.

What people find incredible about the breeder is that it produces more fissionable fuel than it consumes. Such reactors can use fast neutrons with uranium-238 or with thorium-232 to implement the breeding process. Various coolants can be used, such as water, sodium, helium, or steam. Three fissionable fuels can be considered: uranium-233, uranium-235, or plutonium-239, in metallic or oxide form.

From the early days of the U.S. civilian nuclear power program, the breeder reactor has been regarded as critical to realizing the full potential of the nation's uranium resources. Breeders are capable of using uranium resources 30 to 60 times more efficiently than light water reactors and can extend uranium resources from decades to centuries. The energy content of uranium tailings already stored at the government's gaseous diffusion plants, if used in breeders, would produce energy equivalent to over three times the world's oil resources plus that contained in all the recoverable coal resources in the United States.

In the mid-1960s, the number of light water reactors sold and projected was such as to indicate a real need for a breeder reactor. We were optimistic that it could be economically competitive. I set up several teams to evaluate various breeder options. A sodium-cooled reactor team was headed by Bob Creagan, a steam-cooled team by Paul Cohen, and the helium-cooled team by Jim Wright.

After the three studies were completed, we reviewed them and selected the first—a sodium-cooled fast breeder. Not only did liquid sodium have favorable inherent properties, but there was considerable experience with this concept. Since the 1940s, a number of experimental breeder reactors had been under development in the United States and elsewhere.

Included in these was the first attempt by GE to develop a sodium-cooled breeder reactor using an intermediate-energy neutron spectrum. This endeavor led to the GE liquid metal submarine intermediate reactor project and the *Sea Wolf* nuclear submarine, which operated from 1957 to 1959. You will recall that Rickover replaced that propulsion plant with a Westinghouse PWR.

The first breeder reactor, Experimental Breeder Reactor No. 1 (EBR-I), was built by the Argonne National Laboratory (ANL) and operated in Idaho. It reached its full power of 1.1 MW(th) in late 1951. That small but historic reactor provided the first demonstration of nuclear electricity on December 20, 1951.

EBR-II was also built at the National Reactor Testing Station, now called the Idaho National Engineering Laboratory, as a power reactor producing 19 MW(e). Full-scale operation began in 1964 and continues today. Between 1964 and 1969, the EBR-II metallic fuel was reprocessed and fabricated in an adjacent fuel cycle facility and returned to the reactor. Despite being an experimental facility EBR-II has maintained a high-capacity factor comparable to commercial nuclear power plants.

A group of American and Japanese companies, led by Walker Cisler, CEO of Detroit Edison, formed the Power Reactor Development Corporation (PRDC). As part of the AEC power reactor demonstration program, it built the 61-MW(e) Enrico Fermi 1 sodium-cooled reactor plant in 1963. In 1966, flow blockage of a fuel assembly resulted in some local fuel melting. Following modifications, the plant was put back in service in 1970 and then shut down permanently, when its core load of fuel was depleted in 1971, when the AEC refused to fund another core due to budgetary conflict with money for the Fast Flux Test Facility (FFTF).

Our review of detailed experience with these previous reactors led to the conclusion that the feasibility of sodium as a coolant had been demonstrated. It is an excellent heat-transfer agent and its high boiling point allows high-temperature and low-pressure operation. The use of sodium also allows the system to be cooled by natural circulation in the event of an emergency shutdown.

Although sodium has excellent heat-transfer properties, unfortunately, because its specific heat is one-third that of water, a higher coolant flow rate is required. With the same temperature rise as with water, it takes three times as much sodium to transport the same amount of heat. This accentuates the importance of hot-channel factors because of higher temperature rise through the reactor and hence higher temperature differential (ΔT) after applying hot-channel factors.

From our review of these various reactor types, the concept emerged of using mixed oxide (about 20% PuO^2) as fuel instead of more highly enriched metal fuel. Liquid sodium and mixed-oxide fuel in a loop-type reactor configuration thus became the focus of the Westinghouse approach to fast breeders. Our reasons for not selecting the other breeder reactor types are discussed in the technical chapters of this book.

Attempts to Buy Atomics International

Having decided on liquid metal technology, Westinghouse looked around for means of augmenting our capability in that technology. We thought we had found it in Atomics International (AI), a subsidiary of North American Aviation (NAA). That firm had a major program dealing with liquid metals and would have been a desirable acquisition. At that time, NAA was being acquired by Rockwell and AI didn't appear to fit into Rockwell's plans, so it seemed to be an appropriate time to buy it.

We approached North American and opened negotiations for AI. Things seemed to be going well and we had just about reached an agreement, when the whole thing fell apart. Al Rockwell was meeting with NAA in Los Angeles and called my boss, George Wilcox. George was tied up at the moment and did not return the call promptly. This infuriated Rockwell and he called the deal off. As it turned out, that was lucky for Westinghouse because we were able to proceed on our own to develop a good liquid metal capability.

Debate Over the AEC Breeder Program

The AEC's liquid metal fast breeder program (LMFBR) was being criticized in the early 1970s by many people outside the program. Each claimed that his pet scheme was preferable. From time to time, claims would be made for different concepts or fuels that might produce higher breeding ratios. The leaders of the JCAE—Chet Holifield, Melvin Price, and Craig Hosmer—asked me to come to Washington for a private meeting to discuss the claim of higher breeding rates for metal fuel made by the energy group at Cornell.

I reviewed the positive qualities of the liquid sodium and mixed-oxide fuel concept that Westinghouse had chosen and why I believed it gave the highest chance of success. Fortunately they trusted me. They said that was enough for them and they would back the program wholeheartedly.

Later, we developed a core configuration for LMFBRs in which blanket assemblies containing the fertile uranium isotopes were dispersed heterogeneously throughout the active core region. This configuration improved plant breeding performance and for large LMFBRs leads to high breeding ratios in the range of 1.40 even though it increases the inventory of plutonium required per kilowatt-electric.

The Fast Flux Test Facility[1]

The FFTF is a 400-MW(th) reactor, cooled by liquid sodium and designed to operate at temperatures between 400 and 1000°F. The reactor vessel is one of the largest and heaviest stainless steel vessels ever built. It measures 43 feet high, about 21 feet in diameter and weighs 239 tons. The plant contains 25 miles of piping and 1200 miles of electrical cable.

In the liquid-sodium, mixed-oxide fuel, loop reactor configuration concept developed by us and others, the major goal is to achieve low fuel cycle cost. To do that, it is necessary to obtain about 100,000 MWD/T average burnup and a peak burnup of about 150,000 MWD/T in the fuel. Testing of full size fast breeder fuel assemblies and materials in an environment like that found in commercial LMFBRs was, therefore, essential. The FFTF made this possible. It was critical to the success of LMFBRs.

Our advanced reactors division (ARD) was chosen in 1968 to design the reactor for the FFTF and, in 1970, Westinghouse was given responsibility for the complete FFTF project. FFTF was built on the AEC's Hanford site at Richland, Washington. This was the first major step toward commercialization of the liquid metal fast breeder.

More than 2000 Westinghouse engineers, scientists, and technicians participated in design, construction, and operation of FFTF, of which 800 were at ARD. Also more than 300 engineers at Waltz Mill conducted liquid metal tests and designed fast breeder systems and components.

In addition, ARD engineers working on the FFTF project developed reactor development and technology (RDT) standards, which formed the foundation for the LMFBR program. Technical service laboratories performed special chemical analyses of metal alloys, gases, liquid metals, and other materials. A hybrid computer laboratory mathematically simulated behavior of a fast breeder under a wide variety of operating conditions.

A major activity was testing of future fast breeder materials and components in a sodium environment. Sodium was circulated in test loops at

---

[1]"Westinghouse People Make It Happen at Large and Waltz Mill," Westinghouse Electric Corporation, p. 61, 1986.

rates up to 2000 gallons per minute. Eighteen test loops were in operation and in 1970 more than 100 million pounds of sodium was pumped. The materials development and test program was directed for many years by Peter Murray.

The original purpose of FFTF was to provide the test bed for the development and testing of high-burnup mixed-oxide fuel rods. With the discovery of the phenomenon of irradiation, swelling of stainless steel and its impact on LMFBR fuel and core design, it became clear that irradiation testing of full-size, prototype fuel assemblies in the correct environment of flux and temperature was essential. Milton Shaw, head of the Reactor Design and Technology Division of the AEC, therefore, made the decision to proceed with this larger objective and to build the 400-MW(th) facility.

Many things caused major increases in the development and construction costs. Among these were that the design of the core was changed from a split conical to a vertical design, which required a major redesign; the decision to use the very stringent and costly RDT standards; extremely high inflation during this period; and many delays in the construction. Some changes were made to reduce costs. For example, there were originally supposed to be seven closed loops, but these were not included in the final design due to budget limitations.

While these resulted in major cost increases over the original estimate, the fuel assembly data obtained, the high-temperature design codes developed, and the overall performance of the FFTF itself provided the high degree of confidence necessary for the success of the technology. Information gained from this facility was useful in the development of high-temperature boiler and pressure vessel codes. This work was guided by Bill Pennell.

The reactor went critical in February 1980 and started full operation in April 1982. The FFTF has demonstrated the high burnup achievable with mixed-oxide fuel for LMFBRs and the ease of maintenance of low-pressure sodium systems. Under Westinghouse Hanford Company operation, the FFTF provides the controlled, instrumented fast flux environment for full-size testing of fuel assemblies and its value has been recognized internationally with the carrying out of joint fuel irradiation programs with other countries, particularly Japan.

Interest in liquid-metal-cooled reactors has diminished considerably throughout the world since the construction of the FFTF. In addition the reference fuel for the U.S. liquid metal reactor program has changed from mixed-oxide to metallic fuel. Testing of that type of fuel can be done more economically and efficiently in the EBR-II at ANL-West in Idaho.

The Independent Review Team (IRT) chaired by John Landis commended "the Westinghouse Hanford Company, prime contractor for the

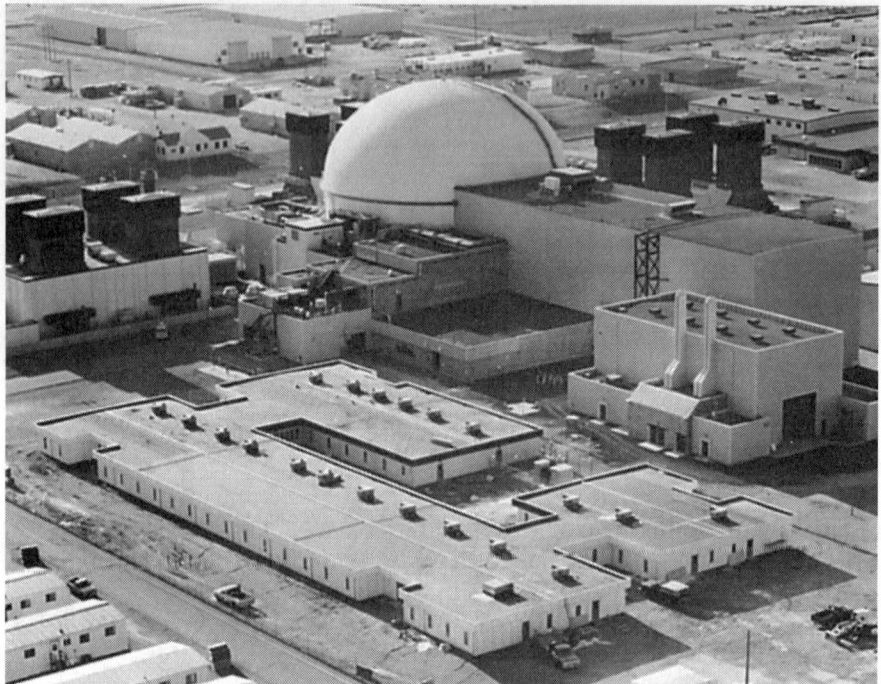

Figure 9.4 The Fast Flux Test Facility. (Courtesy of Westinghouse Hanford Company)

facility since 1970, on its superlative performance in all aspects of the assignment. The IRT believes that the FFTF ranks in the top one or two per cent of the approximately 200 nuclear complexes in the United States, both in performance and in safety/cleanliness."

The FFTF was originally funded for $87.5 million. The capital cost in the end was $167 million in addition to the development costs. The cost of the development program associated with the FFTF has been more than $500 million.

The facility was put in a cold standby mode by former Energy Secretary James Watkins on January 11, 1993. This process is expected to take about five years, in part because of the handling requirements for the liquid sodium coolant. Watkins decided that FFTF would not be needed to produce plutonium-238 for spacecraft reactors, since it could be acquired from Russia. However, on June 15, 1993, Energy Secretary Hazel O'Leary announced that John Landis, senior vice president of Stone & Webster Engineering Corporation, would head an independent study of potential new missions for FFTF. This study has now been completed and determined that there

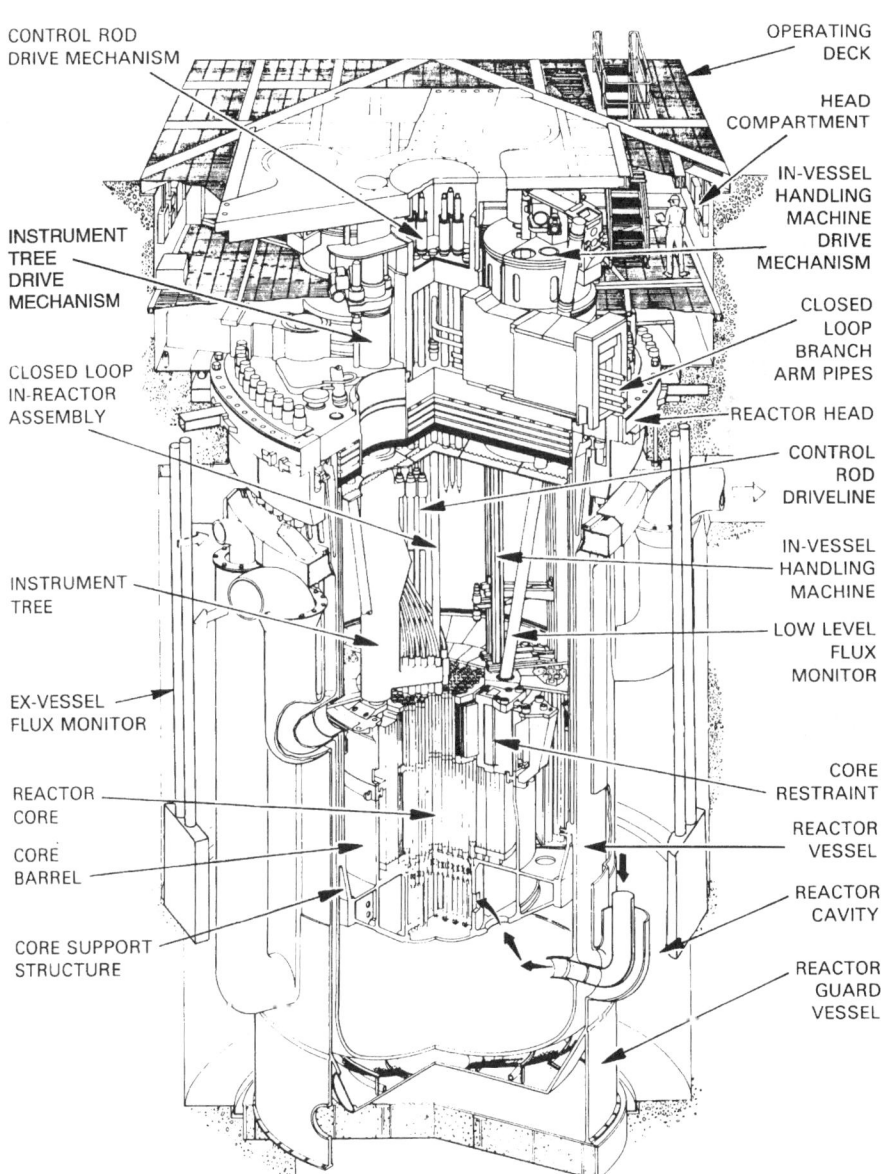

Figure 9.5 Fast Flux Test Facility reactor cutaway.

was no mission for FFTF that would justify the estimated $500 million of continued operation cost in view of the lessened interest in the liquid metals breeder program. Therefore in December 1993, Energy Secretary Hazel O'Leary ordered that the process be started to shut it down.

Clinch River Breeder Reactor Plant

This was it. The time was 1972 and the objective was to win the first contract for a fast breeder nuclear power plant. The AEC had asked for bids for the Clinch River (Tennessee) Breeder Reactor (CRBR), which would produce power for the Tennessee Valley Authority (TVA). It was to be a joint effort between the AEC and the nuclear utilities on a 380-MW(e) breeder demonstration plant.

We submitted our proposal based on the loop design that provides reactor cooling via primary and intermediate sodium filled loops. Our major competition proposed the integrated pool design. We were convinced that our design offered greater reliability and ease of maintenance.

After the proposals were evaluated, the AEC and nuclear utilities asked each contender for the contract to make a presentation. The evaluation board that would hear the presentations consisted of all the AEC commissioners, the CEO and the president of Commonwealth Edison, and the head of the TVA. James Schlesinger, chairman of the AEC, chaired the meeting.

We had heard that GE was having their CEO, Reg Jones, make a statement at the meeting. We thought that demonstrating corporate backing might be important. Unfortunately our CEO had to attend an important meeting of the executive committee that same day, which left us without a designated hitter. To beef up our presentation, Frank Cotter asked me to invite Hobart Taylor, the youngest member of the Westinghouse board, to make a statement for our side. At least we knew that Taylor was known to the CEO of Commonwealth Edison. Taylor was in Sweden, but we persuaded him to return for the meeting. He read a resolution of the board stating a firm backing for the Clinch River project, then he said that as the youngest member of the Westinghouse board, he would be around to see that Westinghouse lived up to its commitment.

John Taylor carried the ball for me on most of our technical presentation on the loop design. Convinced that building the reactor quickly was important, we presented a day-by-day schedule for the first few months to assure getting a fast start.

Toward the end of the meeting, Jim Schlesinger, chairman of the AEC, said to me, "One of the participants in this selection indicated that they would approach this project with humility. Will you approach this project with humility?"

"No," I said, "We will approach it with confidence. That does not indicate that we consider ourselves superhuman or that we know everything, but it does mean that we will not approach the project with humility."

It was the Wednesday before Thanksgiving and I left for Hilton Head immediately after the meeting. Shortly after I arrived the phone rang and it was Schlesinger. He related the story of the man with an inferiority complex, whose psychiatrist told him that his problem was that he *really* was inferior. Then Schlesinger said, "Westinghouse does not have a superiority complex, it *is* superior. Mr. Simpson you have the contract."

The following paragraphs summarize the conclusions of the selection board in awarding to Westinghouse the contract as Lead Reactor Manufacturer for the LMFBR.[2]

Among the factors which led the Project Management Corporation (PMC) to conclude that the proposal submitted by Westinghouse offered the greatest promise of successfully achieving the project objectives were the following:

First, PMC felt that Westinghouse exhibited the best rounded overall technical capability for carrying out the breeder project. In terms of proposed plant design, it was a standoff between Westinghouse and one competitor and with respect to experience on large research and development projects, it was a standoff between Westinghouse and the other competitor. Westinghouse was judged ahead on experience with nuclear quality assurance systems and on depth of technical management experienced in nuclear power.

Second, the Westinghouse management and organizational approach appeared to offer a better chance of achieving the technical objectives of the project within the time schedule and cost targets. Westinghouse was a standoff with one competitor on the degree of advanced preparation for moving ahead promptly and led both in the application of advanced management systems for controlling the breeder project. In addition, the project organization Westinghouse proposed offered greater clarity with respect to assignment of responsibility and authority to key project personnel. Finally, the proposal contained a recognition of the boundaries between the lead manufacturer's assignment and the responsibilities of other parties, such as AEC's civilian base LMFBR research and development program.

Third, the Westinghouse proposal appeared to strike the best balance between the objective of stimulating a future commercially competitive domestic and industrial base in breeder reactor technology and the objective of timely and successful completion of the project at the lowest cost. Westinghouse recognized the need to provide an opportunity for the two other reactor manufacturers to assume responsibility for design and development of significant nuclear systems of the demonstration plant. At the same time, Westinghouse recognized that the Lead Reactor Manufacturer should retain control over the most essential systems to assure adequate coordination and integration of this technically complex project.

Finally, each bidder offered an economic contribution to the project which varied in form and amount. Evaluation led to a judgment that the differences in the value of the contributions from the proposers was not of great significance when weighed in the context of the entire project objectives and should not be determinative in the selection.

---

[2]Memorandum on the Selection of a Lead Reactor Manufacturer for the LMFBR Demonstration Plant, December 5, 1972; attached to a letter from W.B. Behnke, dated December 7, 1972.

The selection of a contractor for a development project, such as the LMFBR, is a trying experience for both the selection board and the proposers. It is nearly impossible for a clearly incompetent bidder to prepare a winning proposal, but it is very easy for a highly competent bidder to make a poor showing. The selection is usually not based as much on the overall competence of the entire company or organization bidding as it is on demonstrating the competence that will be brought to bear on the project. This point is often missed by bidders.

Later the AEC was requested by one of our competitors to conduct an investigation into the award of the Clinch River contract. The report of this investigation later became available to me. It showed that, on reexamination, we were even farther ahead of the others than the original evaluation had indicated.

In this case, the AEC probably set an all-time speed record in making a major contract award. But then came substantial delays. Instead of being able to make the fast start we had proposed, there were several congressional hearings and much discussion over how the contract would be managed by Commonwealth Edison and the government. The utilities formed the Breeder Reactor Corporation (BRC) and the Project Management Corporation (PMC). BRC was to obtain financial and other participation of the 973 utilities contributing to the CRBRP. Congress was refusing to go along with the AEC being responsible for all cost overruns. The contract was not formally signed until almost exactly two years later.

In the interim a major study—the International Nuclear Fuel Cycle Evaluation (INFCE)—was conducted to consider the nuclear proliferation risks associated with civilian nuclear power. This study was intended by the Carter administration to be used to stop the breeder program, but other countries that were involved disagreed. In 1980, INFCE concluded that fast breeder deployment was feasible by 2000 with the timing dependent on the needs of individual countries. INFCE also concluded that international initiatives, with emphasis on institutional arrangements for the fuel cycle, would contribute to minimizing proliferation risk while assuring fuel supplies.

The original cost estimate and the basis for the authorization was $699 million. However it was recognized at the time that there was no reference design as a basis for the estimate and that the estimated cost could be changed when a more definitive basis for revision was available.

The reference design was completed in June 1974 and the cost was then estimated at $1,736 million with a reference schedule for initial criticality in June 1982, based on site clearing commencing in September 1975. These cost estimates included construction costs and the cost of five years of operation.

Figure 9.6 Artist's conception of the Clinch River Breeder Reactor. (Courtesy of Oak Ridge National Laboratory)

Due to the increased cost estimate, in 1975 Congress changed the arrangement among the four parties to the contract (AEC, TVA, Commonwealth Edison, and PMC). The federal government was to manage the project with support and assistance from PMC. There was to be an integrated Project Office composed of representatives of the federal government and the utility industry. This change went in effect May 1, 1976.

In 1975 the cost estimate was again increased, this time to $1.95 billion, with a delay in initial criticality to October 1983. This increase was partly due to delays caused by budget reductions. This cost estimate remained in effect until 1977 when the Carter administration suspended licensing activities and started termination of the project. However, continued congressional support enabled design activities to continue. By 1983 the cost estimate had increased to $3.6 billion with initial criticality delayed until September 1989.

Congress passed a continuing resolution in 1983 to "explore proposals including a reconsideration of the original cost-sharing arrangement, that would reduce the Federal budget requirements for the Clinch River Project or project alternative, and secure greater participation from the private sector."[3]

---
[3]Report to the Congress on Alternative Financing of the CRBRP, U.S. Department of Energy, March 1983.

Since the development of the initial CRBRP baseline cost estimate of $1,736 million in 1974 it was estimated by DOE that 68% of the subsequent cost increases were beyond the control of the project. The biggest part, 56%, was caused by government actions which have "stretched out the Government's funding profile below that required to support the Project schedule, delayed completion, or threatened termination of the project. Another 12% were cost increases due to requirements imposed by the Nuclear Regulatory Commission. Only 32% of the estimated Project cost increases were theoretically within the control of the Project, and resulted from the developmental nature of the Project."[4]

For the first few years, President Nixon wanted to keep the project going and Congress wanted to kill it. Then it shifted with Jimmy Carter. He wanted to kill it and Congress wanted to keep it going. If we could have proceeded according to my schedule when we won the contract, we would have been so far along that it would have been about as cheap to complete it as to cancel. When the project was finally canceled in 1983, the design was complete, 70% of the equipment delivered or ordered, and site preparation had begun.

That was the last of a series of nuclear power milestones for me: the first nuclear-powered submarine, cruiser and carrier; the first U.S. nuclear central station power plant; the first and only nuclear rocket engine; the first fast flux test facility; the creation of a monopoly for almost all French nuclear plants; becoming the world leader in nuclear power; and the first contract for a fast-breeder nuclear power plant. Not a bad run.

Fusion

As you know fusion is the reverse of fission. Instead of splitting the nucleus of the atom in such a heavy element as uranium to release energy, the fusion process combines two nuclei of a light element such as hydrogen to form a heavier nucleus to release energy. A big advantage for fusion in the future, if target fuel temperatures are increased by a factor of about two is that its fuel could be deuterium, an isotope of hydrogen found in almost limitless supply in ordinary water. Otherwise lithium availability for production of tritium limits the energy available.

Fusion also has other advantages over fission, such as less production of radioactive waste; there is, however, the possibility of tritium releases. There is one big catch. We haven't yet accomplished sustained and controlled fusion. We haven't yet succeeded in releasing its tremendous potential for the generation of civilian power.

---

[4]Report to the Congress on Alternative Financing of the CRBRP, U.S. Department of Energy, March 1983.

But that doesn't mean we aren't trying. While most of our energies during my tenure at Westinghouse were concentrated on fission reactors, we also kept a close eye on the long-range fusion research program and participated wherever possible when our expertise matched the program requirements.

Westinghouse involvement began early with Woody Johnson and Don Grove taking part in the Stellarator program in the mid-1950s and later developments at Princeton. The pioneering work of John Hulm and his R&D center colleagues on niobium-based alloys for superconducting magnets led to our building niobium-based superconducting coils at our East Pittsburgh Works. These were successfully tested at Oak Ridge as part of the large coil program. We also carried out several tasks for the Princeton Tokomak program including the effects of irradiation on materials.

The scientific and engineering challenges of developing fusion power make it essential to cooperate and develop projects on an international basis. To this end, the International Thermonuclear Experimental Reactor (ITER) is a collaborative effort by the United States, Europe, Japan, and Russia to design and build a 1000-MW(th) power tokomak. The conceptual design has been completed, the engineering design work started in 1992, and the facility is expected to be built by 2005 at an estimated cost of $7.5 billion. Westinghouse is involved in the program as a member of the McDonnell Douglas team, carrying out R&D on the plasma facing components.

Uranium is in plentiful supply at present and will probably not be sufficiently scarce for decades, thus avoiding abnormal cost increases. It is possible that at some point in the future either an additional energy source or a way to increase fissionable material supplies will be required. Two possible new energy sources are those we have just discussed, breeder reactors and fusion.

It is unlikely that breeder reactors will play a significant part in the U.S. energy picture for at least several decades. Currently, the United States has only a development effort with regard to breeder reactors, while Japan is already building a breeder. It is going to be very difficult to get another reactor of any type ordered in the United States. It appears to me to be unlikely that the next nuclear plant ordered would be a breeder or any new type of reactor. Even if a breeder is ordered in this decade, it would take several decades for the breeder to play a significant role, even if successful. Unfortunately the cost will only come down after a number of breeder reactors are built and operating. It is hard to see how this can come about any time soon.

On the other hand, significant development effort has been aimed at fusion. Even so, economically competitive fusion is a long way off. Since that technology is relatively new, there could be a major technical breakthrough that would substantially reduce cost. The great publicity given to "cold fusion" has turned out to be publicity without performance data to support it, but it is an example of a "not thought of before" idea that may someday produce a wholly unexpected breakthrough. Someday there may be competition between breeders and a fusion-fission hybrid reactor to supply fuel for light water reactors.

Westinghouse has also kept in touch with the fusion program by participating in various committees. Peter Murray has been a member of the Magnetic Confinement Fusion Advisory Committee. I have been a member of several DOE committees evaluating the fusion program and I chaired one of them. I was also chairman of the National Research Council committee evaluating the fusion-fission hybrid reactor program.

Westinghouse is monitoring the situation on both fusion and breeders and will be prepared to participate at the right time.

Long-range, high-technology development programs for the U.S. government have proved to be very frustrating and often unrewarding. During my career we succeeded in getting contracts for the nuclear rocket engine and carried it to successful ground test, only to have the program canceled. Then we were technically successful with the FFTF and power breeder reactors, again only to have the government discontinue the effort. We never went as far in the development of fusion, primarily because of the very long term nature of fusion development, with a payoff—if any—years or decades into the future.

CHAPTER 10

The GOCOs[1]

If you were to select 10 reasonably well-informed citizens at random in this country and ask them what Westinghouse has done in the field of atomic energy, 8 out of 10 probably would mention nuclear power plants, 6 or 7 might mention nuclear submarines. But the third area of the Westinghouse nuclear business probably would not be known by any of them.

That segment, however, employs about 30,000 people at seven U.S. Department of Energy (DOE) plants or facilities. It's a business we got into by chance, but grew in by design.

The business is the operation of government facilities. That type of enterprise began in the period after World War II, when the U.S. Atomic Energy Commission (AEC) decided to use operating contractors at essentially all of its reservations, manufacturing facilities, and laboratories. These government-owned and contractor-operated facilities are called *GOCOs*. Thus, for many years General Electric held the operating contract at the Hanford Reservation and Union Carbide at Oak Ridge. Typically, these contracts would be of five-years duration and would be renewable at the option of the government, if performance was satisfactory. Many of the large operating contracts were quite stable and were renewed several times. At the height of the weapons program, DOE operating contractors had almost 100,000 employees staffing its facilities.

Fees on these operating contracts tend to be rather small—typically on the order of 1% to 2% of the value of the contract, depending on size. On

---

[1] Much of this chapter is drawn from a private communication with W.M. Jacobi.

the other hand, the operating contractor has no significant investment and receives typical government indemnities. Thus, at the larger facilities some relatively attractive fees can be earned.

Unfortunately in the early years before Westinghouse was involved in many of the GOCOs, much radioactive and other toxic waste of all types was apparently not well handled. The cleanup of all these operations will cost many billions.

Bettis

Westinghouse entered the GOCO contract area almost by accident. To set up a facility for development of naval reactor propulsion plants, Westinghouse purchased the abandoned Bettis Airport in the Pittsburgh suburb of West Mifflin. The only buildings on this site were the administration building and a few old airplane hangars.

The government agreed to pay for all the facilities to be constructed. Also, during negotiation of the basic contract for development of naval nuclear propulsion, Westinghouse requested and the government agreed that, at the time of contract expiration, Westinghouse would have the right to purchase facilities the government had built. Almost as an afterthought, the government requested reciprocal rights. It turned out that the contract was to run for a very long term and eventually the government exercised its rights and purchased the land from Westinghouse. The Bettis contract was only the first of many GOCO contracts.

Hanford

In 1970, Westinghouse expanded its role in management of AEC facilities. The situation this time revolved around the Fast Flux Test Facility (FFTF). The FFTF was intended to be a 400-MW(th) high-temperature sodium-cooled reactor. The reactor was designed to be capable of testing the fuels, materials, and components of a new generation of fast breeder reactors and was the centerpiece of an aggressive AEC breeder technology development program. Pacific Northwest Laboratories at Hanford, Washington, a subsidiary of the Battelle Memorial Institute, had been designated by the AEC as a center of breeder reactor technology excellence; it was the overall program manager for FFTF design, development, and construction.

In a 1967 competition that included General Electric and Atomics International, Westinghouse Advanced Reactors Division (ARD) was selected to design the FFTF reactor plant. In a relatively brief period, a design team of almost 300 people was assembled and working on FFTF design. Many of these were recruited from the United Kingdom breeder reactor team as part of the "brain drain" of 1967–68; this included the famous Dr. Peter Murray, who assumed responsibility for Westinghouse fuels and materials development.

Despite the world-class design team that had been assembled, the work proceeded slowly. Breeder reactor design posed some unique issues that had not been successfully resolved before. These included ultra-high fuel burnup; operating temperatures in the creep range of some of the structural materials; and very high neutron fluences, which caused swelling and creep of structures. ARD believed that credible solutions to these problems required both innovation and divergence from Battelle's conceptual design.

There was a great deal of squabbling about the design, and ARD believed it was being micromanaged. The AEC also was aware of these difficulties and, in early 1970, concluded that a simplified and more responsive management structure was required. The official reason given for Battelle not continuing at Hanford was the tax problem it had with the operation at Hanford concerning its tax-free status.

On February 14, 1970, Milt Shaw and Bob Hollingsworth, the AEC general manager, came to see me. They asked whether Westinghouse would be interested in assuming the operating contract for the Hanford Engineering Development Laboratory (HEDL), the nation's leading breeder reactor development facility. Hollingsworth was very surprised when I agreed, on the spot, to take the contract. He had expected that a large company like Westinghouse would need weeks to obtain approvals and enter into a contract.

We were expecting to bid on design and hardware contracts for the FFTF reactor plant and the fast breeder reactor demonstration plant contract. My only stipulation was that we not be hurt in competing for those contracts as a result of running the Hanford Engineering Development Laboratory. The AEC agreed and the contract was quickly awarded.

A Westinghouse takeover team of approximately 20 individuals, headed by Walt Esselman, left for Hanford almost immediately. The Westinghouse Electromechanical Division competed for, and was awarded, contracts for FFTF primary and secondary cooling pumps. Also after competition, ARD was awarded a contract for the Clinch River Breeder Reactor (CRBR) plant; clearly, there was no penalty for operating HEDL.

The FFTF, a sodium-cooled test reactor designed to duplicate the breeder reactor environment, was successfully completed and dedicated in 1982, but it is now being shut down.

The West Valley Project

The next Westinghouse move into DOE facility operations occurred in 1981. The early 1980s were not good times in the nuclear business. The last domestic nuclear plant had been ordered in mid-1978, and since that time many plants had been canceled. Because of long lead times associated with these plants our engineers were still busy and the shops active; nevertheless, backlogs were clearly shrinking, and the handwriting was on the wall—at least for the big-iron part of the commercial nuclear business.

Our commercial nuclear activities were maintaining their traditional profit margins, but sagging personnel levels were becoming an issue. Even more troubling, there were no longer many growth opportunities within the organizations, and it became more and more difficult to provide advancements for our better people.

The situation was not much better on the government-funded side. Bettis was stable, but ARD was shrinking rapidly as the Clinch River Breeder Reactor program was being closed down. Astronuclear was still alive under the leadership of Max Johnson, but it was a shadow of its former self, with fewer than 400 people and a scattering of small development projects. Howard Arnold, as head of ARD, had the lead for John Yasinksky's advanced power systems business unit in support of marketing and strategic planning. He and Johnson collaborated in the search for opportunities in what was now recognized as a new business opportunity for Westinghouse—the GOCOs.

Johnson perceived that the packaging, handling, treatment, and shipment of nuclear waste were going to be a major growth area of the 1980s and 1990s. Further, he observed that DOE facilities, almost without exception, had major waste problems that would require substantial attention in the future. After an examination of upcoming DOE opportunities, Astronuclear Laboratory decided to compete for operation of DOE's West Valley Project.

Much of the capability and perspective had come from the post-NERVA work at the E-MAD facility at Jackass Flats on the Nevada test site. There, Johnson's people had managed to secure a DOE contract to demonstrate the ability to store spent civilian reactor fuel in air-cooled above-ground and below-surface containers. This facility had been built to disassemble and examine spent nuclear rocket reactors after testing and had been manned on a skeleton basis by Westinghouse people continually since the end of the NERVA project. Many of them eventually became part of the Waste Isolation Pilot Project (WIPP) mentioned later in this chapter.

West Valley was a commercial nuclear fuel reprocessing site. It was built by Nuclear Fuel Services Corporation (NFS) during the early 1960s on a site owned by the Empire State Atomic Development Authority (ESADA) and under a contract that permitted NFS to walk away from the project under certain conditions. The plant started up in 1966 and successfully processed a number of core regions of commercial fuel. Indeed, the facility was such a commercial success that NFS initiated a temporary shutdown in order to expand the plant capacity. Unfortunately, the timing of its application for an amendment to the license coincided with the period when regulatory requirements were proliferating and when intervenors were learning how to inflict endless delays. In 1976, after four years of trying, NFS decided that the plant could not be economically expanded under the current state of nuclear regulation and probably could never be restarted in its present configuration.

Exercising its contractual rights, it turned the entire facility over to ESADA. This left New York State with a totally contaminated major facility—almost 750,000 gallons of high-level waste in tanks—and no place to turn. The issue became a major political football in New York. Estimates were prepared showing that cleanup of the West Valley site would cost approximately $2 billion—and the state had neither the money nor the technical or management capabilities to undertake the job. Eventually, the New York delegation convinced Congress to fund and undertake the project as a DOE demonstration project.

As DOE prepared the RFP, Johnson organized so he could put on a full-court press for the job. Ray Mairson, who had a long and successful career at Bettis and the naval reactor facility, was selected as project manager. Bruce Boswell, who was then Bettis manager of the expended core facility in Idaho, became deputy project manager. Other experienced radioactive operations and safety people—such as Jim Krauss, John Knabenschuh, and Hugh Daugherty—were also made part of the team. And last but not least, it included a young control engineer from ARD named Phil Woods, who would be serving in his initial assignment as public relations manager.

Westinghouse won the competition, and in 1983 began assembling the West Valley team on the site. As they arrived, they found a western New York public that was almost in a state of hysteria over the radioactive dangers supposedly emanating from West Valley. The site was featured in Buffalo papers and on TV almost daily. The reports frequently featured the views of antinuclear activists.

One of the first decisions made by Ray Mairson and his young public relations manager was to try to take the mystery out of the project by running a completely open operation. As project plans crystallized and progress was made, the news media was invited to come in and observe. So were local government officials, service clubs, and neighbors. All questions were answered honestly and straightforwardly. A special effort was made to bring in schoolteachers and high school science students for tours; in many cases, they left as disciples.

The bottom line of this policy was that, during a two-year period, fear and hysteria swirling around the project were replaced by public acceptance of our actions—and a belief that DOE and Westinghouse were not part of the problem but part of the solution. From both a DOE and a Westinghouse perspective, the West Valley program can be considered a success.

Oak Ridge and the Idaho Chemical Processing Plant

Very shortly after award of the West Valley contract, DOE reached separate decisions to open for competition the operating contracts for the entire Oak Ridge complex and the Idaho Chemical Processing Plant. Both contracts were of interest to Westinghouse. John Yasinsky, vice president of the advanced power systems business unit, commenced the evaluation of both opportunities. The contract for operation of the Oak Ridge complex was by far the larger of the two. The complex included Oak Ridge National Laboratory (ORNL), the Y-12 weapons components facility, and the K-25 and Paducah, Kentucky, gaseous diffusion plants. More than 20,000 people were employed in these endeavors.

The operating contract had been held by Union Carbide since 1948, and DOE wanted new management. Thus, both because of its size and the need for fresh management faces, the Oak Ridge competition would require an unusually large and well-qualified takeover team.

In contrast, the Idaho Chemical Processing Plant (ICPP) employed about 1,200 people. Its mission was to dissolve and reprocess spent naval reactor cores. The recovered highly enriched uranium was shipped to Savannah River for use as fuel in production reactors. The remaining high-level waste, a sand-like gray powder, was calcined and put in long-term storage in stainless steel bins encased in concrete.

The plant was originally built in the early 1950s, but DOE had initiated and partially completed a major rebuilding. The plant had several operating contractors over its lifetime but was currently operated by Exxon Nuclear Corporation. Exxon was considered to be a very qualified operator, but the need to reopen the contract was created by Exxon's decision to sell its nuclear subsidiary to a West German consortium. DOE did not like the idea that a significant part of the defense nuclear complex would be operated by a foreign corporation.

ORNL was thought to be a potential source of a major increase in our overall technological capability. The uranium enrichment complex was closely related to our own profitable nuclear fuel business, and the experience of operating the complex would be invaluable to Westinghouse.

The ICPP would provide substantial nuclear waste-handling experience, but we were already obtaining that at West Valley and at the naval expended core facility. It would also offer reprocessing experience, but since the Carter presidency, national policy had forbidden the reprocessing of spent commercial nuclear fuel. Thus, the value of this contract was thought to be less than that of Oak Ridge.

The final decision on both contracts would be made in DOE headquarters, and the advice of Frank Cotter and Leo Wright from our Washington office was that we might get one contract but could not possibly be awarded both. Which to choose? Oak Ridge was the more desirable, but there was no way to confirm whether DOE would look favorably on a Westinghouse bid. It was finally decided that the safest course of action was to bid on both and let DOE make the decisions.

We put our first team on the Oak Ridge proposal. Yasinsky was team leader and Howard Arnold was proposed as director of ORNL. Arnold directed preparation of the proposal and chose Eileen Massaro as proposal manager. An excellent supporting cast was provided. The proposals were evaluated by an Oak Ridge-based Source Evaluation Board (SEB), and its recommendations forwarded to the Source Selection Official (SSO) at DOE headquarters.

They evaluated Westinghouse as the most qualified bidder, Rockwell International ranked second, and Martin-Marietta third. The SSO was Martha Hesse, assistant secretary for administration, and she took the unusual step of requesting presentations at headquarters. Westinghouse was represented at the presentation by Tom Murrin, president of the energy and advanced technologies group and Yasinsky, then vice president, energy and advanced power systems business unit.

After the presentation, Martin-Marietta got the nod. The only reason cited by the SSO was that she found the Martin-Marietta management team to be "more impressive." We went to an official debriefing with the SEB, but their only comment was, "We already said in writing that we thought you were the best qualified." Westinghouse was ranked first and Martin-Marietta third. We never could get any reliable information on the reasons for the selection. It seemed strange that if everything was done according to the rules we could not have been given more information.

The Idaho Chemical Processing Plant

Almost in parallel with the Oak Ridge competition, the ICPP proposal was prepared. Max Johnson was our team leader. He was a native of Idaho, had worked for a number of years at the Idaho naval reactors facility, had extensive nuclear operations experience, and was a capable general manager. He also was a well-known quantity to the Idaho customer. Recognizing that he was within a few years of retirement, we backed him up with Ed Pottmeyer, an experienced senior manager from the plant apparatus division and Hanford, as his deputy.

In late 1983, before the decision was reached on the Oak Ridge contract, we won the ICPP operations contract. The transition team went to the site where, in May 1984, we officially assumed full responsibility for ICPP operations. The experience has been a positive one. We have received steadily increasing award fees and have a generally happy customer. The contract was renewed in 1989. As we went on to other competitions of the late 1980s, ICPP began to supply some of the management talent.

Waste Isolation Pilot Plant

The Waste Isolation Pilot Plant (WIPP) has gotten a relatively large amount of national attention because its mission is dramatic and controversial. It is designed to be a permanent repository for transuranic defense waste. Transuranic waste is waste having more than a specified quantity of isotopes of elements with a higher atomic number than uranium—such as neptunium, plutonium, or curium. Some of these isotopes have extremely long half-lives and are not suitable for shallow land burial. WIPP is located near Carlsbad, New Mexico, in a 2000-foot-thick salt bed.

The salt bed is a remnant of the ancient Permian Sea that once flooded much of the western United States. The bed was laid down more than 225 million years ago as land rose and a remnant of the sea dried up. It has been stable since then, and is now covered with 1000 feet of overburden. The actual repository site is 2250 feet below the surface, approximately in the middle of the salt bed. It is considered a suitable site for a permanent repository because of its demonstrated geologic stability—salt is an excellent heat conductor that flows under the static pressure of the earth above it to tightly enclose buried waste.

Astronuclear was awarded a small technical support contract by the Albuquerque Operations Office to assist in the design of the site and its facilities in 1977. The Waste Technology Services Division, headed by Leo Duffy, was formed from the relevant portions of the old Astronuclear Laboratory (earlier renamed Advanced Energy Systems) when Johnson left to head the ICPP. The other parts of Advanced Energy Systems, essentially successors of the old non-NERVA programs, were merged under Arnold with the Advanced Reactor Division under the Advanced Energy Systems name.

As mentioned before, many of the people who had supported the E-MAD post-NERVA work on spent fuel storage became involved in this effort. A team was sent to Albuquerque for that purpose. As time passed, that team grew substantially as more and more tasks were assigned to it, and it moved to the site near Carlsbad. As construction neared completion in the late 1980s, DOE agreed to convert the technical support contract to a management and operations contract for the finished facility.

That contract was entered into and WIPP has become another Westinghouse GOCO. At this writing, there are more than 750 Westinghouse employees on the site; the site and its people have passed all readiness reviews for emplacement of waste. Actual emplacement of that waste is currently on hold because of a pending case in the federal courts.

Feed Materials Production Center

When we made the decision in 1981 to seek more opportunities in the GOCO business we were aware, of course, that such activities almost always offer difficult challenges. We didn't promise anybody a rose garden. Well, we surely didn't find a rose garden at Fernald, Ohio.

This turned out to be the only GOCO job we ever started that we didn't finish. It was, to use the common phrase, a real can of worms.

Early in 1985, the DOE Oak Ridge Operations Office requested proposals for the management and operation of the Feed Materials Production Center (FMPC) at Fernald. FMPC was a large chemical complex dedicated to processing a variety of uranium feedstocks (from yellowcake to UF_6) into metallic uranium forms and shapes.

The plant had been built in the early 1950s, and had been operated since the start by a subsidiary of National Lead. There had been no substantial modernization of the plant since it had been built, and much of the equipment was no longer used and was marked "abandoned in place." The plant supplied uranium to be used for fuel element manufacture for Savannah River and Hanford production reactors; however, by 1985 most of those had been shut down, and FMPC was operating at a fraction of its capacity.

In addition to the poor condition of its equipment and light manufacturing load, the plant had a host of environmental problems. The site was very large physically, and had been used for years as a dumping ground for mildly radioactive waste generated by parts of the Oak Ridge complex. For example, behind the plant was a small mountain of contaminated copper windings that were scrapped when the diffusion plants were upgraded in the 1970s.

In addition to waste produced elsewhere, FMPC itself had generated well over 100,000 steel barrels of low-level waste that sat in the open in many rusting rows. Substantial chemical wastes had also been created, and were retained in several open lagoons. Finally, ventilation filters in a building used as a uranium foundry had burst earlier in the 1980s. Unknowingly, the ventilation system was used in that condition for the best part of a year. In the process, substantial quantities of uranium dust were released that contaminated much of the site and some of the neighboring properties. As a result of these problems, local residents were up in arms. Open warfare practically existed with the State of Ohio, and relations with the cognizant U.S. Environmental Protection Agency (EPA) field office were very strained.

We believed there was a very limited manufacturing job to perform, and it was not of great interest to Westinghouse. However, there was a truly world-class environmental remediation job to be performed, and we believed those skills would be a very valuable addition to the Westinghouse portfolio. Despite the grim situation, we decided to bid the job. It's probably one call we would like to have back.

Bruce Boswell headed the Westinghouse team. He reported to Bill Jacobi, who by now had replaced Yasinsky as head of the business unit when Yasinsky moved into Westinghouse International. Boswell had done a great job at West Valley and had dealt very successfully with the same kind of public concerns that existed at Fernald. He was supported by Phil Woods as public relations manager, filling the same role as at West Valley. Bill Britton, who had previously been plant manager of the Westinghouse nuclear fuel plant in Columbia, South Carolina, was proposed as manufacturing manager. At Columbia, and before that at Bettis, he had developed an excellent knowledge of uranium chemical operations, as well as the accountability and radiation protection disciplines necessary to work with uranium.

There were several competing proposals submitted, but the Oak Ridge SEB decided that Westinghouse and Babcock & Wilcox were the most qualified. Both were invited to Oak Ridge for presentations, and we were each given a full day with the SEB. The Westinghouse delegation was headed by Ted Stern, Jacobi, and Boswell, all of whom made presentations. Within a few days, we were notified that we had been selected.

The takeover team arrived at the site on October 30, 1985. In parallel we began definitive contract negotiations. An agreement on the definitive contract was reached by an exchange of faxes at 7:00 p.m. on Christmas Eve of 1985, but Peace on Earth was not the way of things at Fernald.

Right from the beginning, Fernald proved to be anything but peaceful. National Lead was difficult to work with during the transition, and our team was assigned office space in remote trailers from which it was difficult to observe operations. After we assumed full responsibility, we slowly began to learn that environmental issues were substantially larger and more complex than we had perceived. Also, it was made clear that DOE, and not Westinghouse, was in charge of contacts with the environmental regulators. At that time, DOE was still contesting some elements of EPA jurisdiction, and relationships were very antagonistic.

Following the West Valley playbook, Westinghouse moved aggressively to open communications with the local public. However, those efforts were not as well received as at West Valley. It took almost five years to gain credibility with our local critics.

One clear area of success has been in cleanup of the plant itself. Much of the plant has been decontaminated and is now open to unrestricted access. Contamination levels have been substantially reduced in other areas of the plant, and rigorous radiation protection controls are in place to protect employees. A great majority of the 100,000-odd barrels of radioactive waste have been shipped from the site for permanent disposal in Nevada. At present, the production mission of the plant has been completely phased out and the plant is in a cleanup mode.

In 1990, DOE headquarters conceived the concept of an Environmental Remediation Management Contractor (ERMC), which would be responsible for management of such activities at a given site. The next year, DOE decided to use that concept at Fernald, and shortly before Christmas 1991 an RFP was issued.

Westinghouse had never yet left a GOCO site once it assumed management responsibility. However, we decided not to make a bid to continue at Fernald and a consortium headed by Fluor-Daniel has been awarded the Fernald remediation contract.

The Hanford Consolidation

In 1985, DOE Richland Operations Office announced its intention to consolidate the four major Hanford operations contracts into a single contract. At that time, Westinghouse was still responsible for HEDL, and had approximately 1800 employees operating FFTF and pursuing liquid metal reactor development programs.

Rockwell had overall site-landlord responsibilities (buses, laundries, patrol, fire department, etc.) and was responsible for the 200-Area with its chemical separations and waste management activities. It also managed the Basalt Waste Isolation Project (BWIP), which was later terminated when DOE focused on Yucca Mountain as a repository site. Rockwell had the largest of the operating contracts, with more than 4000 employees. United Nuclear had contractual responsibility for operation of the N-Reactor and for environmental monitoring and remediation of several long-closed production reactors along the Columbia River. United Nuclear's activities were primarily in the 100-Area, and it had more than 2000 employees.

The fourth and smallest of the operating contractors was Boeing Computer Services Richland (BCSR). It provided "white-collar" services (word processing, graphics, photography, printing, etc.) for the site, and had somewhat fewer than 1000 employees.

The RFP for consolidated Hanford Operations was issued early in 1986. At that time, Westinghouse Hanford Company had an unbroken string of six consecutive excellent-award fee ratings, and the FFTF was widely considered to be a first-class operation. Of the other activities on site, we believed that the waste management, environmental remediation, and chemical operations would all be valuable experience for Westinghouse. At the business unit level, it was decided to compete for the job. Because of its size, and because the Hanford reservation had the potential for adverse publicity, we went up the management chain for approval. Within a month, Paul Lego, Doug Danforth, and the board of directors had all concurred.

Intelligence soon indicated that Bechtel and Rockwell were the only competitors. It also became apparent that United Nuclear had agreed to support Rockwell in its bid. On our part, we thought that BCSR could strengthen our proposal. It received excellent-award fee ratings, and we had no notable experience in the area of office services. Tom Murrin knew Boeing senior management well from working with them on the AWACS program, and he personally closed a deal with "T" Wilson (then the Boeing chairman) for Boeing to join with us as a subcontractor in its existing role.

Proposals were submitted in the second quarter of 1986. The Westinghouse team included Jacobi as president, John Nolan (who headed our existing operations at Hanford) as deputy, Les Weber as vice president for defense operations, and Arnold as vice president for development activities. Essentially the same team that had prepared the Oak Ridge proposal prepared this one. The quality of this proposal was even better, since we had learned from the Oak Ridge experience.

The SEB found all three proposers were qualified, and it scheduled oral briefings in Pittsburgh in September. Our entire team participated, and both Danforth and Wilson attended to express the corporate commitments. There was still no word as to who was on top. After a hiatus of several weeks, all three competitors were invited to give oral presentations at DOE headquarters on a Thursday in late December. On the following day, Danforth received a call notifying him that Westinghouse had been selected. Several of the team were at Richland the following Monday to negotiate the transition contract, which was quickly accomplished. On January the second, the transition team moved into temporary quarters at Hanford.

Fortunately, the transition proceeded much more smoothly than at Fernald. Both Rockwell and United Nuclear cooperated fully, and we had the advantage of a cadre of managers who had worked on the site for years and knew it intimately. On June 29, 1987, the new Westinghouse Hanford Company took over full operating responsibility.

Our first priorities were to tighten discipline of nuclear operations, particularly in chemical operations and tank farms (200-Area), and to complete the safety upgrades and restart the N-Reactor. We made significant improvements in the discipline and quality of the nuclear operations, although not as rapidly as DOE would have liked. On the N-Reactor, we largely completed safety modifications and personnel retraining only to have DOE determine that it should be mothballed in light of changing defense needs.

A "Blue Ribbon Committee" evaluated the N-Reactor plant and gave it an okay for operation, except for one committee member, Louis Roddis, who wrote a dissenting opinion. He said, "The N-Reactor should be restarted only if it is mandatory for national defense." The DOE finally decided against restarting the reactor.

In early 1988, DOE announced a decision to terminate BWIP and focus all its repository efforts on the Yucca Mountain site. Still later, the department decided to totally phase out the defense production mission at Hanford. Today, the mission at Hanford is twofold: closing down of the FFTF and the challenging environmental effort to safely maintain and remediate defense production and waste sites.

It is in this latter mission that we have had both our greatest achievement and our largest problem. Our achievement was in successfully playing the midwife role in the tripartite agreement among DOE, EPA, and the State of Washington—spelling out the direction and timetable for overall cleanup of the site. Although Westinghouse is not a signatory to the agreement, it was a major contributor to the technical content, a participant in negotiations, and, on many occasions, the facilitator when the parties seemed to be deadlocked.

On the negative side, we were affected by the 1989 furor over the single shell-waste tanks and our inability to prove that an explosion could not occur. Although much scientific detective work has been done, and the risk shown to be less that had been believed, that controversy persists to this day.

The original Hanford consolidation contract expired on September 30, 1992. The mission is now considerably smaller than when we entered into the contract, however, DOE has extended the contract, and Westinghouse is glad to remain on the site.

Savannah River Site

If you were to walk down executive hallways at the Du Pont corporate headquarters in Wilmington, Delaware, you would come upon a framed copy of a letter from President Truman to then Du Pont chairman, Crawford Greenewalt. The letter requests Du Pont to undertake the design, construction, and operation of a plutonium production complex in the Savannah River area of South Carolina. Immediately adjacent to that letter hangs a copy of the response. In it, Chairman Greenewalt commits Du Pont to undertake the requested tasks "for the fee of one dollar for the duration of the contract."

That initiated Du Pont's association with the Savannah River site (SRS), an association that was to last for more than 35 years. The complex was built and operated safely and well. For more than 30 years, it attracted very little public attention, except from the local community—which was highly supportive. That situation began to change in the mid-1980s, when there was increasing and high-profile media coverage of safety and environmental concerns at various sites of the DOE weapons complex. Criticism was levied against both DOE and its contractors. Several congressional committees held hearings on conditions at the DOE sites, and there began to be congressional pressure to expose contractors to liability for environmental violations. In this climate of mounting public criticism and potential future liabilities, Du Pont notified DOE that it did not wish to extend its Savannah River contract.

Here was an obvious opportunity for Westinghouse.

In 1988, Westinghouse began to assemble a Savannah River team and prepared to compete for the contract. At that time, the commercial nuclear business was beginning to contract sharply as the last of the nuclear steam supply system (our part of a nuclear plant) contracts were completed, and annual layoffs were necessary.

We decided to put our best team on the Savannah River proposal. Jim Moore, who had been vice president of the water reactor division, was proposed as president of the Westinghouse Savannah River Company. Ambrose Schwallie from the Advanced Reactor Division, who had been part of the proposal teams for Oak Ridge and Hanford and now headed this effort, was his deputy. Jim Gallagher, who was general manager of the nuclear and advanced technology division, was proposed as vice president of engineering and projects with Les Weber from Hanford as vice president of operations. Dick Begley, formerly at the R&D Center and head of our steam generator activities, was to be director of the Savannah River Laboratory.

Tom Anderson, who had been in both PWR nuclear safety and nuclear services, was appointed manager of nuclear safety. He has since moved to Hanford as president of the Westinghouse Hanford Company. Bechtel National agreed to support Westinghouse as a subcontractor for architect/engineering services and for construction management.

As with the Oak Ridge contract, Westinghouse's competitor for Savannah River was Martin-Marietta. From the outset, the contest was close; however, an incident late in the competition highlighted to DOE the value of Westinghouse's extensive nuclear industry experience. Although we were not asked to assist at this time, DOE gave Westinghouse a decisive edge.

In August 1988, during an attempt to restart P-Reactor after a long outage, the 34-year-old reactor did not reach criticality when expected because of helium-3 in the core, which acted as a neutron poison. The helium formed in the reactor tank from decaying tritium during the outage. This problem was not anticipated by the reactor operators and drew attention to shortcomings in the site's reactor operations—a challenge we were uniquely suited to address.

Exactly one month later, the Westinghouse/Bechtel team won the contract. In September 1988, key managers went to the site to begin a seven-month transition to assume the contract on April 1, 1989. In addition, several hundred Westinghouse and Bechtel engineers transferred to the site to undertake some of the engineering duties Du Pont had performed at Wilmington. Westinghouse had emphasized in its proposal its strengths in both the current licensing and the safety practices of commercial reactors; also stressed was its outstanding record of disciplined operation at other GOCO sites, particularly at FFTF.

As Westinghouse assumed responsibility, the key priorities were, first, to install modern nuclear safety practices and disciplined nuclear operations throughout the site and, second, to complete whatever hardware upgrades and training were necessary to safely restart the production reactors. Major progress has been made in instilling a disciplined nuclear operations culture across the site. K-Reactor operating crews have been thoroughly trained to today's standards and have passed all Westinghouse and DOE readiness reviews.

K-Reactor was restarted and successfully completed a demonstration run in the summer of 1992. However, as the nation adjusted to the end of the Cold War and growing federal budget pressures, the mission of the SRS began to shift. One by one, the site's three reactors were taken out of service as Westinghouse helped DOE move the site's primary focus from defense production to greater emphasis on environmental cleanup and resolution of nuclear waste issues.

Today, P- and L-Reactors are permanently shutdown and are slated for decommissioning. With no projected need for new plutonium and virgin tritium production not required until after the year 2006, K-Reactor is being placed in standby condition in case of a national emergency until a new tritium source can be built.

Westinghouse's major thrusts today at the Savannah River Site are (1) the application of commercial nuclear operational standards to nonreactor nuclear materials processing facilities; (2) startup of the nation's first large-scale vitrification facility (Defense Waste Processing Facility) to glassify 34 million gallons of high-level nuclear waste from 35 years of nuclear materials production; (3) consolidation of DOE's tritium handling and recycling facilities at one site; and (4) environmental remediation.

The mission at SRS is currently undergoing the most dramatic change in its history, and will continue to shift profoundly as the nation struggles to adjust to the end of the Cold War.

The Value of the GOCO Business

What has been the value of this GOCO business?

In addition to earning a reasonable profit without a major corporate investment, there exists an important benefit to Westinghouse that should not be overlooked. Among the some 30,000 employees working in the seven DOE plants or sites, there are hundreds of Westinghouse managers and key professionals handling some of the most difficult challenges DOE has to offer. This is excellent preparation for the future resurgence of the commercial nuclear power industry, and the GOCO business holds this core of expertise together. It is also enabling Westinghouse to forge strong credentials in the field of environmental remediation.

On a similar theme, perhaps this is the appropriate place to note that the transfer of experience within the nuclear division extended far beyond the GOCOs. That "there is a tide in the affairs of men" certainly applied to the company's nuclear divisions; while one division was experiencing funding problems or perhaps even termination of a business, other divisions were expanding and able to absorb the personnel at most seniority levels to everybody's mutual benefit. Personnel remained employed and were able to continue to use much of their experience, often without relocating. The expanding division obtained engineers, managers, and even technicians who had experience on related problems and did not need training, thus allowing for more rapid program implementation, and the company reduced its financial layoff exposure.

The principle is illustrated by the major movements at Waltz Mill. Engineers from the Kansas City gas turbine division, which closed in 1960, provided valuable manpower at the early nuclear sites, including Waltz Mill and Astronuclear as well as other non-nuclear divisions. When Bettis had a fiscal downturn in the early 1970s, many people went to Waltz Mill, which was working on the FFTF as well as preparing the project definition phase (PDP) documentation to support our bid for the first demonstration reactor, the CRBR. Shortly thereafter, as the NERVA project wound down and was terminated, engineers and managers flooded into Waltz Mill just as the CRBR project took off, and others arrived in the commercial divisions as their order books began to fill. In the mid-1970s, as the FFTF design activities began to wane at Waltz Mill, construction, commissioning, and early operation were under way at the Hanford Site, and several engineers and managers were transferred to that site.

About 1979, when the offshore power systems division was cutting back, stacks several feel high of resumes of the Jacksonville staff were passed through the other nuclear divisions, and many engineers returned north to a corner of Pennsylvania where many had gained their first nuclear experience. In 1981, as the CRBR project started to feel the impact of congressional pressure on project funding, some of the engineers on that project transferred to the new and growing project at West Valley. Two years later, in 1983, when the CRBR project was terminated, even though the commercial divisions were not growing much, a significant fraction of the staff found at least medium-term assignments in those divisions. Others moved out to the Feed Materials Production Center at Fernald, Ohio, or to the WIPP project at Carlsbad, New Mexico. In the late 1980s, when Westinghouse took over management of the Savannah River Site, this provided an opportunity to relieve the pressure on the commercial divisions and Waltz Mill.

Clearly, although unsettling while the events are occurring, the transfer of personnel following the growth and decline of various projects in the nuclear enterprise provided an opportunity for maintaining and developing the nuclear experience base within the company.

CHAPTER 11

Other Nuclear Activities

The Commercial Nuclear Fuel Business

As was pointed out in Chapter 8, nuclear fuel was recognized from the outset as one of those evergreen businesses. As film is to the camera manufacturer, nuclear fuel is to the reactor builder—a product necessary for continuing operation of a nuclear power plant. If nuclear fuel is of top quality and competitively priced, it comes as close to "selling itself" as any product I have ever known.

The utility equipment business depends on the growth of the utility. If growth of capacity drops, say, from 7% to 2%, the manufacturer's market drops by more than 70%. In the 1980s with almost no growth in utility capacity, the market for equipment simply dried up.

To the credit of Westinghouse, the nuclear fuel business has been well managed from the start. It is produced today in a blue ribbon plant, which is the world leader in the production of light water reactor fuel. There aren't that many success stories around, so it is worth taking a closer look at this one.

Westinghouse got into this business 34 years ago with the development and manufacture of a complement of fuel assemblies for the Yankee Rowe power plant. That 1959 design called for low-enriched uranium dioxide pressed into cylindrical pellets. These were loaded into stainless steel tubes, which were then seal welded with plugs at both ends. These rods were then assembled with stainless steel ferrules into fuel bundles. The bundles were furnace brazed to form the completed fuel structure.

The design was good, but where were we to carry out all the operations necessary? Westinghouse had no integrated facility so we had to look both inside and outside the company.

For the first Yankee core loading, we obtained the pellets from Mallinckrodt Chemical; the fuel rods and assemblies were then fabricated at the former Westinghouse research laboratories located in Forest Hills, a Pittsburgh suburb. Fabrication of the second-core loading directly followed fabrication of the first, but this time, we assigned half of the fuel pellet job to the Westinghouse metals plant at Blairsville, Pennsylvania. This plant also provided most of the zirconium tubing for our reactors.

Anticipating the growth of commercial nuclear power and recognizing the benefits of manufacturing fuel in-house, Westinghouse laid plans for a complete fuel plant at its Cheswick, Pennsylvania, site. This location was adjacent to the Westinghouse-owned Navy fuels manufacturing facility and benefited significantly from sharing of many common facilities, such as the chemistry lab, health physics, administrative services, and works engineering. Construction of this integrated, pilot-production fuels plant began early in 1960.

Again our breadth of experience paid off. John Theilacker provided much engineering design information for the building and equipment, factoring in a wealth of information he had gained while associated with the Shippingport effort at Bettis. Ron Bish directed plant construction and equipment installation efforts. Development didn't stop with completion of the plant. It continued with a manufacturing process that minimized both labor and hands-on materials handling.

Getting the specification density for the fuel pellets was always a concern. With each shipment of feed material the grain size and other characteristics varied and this affected the final density of the pellets. Therefore, each time we received a new shipment we had to make a pilot run to determine the proper procedure.

Fuel and fuel rods for the third and fourth core loadings for Yankee Rowe were fabricated in late 1960 and 1961 in these Cheswick facilities. As with the initial two loadings, finished fuel assemblies were fabricated at Forest Hills.

In the time frame—1962 and early 1963—equipment was installed and processes established for manufacture of larger fuel assemblies. The business was growing in size and volume. Then a change in fuel construction—the alloy Zircaloy replaced stainless steel for fuel cladding—required an all-out development effort at Cheswick to accommodate this new material.

By mid-1963, Cheswick could carry out all fuel-component manufacturing operations, from receipt of the UF_6 gas to final accepted fuel assemblies ready for shipment to the power plant and insertion into the core structure.

This capability was demonstrated in 1963 with the successful manufacture of the initial fuel for the Carolina-Virginia Tube Reactor.

Our problem at this point became one of capacity. We couldn't carry out development of improved processes and at the same time produce components for core applications. The need for greater capacity was obvious, particularly since the demand for fuel was continuing to grow. Thus, in 1965-66 additional buildings were built and equipment installed to achieve a twofold increase in manufacturing capability. We went from 125 tons of uranium dioxide per year to 240 tons. Even with this greater plant capacity, the demand for fuel by 1968 required operation on a two-shift, 16-hour schedule.

When forecasts for the period beyond 1970 indicated still greater demand, the decision was made to construct an entirely new plant in Columbia, South Carolina. This would be a plant with at least three times the capacity of Cheswick, featuring all the latest manufacturing and inspection techniques. The goal was to emphasize automation and quality control. The new Columbia plant went into operation in 1970 with a rated capacity of 850 tons of uranium dioxide per year—becoming the world's largest producer of light water fuel assemblies for the pressurized water reactor system.

With this emphasis on quality and volume, management faced a major challenge: how to recruit and train a new work force to handle the many functions—operations, maintenance, quality control and assurance, and administration. Only a small number of supervisory and technical personnel had been transferred from Cheswick and the job of preparing a new cadre of people was squarely on their shoulders.

The state cooperated by providing a grant for training. Some 300 inexperienced workers were hired and trained. The soundness of the training program is attested to by the results. These people eventually performed varied and complex operations, including the handling of radioactive components, and met the stringent specifications required in the manufacture of fuel assemblies. By 1974, the Columbia plant was producing fuel assemblies at roughly 50% capacity; three years later its output was nearing 80%. But challenges lay just ahead.

Forecasts *again* indicated that more manufacturing capacity would be needed in the early 1980s. At the same time it was becoming clear that overseas competitors were thirsting for this business. Both Framatome of France and Mitsubishi of Japan were making strong bids to become major suppliers of reactor core fuel. In the near future, they would challenge Columbia's dominant position in the fuels industry.

Westinghouse lost no time in responding to these challenges. Utilizing the technical resources at Pittsburgh headquarters, the research and development center, the automation division and at the Columbia plant itself, plans were laid for a 50% expansion in output.

The plant had three processing lines and now a fourth was to be added, but it would be different. This new line would be completely automated and computer controlled with respect to production, process control, and inspection. The number of personnel needed to staff and operate the line would be only one-third that of the other three lines. Personnel would work as a team, being collectively responsible for quantity and quality of product. There would be limited supervision by management personnel; each person would be capable of performing any administrative, production, or quality job on the line.

This was a complex challenge. While the new line went into operation in 1984, it took another 12 to 15 months to work out the bugs and prove the new concepts. Once the new line was completely operational, retrofitting of the older lines began. So a complete upgrading of the plant was not finished until 1988.

How good was it? In 1989, the Westinghouse Columbia Plant received the first Malcolm Baldridge Quality Award—a national honor carrying great prestige. This plant today is maintaining a production rate consistent with demand for 1000 to 1200 tons of uranium dioxide per year. It is simply the best in the world.

Why Does the PWR Reactor Lead All Others?

Although we would like to attribute world leadership of the PWR reactor type to the excellent job done by those who have developed, built, and sold it, we also have to be honest: The PWR has major inherent technical advantages.

Its major advantage is its dual cycle. PWRs have a radioactive primary side consisting of reactor, nuclear fuel, and reactor support systems. There was nucleate boiling on fuel element surfaces, but bulk boiling was limited to 2% per outlet channel. It has a normally nonradioactive secondary side out to the turbine. Steam generator tubes separate the two coolant streams.

This division of the plant into two parts provides advantages in several major areas. The first is in operation and maintenance. Since the steam cycle is nonradioactive, simpler fossil plant procedures can essentially be used in the construction, operation, and maintenance of the turbine island, permitting unrestricted equipment access and on-line maintenance.

A second area of advantage is in occupational radiation exposure. This has been lower in PWRs on the average and has shown a smaller increase with plant age. The reason: Radioactivity is restricted to the reactor and reactor auxiliary systems. Reactor coolant and steam cycle water chemistries each can be optimized individually for minimum corrosion and crud formation. This means a cleaner plant. For the same reason, PWRs have produced less low-level radioactive waste.

Also important in the superiority of PWRs is simply their large number. This has meant more experience and higher quality of development and design. These advantages not only contribute to market dominance by the PWR, but also provide for higher availability and capacity factors compared with the BWR.

Nuclear power supplies a much greater percentage of the total electrical supply overseas than it does in the United States. Figure 11.1 shows that percentage for 1992.

The PWR has had a slightly higher capacity factor than the BWR worldwide as shown in Figure 11.2 for 1980 to 1992. Figure 11.3 shows that

Figure 11.1 Percentage of electricity from nuclear energy in various countries. (From "1992 International Survey," U.S. Committee for Energy Awareness, June 14, 1993.)

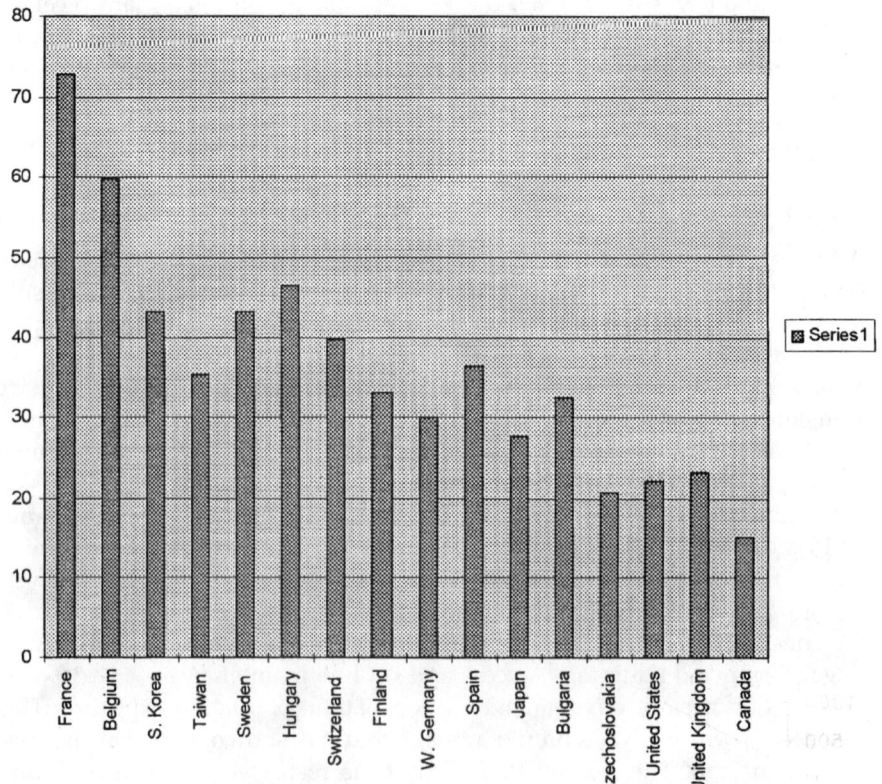

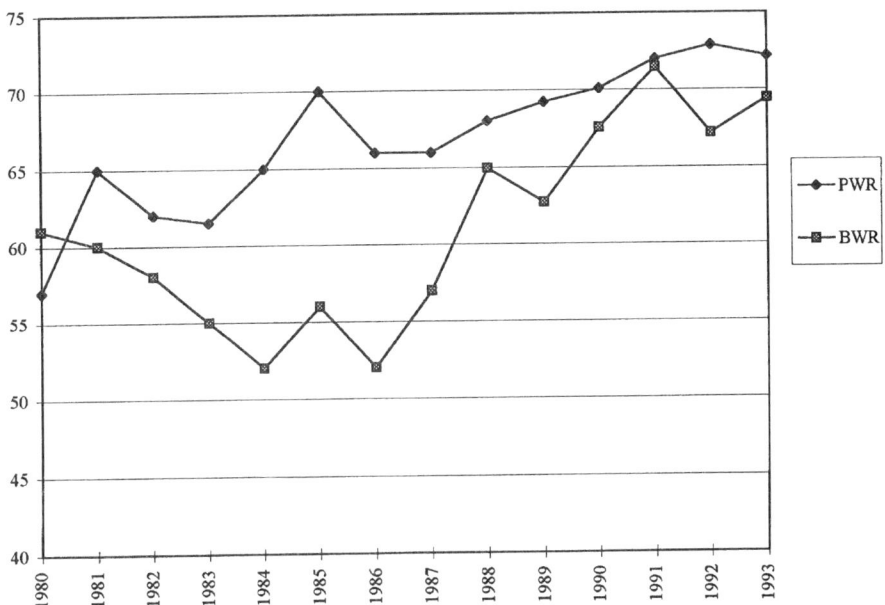

Figure 11.2 PWR and BWR capacity factors, 1980–1992.

construction costs in France were slightly higher than that in the United states in 1970, but the increase has not been nearly so drastic. By 1990, construction costs in France were about half of that in the United States.

Figure 11.3 Construction costs per kilowatt-electric. (From Japan Nuclear Information Center, 1985.)

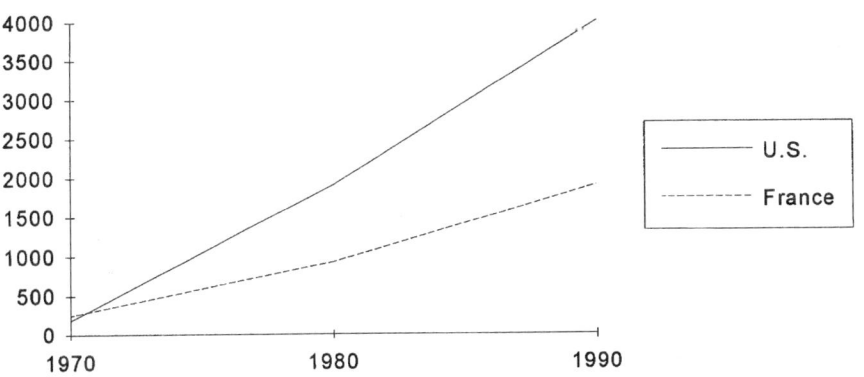

Plant Performance

For many years Connecticut Yankee had the largest total number of kilowatt hours of any nuclear plant, despite its relatively small size. Yankee Rowe has a cumulative capacity factor of 77.03% and Connecticut Yankee 66.68%.

Table 11.1 shows the worldwide gross generation and capacity factor for various commercial reactor types for 1992.

The worldwide capacity factor for water reactors is roughly the same, while that of gas-cooled reactors is considerably lower. U.S. nuclear plants operated at a record 71.3% average net capacity factor in 1992, according to data from the U.S. Committee for Energy Awareness. During 1993 the capacity factor was considerably higher with many plants exceeding 90%.

Technological Improvements

Westinghouse has contributed significantly to commercial development of the PWR over several decades. Large-scale research and development programs—incorporating critical facilities, loop tests, scale models, and prototype plants—were used to verify design techniques and methods currently in use. In some cases development programs were funded through joint ventures between Westinghouse and the AEC.

In others, joint ventures were formed with groups of utilities. Examples are the Westinghouse–AEC–Yankee Atomic Power Company, Yankee-Rowe; Westinghouse–AEC–the utility companies for Carolina-Virginia Tube Reactor (CVTR); and Westinghouse–AEC–Southern California Edison for the San Onofre power plants. On the basis of the results of these and many other experimental programs, such as critical facility measurements, significant improvements were made in initial design concepts.

Table 11.1
Comparative Performance of Different Reactor Types

Type	Gross Megawatts	Generation (MW hours)	Average Capacity Factor
PWR	192,394	1,232,546,710	72.92
BWR	76,104	453,801,437	67.18
GAS	15,969	80,708,932	56.80
PHWR	18,042	98,418,622	62.62

Source: "Status Report on U.S. Nuclear Power Plants," U.S. Committee for Energy Awareness, October 1, 1993.

Some of these improvements were so significant that they actually revolutionized the PWR concept, resulting in major gains in commercial applications. An example was chemical shim control, which was first tried on an experimental basis at the BORAX reactor at NRTS and later at the Saxton training facility and then incorporated into Yankee Rowe and the Enrico Fermi (ENEL) plant in Italy. Some of these developments are as follows:

- Magnetic-jack control rod mechanisms (Yankee Rowe)
- Canless fuel assemblies (ENEL)
- Chemical shim (Yankee Rowe)
- Shaft seal coolant pumps (Con Yankee)
- Rod cluster control rods (San Onofre)
- Plutonium recycle (Saxon and San Onofre)
- On-line DNB protection (Beznau 1)
- Pressurized fuel rods (Mihama 1)
- In-core flux monitoring.

See Chapters 8 and 15 on the commercial nuclear power story for more information on these developments.

Life Extension of Nuclear Plants[1]

Strong economic motives exist for extension of nuclear-plant-licensed operation periods beyond the government's current 40 years. The functional lifetime of various components is much longer than 40 years, and doubling the period is technically a reasonable extrapolation. By the time the license expires, the replacement cost far exceeds the original cost and the cost of the changes required to extend life. For example, the oldest operating nuclear plant in the United States was Yankee-Rowe when it was shut down. It was built for about $250/kW and was still operating satisfactorily to the end. The various components have proved their reliability over many years.

To extend a license to operate for another 40 years may be economically desirable if it can be done for less than the cost of an equivalent new plant. Of course, one must take into consideration that a new plant would be expected to operate for more than 40 years. Such will be the case; however, the present discounted value of that additional 40 years would be small.

Information is being collected to prove to the U.S. Nuclear Regulatory Commission (NRC) that extending the lifetime for another 40 years is technically sound. If the commission agrees, then it makes economic sense to

---

[1]R.J Creagan, Private Communication dated July 8, 1991.

continue to operate the plant—repairing, upgrading, or replacing components as necessary.

The situation varies with the equipment involved. One problem that has been advanced is that the reactor vessel may have become embrittled by neutrons over the years and, thus, may pose a danger. I believe the economics for continued life are so favorable that even if the old reactor vessel is removed and replaced by a new one, it can be a prudent decision.

Unfortunately such decisions are often made by public utility commissions with an eye on the political situation as much as the technical and economic situations. In the case of the retirement of San Onofre 1, the Integrated Resource Planning (IRP) study covered some 32 different scenarios. Twenty-seven of these showed a positive economic benefit that ranged from $175 million to more than $600 million and only five that showed unfavorable economics. Despite these results the California Public Utility Commission (PUC) decided to endorse the extreme negative scenario, one that assumed that San Onofre 1 would operate at a low capacity factor and require high capital expenditures and that gas prices would remain low into the indefinite future. So despite all this, the PUC ordered SCE to install a large amount of very expensive renewable energy sources on the grounds that the SCE system was overwhelmingly fossil fueled, while at the same time they removed 400-MW(e) of non-fossil-fueled nuclear capacity.

The story of the Portland GE Trojan plant was somewhat different in that different unrealistic assumptions were used to justify closing the plant in 1996 rather than continuing operation for another 20 years. If one had assumed that gas prices might rise just a fraction and that operating and maintenance costs could be reduced a similar fraction then the outcome would have been in favor of continuing. It turns out that Portland GE will have to spend about $650 million to save the $200 million cost of new steam generators.

The study showed a saving of about $100 million out of total planned expenditures of $3 billion during the 20-year period. SCE and Portland GE, however, are going to be allowed to continue to depreciate the asset as though it were still in operation. This money will be collected from their customers. It's a strange world in which utilities must operate.

What About the Future?

For almost a quarter century after the launching of the first nuclear-powered vessel, the atom was widely hailed as the energy source of the future—the answer to the world's need for a reliable, nonpolluting source of power that would free mankind from the increasing limitations of fossil fuels.

But from the late 1970s to the present, a combination of factors raised doubts in the mind of the people and governments about the future of nuclear power. Rising costs, due to a considerable extent to the long construction delays, began to make plants less economical and in a few extreme cases uneconomical. OPEC's oil embargo led to great emphasis on energy conservation in all user nations and a sharp economic decline, thus requiring few capacity additions for many years. Safety concerns were raised by the Three Mile Island accident, which caused a major melting of the nuclear core. However, there was no damaging biological effect to the public, the greatest dose received was about four REM, which is about twice a gastrointestinal x-ray dose. The next highest dose was about two REM to a company employee. Other exposures were all less than the legal limit. Despite all this, the Three Mile Island accident in 1979 gave added ammunition to the nuclear critics and the changes it required caused added costs.

The Chernobyl accident in 1986 further aggravated the situation as it raised worldwide concern about nuclear plant safety. This was a very serious and most regrettable accident. However, the latest assessments indicate that the effect on the health of the public was less than first believed. It is difficult to assess the accuracy of many of the reports on the conse-

Figure 11.4 The damaged Chernobyl power plant before Unit 4 was buried in concrete. (Courtesy of Tass/Sovfoto)

quences of the accident. The accident was the result of a poor design, a lack of safety standards and operating procedures, and poorly trained operators and supervisors. The lack of containment was a major factor in the seriousness of the consequences of the accident. A study[2] of the actual amount of radiation exposure revealed the following:

> Virtually all Russians in the region contaminated by the 1986 Chernobyl accident received harmless doses of radiation and face no health risks as a result, according to a recent study conducted by Germany's Julich research center.
>
> The German-financed study was conducted between May and October 1991 to help people living near Chernobyl determine their actual radiation exposure.
>
> R. Hille of the Julich research center says measurements of food samples, external radiation and whole-body radiation for 160,000 people indicate "that over 99 percent of the cases are within a completely safe range. There is no risk to the health of this proportion of the population from their food or environmental activity."
>
> The Julich measurements agreed with previous results obtained by Ukrainian authorities.
>
> "In all cases, the 1991 pollution was smaller than that permitted for persons professionally exposed to radiation in the Federal Republic of Germany," said Hille. "These values therefore do not represent an acute health risk."
>
> Hille said that by providing objective information on radiation risk, the project helped reduce "unfounded fears" and would improve "the psychological well-being of many [Russian] people."

Richard Wilson,[3] a professor of physics at Harvard, reports that it is very difficult to get an accurate picture of the effects of radiation due to the Chernobyl accident. The 600,000 people involved in the cleanup received considerably higher doses than would be permitted in the United States. In the first year it had been stated that these workers had received as much as 20 rems on the average. This was reduced to more like 10 rems the next year.

Some 200 to 300 workers at the plant when the accident occurred received very high radiation doses. These workers suffered from acute radiation sickness. A number have died, but many recovered.

Nikolai Napalkov,[4] assistant director of the World Health Organization, reported on the seventh anniversary of the accident that there had been no increase in leukemia in Belarus, Ukraine, or Russia. Anatoly Tsyb, of the Medical Research Center at Obninsk, reported no observable rise in leukemia among those who worked on recovery and cleanup operations. Based on the experience of Japanese bomb survivors some would have been

---

[2]*Nuclear Energy*, First Quarter 1993, p. 7, U.S. Committee for Energy Awareness, 1993.
[3]Richard Wilson, Mallinckrodt Professor of Physics, Harvard University, 21st Century Science and Technology, Summer 1993.
[4]*Nuclear News*, Vol. 36, No. 8, p. 87, June 1993.

expected. Since some increases in chromosome abnormalities were reported radiation-induced leukemias may be on the way.

Cases of thyroid cancer in children in Belarus[5] increased from 25 in 1987–89 to 114 in 1990–92. Most of the increase was in the Gomel area where fallout was heaviest. "One of the theories put forward by WHO is of an 'iodine shock' caused by stable iodine tablets, which were wrongly administered in the region two weeks after the accident—long after any usefulness for reducing the effects of iodine-131."

The Case for Nuclear Power

With the end of OPEC's power to maintain unrealistically high prices for oil came a return to greater dependence on Middle East oil production and a lessening in the search for energy alternatives such as nuclear fission and fusion. There also appeared to be more natural gas reserves than had been thought and the decrease in oil prices resulted in the lowering of natural gas prices.

Today it is a fair question to ask, "Does nuclear power have a role in the future, or has it reached its apex and begun an irreversible decline?"

Many factors will affect the decisions by electric utilities on the type of generating plants to build in the years immediately ahead, but there is little doubt that they will need to build them eventually.

Generation Capacity Needed[6]

The total generating capacity in the United States at the end of 1992 was 692 gigawatts,[7] and at the peak demand there was a capacity margin of 19%. The electric utilities consider 20% to be a satisfactory reserve margin. The growth in noncoincident peak demand has grown 87 gigawatts[8] or 2.7% per year since 1986. The demand for electricity—including the impact of demand-side management (DSM) savings—is expected to grow by 1.5% annually during the next decade. Without the inclusion of DSM savings, the annual growth in electricity demand would be 1.8%.

U.S. utilities[9] expect to need an additional 94 GW of generation to meet the country's electricity demand during the next decade. They plan to add

---

[5]*Nuclear News*, Vol. 35, No. 15, p. 48, December 1992.
[6]Data from the Electric Power Research Institute, in a private communication with W.H. Esselman, June 1993.
[7]Data from the Electric Power Research Institute.
[8]Data from Virginia Sultzberger of the North American Reliability Council, quoted by Electric Power Research Institute, in a private communication from W.H. Esselman.
[9]"Electricity Supply and Demand Into the 21st Century," U.S. Committee for Energy Awareness, July 1993.

66.4 GW of generating capacity and purchase 27.6 GW of capacity. New plants represent 66% of the 94 GW. In addition, utilities plan to increase the available capacity of existing generating plants by 4.8 GW.

Half of the new capacity additions will be gas fired or dual fired—able to burn either natural gas or oil. However, new gas-fired capacity will account for just 22% of the additional electricity generated, because most of it will be used for providing peaking electricity. On the other hand, coal and nuclear plants represent only about 10% of new capacity additions, but will account for 45% of the additional electricity generated—coal for 35% and nuclear about 10%.

U.S. utilities intend to purchase an additional 15.6 GW from nonutility generators. They also intend to import 12.4 GW from Canada.

The economy has been growing at a slower rate in the last few years than it is expected to in the future. Also there has been a considerable slowing of the growth in the use of energy as a whole due to serious conservation efforts, but the growth in the use of electricity has been greater than the growth in the overall use of energy, because electricity is a pre-

Figure 11.5 Energy growth versus gross domestic product. (Chart from *DOE/EIA Monthly Energy Review*, "1992 Nuclear Energy Review," U.S. Committee for Energy Awareness, December 11, 1992.)

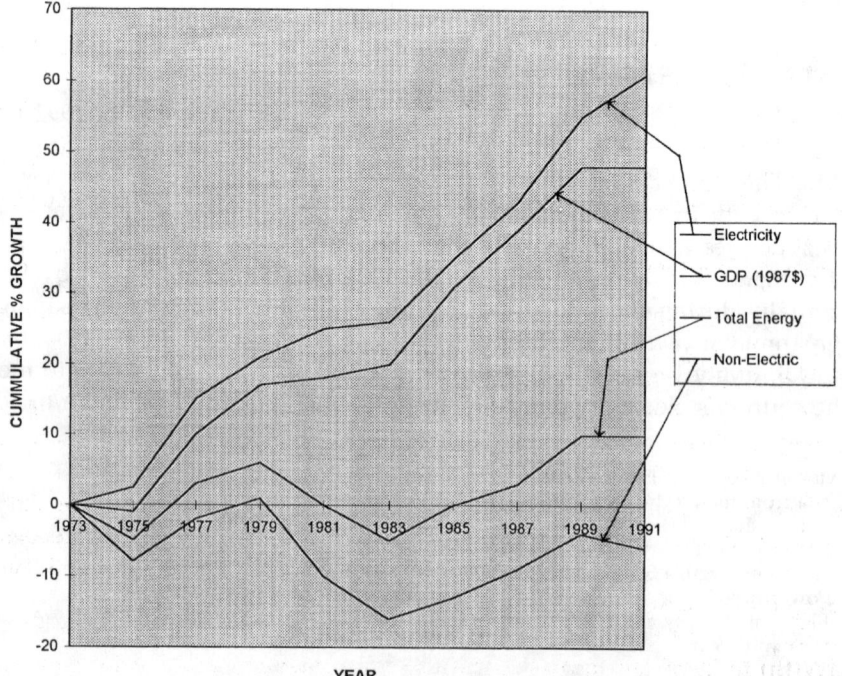

ferred source. This trend is likely to increase in the future. Figure 11.5 shows the cumulative increase in percent compared to 1973.

By the year 2000 40% of today's power plants will be more than 30 years old, and many will need to be replaced. The United States might well be facing a severe energy crunch, which will require the addition of too much uneconomical short-lead-time equipment, such as gas turbines for base load service. There still may be sufficient time left to prevent this unfortunate occurrence, but proper action will have to be taken soon.

Decisions about new plant construction will be affected by many unknowns—the future cost of various fuels, the potential danger of global warming, the problem of acid rain, and the unpredictability of construction costs, to name a few.

What does nuclear power have in its favor?

First, a record of achievement. In the past 30 years, U.S.-designed nuclear plants that emit no particulates, no nitrogen oxides, no sulfur dioxides, and no carbon dioxide have logged more than 1800 reactor-years of safe operation. U.S. nuclear power plants reduce carbon dioxide emitted by all utilities by 20%—saving 420 million tons in just one year.

Second, a development program for future plants. The nuclear plants that can be built in the future will achieve their demonstrated level of safety by simpler more direct means than those in the past and will be built at lower cost.

Third, a growing environmental consciousness that is placing even greater importance on the need to reduce air pollution. Among the major energy sources with prospects for sufficient growth to meet future needs, only nuclear power does not pollute the air.

Fourth, some evidence exists that the stalemate over nuclear waste disposal will be solved as part of the overall waste problem that faces the nation.

Current Developments

Development currently is taking place on two fronts: improved large plants and a new generation of simplified, mid-sized plants that could spearhead the breakthrough for nuclear power in the future.

In a cooperative program with Mitsubishi of Japan, Westinghouse has designed two versions of the advanced pressurized water reactor (APWR), at 1000 MW(e) for international sale and at 1300 MW(e) for domestic sale. The APWR 1300 has received its preliminary design authorization from the NRC.

These designs were the next logical step in development of the basic technology utilized in recent three-loop pressurized water plants. By utilizing the standard four-loop core and vessel in a three-loop piping configuration, the 1000-MW(e) reactor has attained a power increase from 2785 MW(th) to 2910, an improvement providing better operating margins and greater reliability.

Figure 11.6 Westinghouse's AP 600. (Courtesy of Westinghouse Electric Corporation)

By requiring less electric cable, using advanced instrumentation and control systems, and making other improvements, these APWR plants are projected to have capital cost savings as high as 20%. They can achieve a plant availability factor of 90% with a 20% reduction in fuel cycle costs. They can be built in a shorter time at lower cost.

The AP600 Nuclear Plant[10,11]

As good as the new large plant designs are, it is the simplified, 600-MW standardized plant that holds the greatest hopes for the nuclear future, in the opinion of its developers.

---

[10]"Design Certification, Modular Construction, Predictable Economics, and Proven Technology," Westinghouse Electric Corporation, November 15, 1992.
[11]Howard Brushi and Theodore Andersen, "Turning the Key," *Nuclear Engineering International*, November 1991.

This design is being worked on cooperatively with utilities and the government to meet utility requirements for advanced light water reactors that will be the ultimate in safety, flexibility, low-cost and simplified construction.

The Westinghouse AP600 is a 600-MW pressurized water reactor that uses the finest of proven, state-of-the-art components, reconfigured to create a new generation of simplified plants. Ultimate safety is assured by nonactive or passive systems that rely on natural forces such as gravity, evaporation, convection, and physical properties of materials. In the unlikely event of an accident at an AP600, no operator action would be required to protect the public. Of course, active safety systems are also available to the operators.

Because it is based on the world's most used nuclear power technology—more than 57% of the world's reactors are pressurized water—this standardized plant will not require a demonstration plant for NRC licensing.

The mid-sized AP600 gives utilities more flexibility to adjust capacity increases to changes in demand growth. Analysis by the Electric Power Research Institute (EPRI) shows that staged 600-MW new capacity additions will reduce construction time, interest costs, and rate "shock" to the consumer.

The availability of the six Westinghouse two-loop plants was an average of 86.7% from 1984 to 1991, and the goal for the AP600 is 90.0%. This has been accomplished by a design that reduces refueling time and gives a period of 18 to 24 months between refuelings. The best refueling/maintenance outage of a two-loop plant in 1990 was 30 days. The average availability assumes an outage extension for replacement of both steam generators once in the lifetime of the plant.

The 84 model F steam generators placed in service since 1980 have accumulated more than 2.4 million tube-years of service with less than 0.1% of the tubes plugged. Corrosion potential in the AP600 will be reduced even further by the 600°F hot leg temperature, copper-free condensate system, and Inconel 690 tubes.

The fully integral, ruggedized low-pressure turbine rotors have more than 54 rotor-years of nuclear service with no forced outages. The canned motor pumps eliminate seals and their associated systems. AP600's pumps have a 12-year mean time between repairs. Westinghouse distributed instrumentation and control architecture with fiber optic data buses has been proven in new and retrofitted plants around the world, including the United Kingdom's Sizewell B and the Swiss NOK Beznau plants. Also the AP600 has ample design margin to reduce component stress and meet a 60-year design life objective—lower power density (79 kW/L), larger water inventories and a 15% fuel temperature margin during transients.

The AP600 is much simpler in design than the already successful two-loop plants. It has 50% fewer valves, 35% fewer pumps, 80% less safety grade pipe, 80% fewer heating, ventilating & cooling units, 45% less seismic building volume, and 70% less cable.

NRC requirements specify a core melt frequency of less than one in ten thousand per year; EPRI requirements specify one in one hundred thousand. A preliminary probabilistic risk assessment (PRA) submitted to the NRC in 1992 indicates that the AP600 exceeds these requirements.

Modular Construction

The AP600 carries plant standardization and simplification to its natural next step—modular construction. Its components and systems will be built in modules in a factory setting for rail shipment to the final site. Using modular techniques, the entire construction cycle of an AP600 can be completed in five years—from "commitment to build" to startup.

Modular construction is becoming the norm for large capital projects. For more than a decade, ships, aircraft, and petrochemical plants have been built with modular construction techniques. This reduces capital costs and construction time. It permits use of a stable and well-trained factory workforce for better quality. Work goes on regardless of weather. More efficient tooling and materials handling becomes possible. All in all, modular construction can revolutionize the building of power plants.

Once completed, the AP600 modules are designed for shipment by rail or barge. Rail module size is limited to 12 feet (height) by 12 feet (width) by 80 feet (length). Maximum weight is 80 tons. Larger assemblies, such as the ring sections used to assemble the steel containment, are built from submodules at an on-site shop.

This revolutionary plant will meet EPRI's requirement on costs. It will come in at about $1411/kW (in 1990 dollars) vendor's overnight cost. The owner's cost are estimated to be $156/kW and interest during construction to be $198/kW. This gives a total of $1705/kW, making it competitive with other nuclear and non-nuclear generation alternatives. A study made for the USCEA[12] by EPRI and United Engineers & Constructors, Inc., estimated an "overnight" capital cost for the advanced light water reactor (ALWR) to be about $1500 per kW(e) for the third to sixth plant and that the overall generating cost would be competitive with pulverized coal or combined cycle gas turbine plants. But utility executives know that construction costs are only half the picture. They must consider fuel prices and mix, capacity

---

[12]"Advanced Design Nuclear Power Plants: Competitive, Economical Electricity," U.S. Committee for Energy Awareness, June 1992.

factor, plant staffing, regulatory changes, demand growth, and, above all, predictability. AP600 costs are predictable because its technology has an experience-based track record.

AP600 Levelized Generation Cost Advantages

Generation costs (Figure 11.7) are calculated by adding capital, fuel, operation and maintenance, and decommissioning costs over a 30-year period, levelized at a utility cost of money (11.4% nominal, 6.1% inflation-free, constant 1990 dollars). Coal plant costs are based on a 1992 study by the U.S. Council for Energy Awareness, with 1.2% real coal price escalation. Combined cycle natural gas prices are calculated at various price escalations. All capacity factors are assumed to be at 80%.

Schedule for Certification

The AP600 is on schedule for design certification by NRC for 1995. However these schedules seem to have ways of slipping. It will be the first mid-sized, passive, light water reactor of advanced design to be certified by the NRC.

The year 1992 was one of major milestones toward certification, with the complete standard safety analysis report and the PRA being submitted in June. In achieving these early milestones, Westinghouse is on track to meet the industry's objectives for a standardized design by 1995, as defined by the Nuclear Power Oversight Committee. The beauty of certification of a standardized design, of course, is that it only has to be done once. Plants

Figure 11.7 Generation cost comparison levelized over a 30-year period.

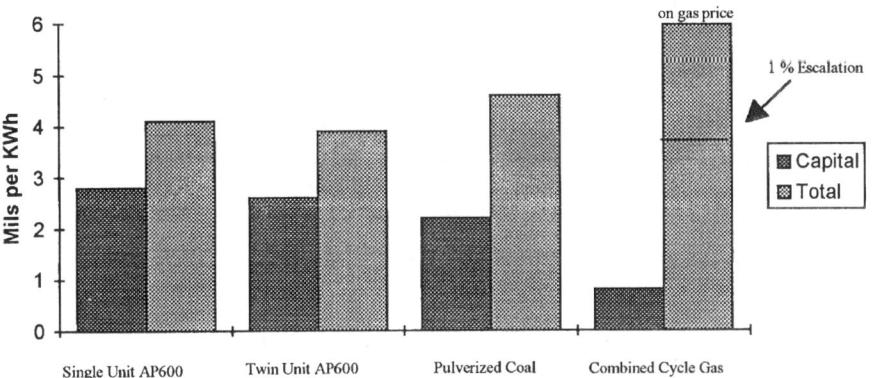

can then be built without going through any further certification process. This will virtually eliminate schedule risks for the utilities, allowing them to plan with confidence on bringing an AP600 on line by the turn of the century.

Team Leaders and Sponsors

The AP600 team is composed of a worldwide network of leaders in the design and construction of nuclear electric generating stations. They bring to the project decades of experience in every aspect of plant design and construction. The sponsors are as follows:

- U.S. Department of Energy, supporting the development and certification of ALWR technology as an integral part of the national energy strategy.
- EPRI, leading the electric utility industry in the development of ALWRs.
- Advanced Light Water Reactor Steering Committee, representing the electric utilities in establishing ALWR criteria that meet their needs.
- Westinghouse Electric Corporation, leading the development and design of the AP600 reactor and demonstrating advancements in passive safety, simplicity, and modular construction.

Westinghouse's major contractors are Bechtel North American Power Corporation, Burns & Roe, Avondale Industries, Inc., MK-Ferguson Company, CBI Services, Inc., Southern Company International, and R.L. Cloud & Associates.

Commitment to Commercialization

The AP600 team is committed not only to technological development but also to commercialization. Its principal members will share the costs and profits. They will offer AP600 at a firm price on a firm schedule. They will work with utility sponsor groups to develop the infrastructure for commercialization. They are working with the NRC to establish a joint AP600 testing program so that design certification will proceed on schedule and the "advanced passive" features of the AP600 design will be preserved.

Message from the Chairman of NPOC

The chairman of the Nuclear Power Oversight Committee (NPOC), Sherwood H. Smith, Jr., said this recently in his annual message:

Developed in cooperation with the U.S. Department of Energy and the Electric Power Research Institute, the AP600 offers the certainty required to revitalize nuclear power in the United States: predictable licensing; a firm construction price and schedule; and competitive capital, fuel, operating and maintenance costs.

He went on to cite the design features that will contribute to those results:

A greatly simplified design that uses far fewer components and equipment and less seismic building volume than conventional reactors . . .
Mid-sized to allow utilities to add capacity in smaller increments and more closely match capacity additions to demand growth.
A fully standardized plant with a design that is completed and certified by the NRC before construction begins.
Modular construction for ease of construction with a planned schedule of three years from ground breaking to initial fuel load—a full year shorter than the 1200 MWe evolutionary plant. Proven technology for lower O&M costs and a high plant availability goal of 90% . . .
Innovative safety systems that rely primarily on dependable natural forces like gravity, instead of operator actions.

It would appear that, by the turn of the century, commercial nuclear power will be ready to justify its promise of 50 years earlier. That may just be in time for an energy hungry world.

Further information can be obtained from the publication by the NPOC entitled *Strategic Plan for Building New Nuclear Power Plants, Second Annual Update,* dated November 1992, from which the preceding quotes are drawn.

Nuclear Waste Management

One of the most important questions to be resolved is that of the management of nuclear waste in the development of nuclear power. One of the basic problems is the need, as expressed by the environmentalists, to prove beyond any doubt that under no possible circumstances could any radioactivity from the storage site reach the public for at least thousands of years. However, they seem oblivious to the fact that spent fuel has been safely stored at plant sites, in some cases, for decades.

Westinghouse recognized at an early stage the importance of demonstrating and establishing the safe handling, transportation, and storage of nuclear fuel—both fresh fuel and also spent fuel after use in the nuclear plants to produce electricity. To this end, we developed the shipping containers, systems, and procedures for the transportation of nuclear fuel from the Columbia manufacturing facility to the utility nuclear sites. In addition, we were directly involved in the Navy nuclear program in the design, handling, transportation, and storage of reactor fuel—both new and spent. These activities involved transportation, storage, and reprocessing of spent fuel in Idaho. All these activities were developed and put in place safely, reliably, and effectively.

During and after the years of the Carter administration, the ban on reprocessing of commercial fuel and the emphasis on the once-through fuel cycle focused our attention on spent nuclear fuel. In a carefully constructed program:

- Under contract to DOE, we carried out definitive experiments at the E-MAD Nevada facility on spent fuel from Florida Power and Light's nuclear plants to determine the parameters for dry storage.

- Working with the Parsons Company under contract to DOE, we developed a conceptual design for a Monitored Retrievable Storage Facility (MRS). Such an MRS would receive spent fuel from the nuclear power plants and store it aboveground until a permanent repository was developed and in operation. The spent fuel would be carefully monitored at all times.

- On our own initiative, at our Pensacola plant we designed and manufactured and obtained the license from the NRC for dry storage containers for spent fuel assemblies.

These were the major elements of a satisfactory system for the handling and storage of spent fuel, which, if proceeded with, would have provided the near-term solution; I outline these activities and approaches in some detail to show that, as a result of these carefully engineered efforts and those of others in the nuclear industry, spent fuel has been handled, transported, and stored for more than 30 years without harm or danger to the workers, the public, or the environment. The technology and techniques and procedures are well proven and well established. In fact, the NRC has determined that spent fuel can be safely stored in dry storage or MRS facilities for up to 100 years.

Storage of spent fuel at nuclear plant sites or in MRS facilities was never intended to be permanent and by adopting the Nuclear Waste Policy Act (NEPA) of 1982, Congress recognized the scientific consensus that deep geological isolation is the preferred solution for ultimate disposal of spent fuel. NWPA directed DOE to evaluate only one site at Yucca Mountain, Nevada, to decide whether it is suitable as a repository. A one mill per kilowatt hour fee was levied on the nuclear utilities to pay for the program; in return for that fee, DOE would accept spent fuel from utilities for disposal by 1998. The Environmental Protection Agency (EPA) would set safety standards for disposal of spent fuel and the NRC would issue regulations to enforce those standards.

Yucca Mountain, from a scientific and technical standpoint, continues to be a promising candidate site for a permanent repository. Moreover, a recent DOE review of the studies conducted to date concluded that none of the conditions that could disqualify Yucca Mountain is present, but that additional scientific studies are needed to make a final suitability determination.

Both the National Academy of Sciences and the Nuclear Waste Technical Review Board (NWTRB), established to provide Congress and the administration with unbiased scientific and technical advice, have concluded that they do not know of any scientific reason to discontinue the studies at Yucca Mountain. But they have also recommended more studies to determine if the site is suitable for a repository, particularly underground studies. The NWTRB has specifically concluded that a final suitability determination for Yucca Mountain can only be made after the program completes underground studies.

Congress, through the NWPA, has determined that the United States should pursue deep geologic disposal at Yucca Mountain. The nuclear industry believes that a timely and comprehensive study of Yucca Mountain as a potential site for a permanent repository is the responsible way to proceed.

However, 11 years after the enactment of the NWPA, no substantive progress has been made. One significant factor has been the inability of DOE to get organized to proceed with the site characterization of Yucca Mountain on a timely basis. Further, the State of Nevada had adamantly opposed as unfair the selection of Yucca Mountain as the single repository site. This has resulted in a stalemate, and the estimated date for operation of the repository has slipped from 1998 to 2010 and could slip further. The repository program can no longer provide the basis for accepting spent fuel on a dependable schedule. The estimated costs for site characterization have also increased dramatically and the present program, if continued, is expected to cost $6 billion. It will be 2001 before a decision is made that the site is suitable and application made for the NRC license. A further $3 billion would have to be spent by 2010 before the decision is made whether to allow operation. This has created critical problems for the program. On the one hand, the nuclear industry sees high and escalating costs with no favorable result in sight; on the other hand, the State of Nevada and environmentalists are afraid that the huge investment, if the program proceeds as currently planned, will create so much pressure that DOE will have to find and declare the site suitable.

So, in 1994, whither spent fuel and what can be done? Clearly, there is an impasse and the program cannot continue in its present form. With licensing reform of nuclear power plants established by the Energy Policy

Act of 1992, spent fuel management is the major issue confronting the nuclear industry. Without a restructured program, both license renewal of existing plants is threatened and the plan to restore the nuclear power option with ALWRs is also jeopardized.

I strongly believe from my experience in large nuclear projects that we have to be entirely realistic and restructure the program along the following lines, together with assigning more responsibility to the stakeholders representing the nuclear industry.

- The Yucca Mountain program should continue but should be revised to reflect the long-term nature of the program and institute a phased, step-by-step approach using experimental fuel as a basis for eventual licensing determinations. DOE should continue to be responsible for the Yucca Mountain Program.
- The main deficiency in the current DOE program is the complete lack of an agreed and accepted near-term interim storage program with realistic objectives and dependable schedules with milestones. Our highest priority must be given to setting up this near-term program to enable spent fuel to be moved from the utility sites to interim storage.

The major elements of this program to be considered are as follows:

- The development of a universal container system to integrate spent fuel storage activities, starting with removal from the reactor spent fuel pool, transportation to an interim storage at a monitored retrievable storage (MRS) site, and eventually to permanent disposal at the repository. As an example, Virginia Power Company, Newport News Shipbuilding & Dry Dock Company, and Westinghouse have developed a design consisting of a basic metal container with three separate overpacks for transportation, interim storage, and permanent disposal. Development and use of a container system will standardize spent fuel handling, reduce overall costs, and provide a robust package for interim storage and eventually for the repository.
- The development of an MRS program as quickly as possible. There have been several false starts on MRS programs during the past 15 years and it is essential that all stakeholders combine to follow through to the design and construction of an MRS with the required capacity to fulfill interim storage needs prior to implementing the repository. The MRS should be integrated with the universal container system; the use of the universal container system approach enables the development of a simplified MRS

since little or no radioactive handling is required at the MRS site. Selection of sites is the key consideration and advantage needs to be taken of the possibility of federal sites following along the lines of the 1992 proposal by Admiral James D. Watkins, Secretary of Energy. The medium-sized military bases being shut down as a result of the downsizing of defense activities could also provide potential sites.

- This near-term storage program is the vital new element that must be developed and managed by the nuclear utilities and the nuclear industry since the future of the industry in the United States is at stake.

Finally, and this would be an essential element of the near-term program, we have to deal with those two pernicious and paralyzing diseases: NIMBY and NIMTOO—"Not in my backyard" and "Not in my term of office." Technologically the spent fuel problem is solved, but we have not been getting our messages across in the political and institutional framework in which we operate. The near-term interim storage initiative that I have outlined and strongly recommend provides the opportunity to regroup and refocus our efforts on public acceptance, emphasizing not only the safety of this technology, but also the advantages and economic benefits in the near term to the site and to the local communities.

Personal Reminiscences

Joint Economic Committee Hearing

The joint economic committee of the House and Senate arranged a panel hearing with nuclear critics and advocates. I was to take one side with my economist; Ralph Nader and an economist were on the other side. Frank Cotter, as usual, was a big help. He arranged for me to meet with Senator Humphrey, the chairman of the committee, just before the hearing and to walk into the hearing room late with the Senator. He felt I needed all the help I could get facing Nader, and showing a relationship with the chairman couldn't hurt with the other members.

The hearing was held during the mid-1970s energy crisis. Nader made his statement first; he raked nuclear energy over the coals and advocated conservation and new nondepletable forms of energy. The only problem with his solution was that the new forms did not exist; they still play little part in the energy picture, and conservation is no complete solution for Third World countries.

My statement was next. I pulled out all the stops with the scenario that we were for the poor and downtrodden of the Third World. We did not want to deny, as some did, all people from being able to look forward to a rising economic standard, with jobs for all. This could not be done without enough energy, and nuclear energy could be made plentiful and safe.

Nader broke into my statement, saying I was attacking him. I told the chairman that Nader was right—I was attacking him—and that was the most correct statement he had made all morning. My people were amazed that anyone was really taking Nader on. The chairman told me to proceed, that I had the floor.

Westinghouse Board Meets in Europe

In 1973, Westinghouse took the Board of Directors to Europe to visit our operations in Spain, Belgium, and France. Since almost all the operations we visited came under my jurisdiction, I was heavily involved. The Westinghouse planes had been sent ahead and met us in Spain. Esther and I had come over earlier, rented a car, and spent a week touring the chateau country and the vineyards of Bordeaux. My daughter, Patty, was in school in Biarritz, preparing for her year of study in Paris. We stopped by to see her and then drove across the Pyrenees and on to Madrid, where we joined the others.

We toured the Westinghouse Electric Spain plants around Madrid and then flew to Bilbao to inspect the plants there. After returning to Madrid, we flew on to Brussels for the meeting. We were the guests of Societe Generale de Belgique, because their chairman was one of our directors. King Baudouin of Belgium attended the reception that evening. He had asked specifically for an opportunity to talk with me about nuclear power. Don Burnham and I spent several minutes in conversation with him, and I gave him my version of the status of nuclear power worldwide.

From Brussels, we went to Paris and from there made a day trip to factories of Creusot Loire, where they were building components for nuclear plants. Burnham and I were scheduled for a meeting in Paris with Premier Messmer of France at five o'clock and we were late leaving for the plane. Creusot Loire arranged for an escort of a squad of motorcycle police. They went ahead with sirens screaming to clear traffic, then two would speed up and block traffic coming in from side roads, as others went for the next intersection. It was the wildest ride I ever had.

Messmer, the French prime minister, wanted to talk to us about ownership of a major interest in Framatome and the monopoly contract for French nuclear power plants. He spoke no English in the meeting, but I am sure he understood most of what I was saying. He answered in French; I understood quite a bit of what he was saying and several times answered without waiting for the interpreter. This was not proper because Burnham understood no French.

We were crowding too much into the day, being scheduled for a reception and dinner Burnham was hosting at seven. The reception was at Ledoyen, a restaurant in the park on the Champs Elysee. Mr. and Mrs. Burnham and my wife and I left the George V late and then ran into a major early evening traffic problem, consequently arriving late.

Many French dignitaries and leading U.S. industrialists who were visiting Paris had been invited. I sat with Andre Giraud, head of the French CEA and later minister of industry. I had gotten to know Giraud quite well because I had been negotiating the Westinghouse French nuclear contract with him and had seen him several times in the United States. This trip was intended to show the flag and show top corporate interest in international business.

President of ANS

I had been vice president of the American Nuclear Society and was elected president in 1973. It was customary for the president to visit as many of the local societies as possible; there were also two conferences a year and many committee meetings. It is hard to see how a man could handle the president's job and an industry job at the same time and do justice to both without a lot of help.

My assistant Geoffrey Keyes arranged the details of local section visits; with the use of the company jet, I could leave Pittsburgh late in the afternoon, make a dinner meeting, and be home that evening in time for bed. Meetings farther away were scheduled to coincide with other business in the area. For the conferences and committee meetings, Keyes always had a complete report for me as to what was coming up and the possible problem areas.

Bomb Threat

In January 1974, I was in Paris negotiating with Framatome for a controlling interest in the company. This involved not only persuading them to agree but obtaining Electricité de France and French government approval. I made several trips to Paris that year to deal with these negotiations. On each of these visits my daughter, Patty, who was at the university in Paris, would have dinner with me.

After we had returned to my suite one evening, my secretary called me. She opened with, "I don't want to scare you, but I think you should know that a group like the SLA called the San Francisco newspaper, telling them that you and the head of Bechtel and Bank of America were on a hit list. If you don't stop all activities in nuclear energy by September first, they have stolen some plutonium and will use it."

The next day I borrowed the company representative's Rolls Royce. It was parked in front of the house where Patty was staying. I was across the street when I saw a man look the car over and then write down the license number. Could the two events be related?

A few weeks later, before taking a trip to Southern California Edison (SCE), we asked them whether I would be welcome. They agreed to the visit, but there were elaborate security measures. The Westinghouse security chief flew with me in the Jetstar. On arriving in Los Angeles, we taxied to the far end of the runway, and the SCE security people met the plane and took us to the hotel. They had checked me in under an assumed name. They took me straight to my room, then they moved me to my security man's room while he took mine. This was all pretty scary, because either they were overcautious or there was real danger.

Later, shortly after someone had taken a shot at President Ford, the ANS/AIF convention was being held in the same hotel. My keynote speech was going to be a strong endorsement of nuclear energy. This caused elaborate security precautions to be taken. Although they were scary times, nothing ever really happened.

Step Down at Sixty

The 1972 Westinghouse Policy Committee, made up of the six top officers of the corporation, met at Frenchman's Cove, Martinique. At that meeting Burnham, our CEO, proposed that we all "step down" at age 60. We would continue to be officers and directors but would have no line executive responsibility. We would work eight months of the year and take a three-week vacation during that time besides the four months off. We would have an office and could do whatever we believed was in the best interest of the corporation, the United States, or the world. Of course, there were some catches. Our pay would be reduced by a third, and we were not eligible for raises or bonuses, which represented a significant fraction of our salary.

I had been working very hard and was away a lot as the kids were growing up. I had also received quite generous stock options and was pretty well set financially. However, I have since regretted voting for the step-down policy. For a couple of years, being chairman of the Atomic Industrial Forum was quite interesting. I represented the entire nuclear industry—miners, manufacturers, and utility users—to the government and the public at large. But as that ended, being a corporate consultant and finding that nobody wanted my services was not the greatest feeling. In addition, I had joined with John Gray to form an engineering consulting firm in Washington, and that had the potential of creating conflicts of interest. So on September 30, 1977, I retired.

Three of the six were to reach 60 within a space of four months. I do not think it was in the best interest of the company to lose the services of that many top executives at as early an age as 60. When Kirby became CEO, he persuaded the board to abandon the step-down policy.

The Advocates

An appearance had been scheduled for me in the winter of 1971 on the nationwide TV program, *The Advocates*. This was a very difficult assignment. I was to represent the nuclear power industry, and Tony Roissman, a brilliant lawyer, was to represent the nuclear critics. He asked at one point, "In case of a major accident at a Westinghouse nuclear plant, caused by the fault of Westinghouse, would the company accept financial responsibility?" The answer was a quick and firm yes! Of course, nothing was being given away because, if it were Westinghouse's fault, we would be liable. But my worry only ended later, after our lawyers agreed with me.

My schedule called for flying to Madrid that night for partridge hunting. The snow was so bad that my flight was canceled. To get to Madrid, I had to backtrack to New York. I was going there for a Westinghouse Electric Spain (WESA) board meeting and to try to sell another nuclear plant.

Those partridge hunts in Spain were something else. Everyone dressed as though he were going to a country villa; there were gun bearers and gun loaders. Lunch was catered by the Madrid Jockey Club, complete with cocktails, two wines, crystal, silver, and china.

Typical Trip Schedule

A typical trip schedule for me might be as follows: meeting with the AEC October 8 in Washington; October 9 to New Orleans to visit a customer; October 10 to San Francisco to visit Pacific Gas & Electric and leave that evening for Tokyo; two days in Tokyo, then on October 13 to Seoul, Korea, for meeting on the nuclear plants we sold them; then October 15 to Taipei to visit Tai Power; October 16 to Hong Kong to work with Jardines on selling a nuclear plant there; October 18 to Bangkok to visit our distributor; and finally October 20 to Sydney to tour our manufacturing plants. In each city, we would be wined and dined and sometimes taken sightseeing. I would return home exhausted after a long West-to-East flight, only to have to be at work the next morning.

There was a visit to Paris in September 1973, then again in January, March, May, and July of 1974. Of course, it wasn't this hectic all the time. However, 150,000 miles a year was about average. The company plane was a godsend.

CHAPTER 12

The Naval Reactors Technical Story

In a development project like that of the naval reactors program, one of the most important facts for consideration is the status of the existing knowledge—whether it is sufficient to give reasonable assurance that a successful result can be achieved. This consideration led to the choice of a water-cooled and moderated, highly enriched, heterogeneous, thermal reactor plant coupled to a steam-driven propulsion system. While steam-driven propulsion systems were normally used for large vessels, this was the first use of steam plants for submarines.

As in all naval vessels, it was necessary to have the plant as small, lightweight, and simple as possible, commensurate with reliability, safety and accessibility.[1]

Basic Principles

Thermal reactors were selected for the naval program. The essential parts of such reactors are the fuel, the moderator, and the coolant. A thermal reactor is so called because the neutrons are slowed down from the very high velocity (2 MeV of energy) at which they are produced by the fission (splitting) of a U-235 atom. When the neutrons are slowed down (below about 2 eV), the probability of splitting the U-235 is greatly increased, thus requiring far less U-235. Any moderator containing low atomic weight at-

---

[1]J.W. Simpson and L.H. Roddis, *Propulsion Plant for the Submarine Nautilus,* Society of Naval Architect and Marine Engineers, New York, 1954. Reviewed and modified by A.F. Henry, W.H. Esselman, and J.J. Taylor.

oms will do, e.g., a hydrogen atom having the same weight as a neutron, so that the neutron is slowed appreciably with each collision.

There have been reactors designed with a great many different moderators—deuterium (heavy water), beryllium, organic liquids, graphite, etc. For the submarine program, plain water was selected, and it served the purpose of both moderator and coolant.

The fuel selected was highly enriched U-235, and the problem was to convert it to a form that would be mechanically suitable. It needed to have a cladding that resisted corrosion by the high-temperature water and could contain the gaseous fission products. Also, a suitable mechanism was needed for controlling the chain reaction, and thus the power level.

To the marine engineer, nuclear fission offers just a new source of heat, but one without the liabilities of massive combustion products. The resulting heat is transferred from the reactor into a steam generator or boiler by ordinary water (called primary coolant) retained under high pressure to prevent boiling. The water is pumped back to the reactor from the boiler in a closed loop.

The boiler, which is essentially a shell-and-tube heat exchanger, contains ordinary boiler feedwater on the shell side at a pressure much lower than that of the primary coolant. This boiler feedwater, when heated by the primary coolant, becomes steam used in a conventional marine turbine-condenser propulsion system to provide energy for ship propulsion and for other purposes.

Of course, you must control the reaction and have instruments to monitor it and provide safety features, in case things don't go as planned. Many materials that would be needed were not available, many components had to be developed, and—to a considerable degree—the theory of the interaction of the various parts was unknown.

The Fission Process[2]

The process of nuclear fission occurs when the nucleus of certain atoms is broken into pieces through a rearrangement of the nuclear binding forces that hold them together. In the *Nautilus*, the light isotope of uranium, U-235, fissions because of the absorption of thermal or low-velocity neutrons into the nuclei. The result is a breaking apart of the U-236 (U-235 plus a neutron) into two major parts known as fission fragments. These fragments are the highly charged nuclei of two elements in the middle of the periodic table, and they move apart with a very high velocity, therefore possessing high kinetic energy.

---

[2]Reviewed by R.J. Creagan and J.J. Taylor.

These fission fragments are stopped through electrical (coulomb) interaction with the nuclei of surrounding materials (the rest of the original uranium, structural materials, etc.) and the kinetic energy is dissipated as heat. Because fragments are stopped and their kinetic energy released as heat practically at the point of fission, fission in effect produces heat directly. Also produced is radioactivity as beta particles, gamma rays, and neutrons emitted at the time of fission and later.

The energy released in U-235 fission in mega electron-volts is as follows[3]:

		Gamma rays	
Fission fragments	169.7	Prompt	7.6
Neutrons	4.7	Delayed	7.1
Beta particles	7.4	Capture	9.3
		TOTAL	204.0 MeV

The fission fragments produce more than half of the total heat generated. About half of the radiation particles are emitted later during the process of radioactive decay of the fission fragments into more stable forms. Some heat is associated with this later release of radioactivity; therefore, even after a nuclear reactor is shut down, a small amount of afterheat continues to be released. If this decay heat is not removed, it can cause overheating and melting of the reactor core.

The process of fission can be studied and understood best on a statistical basis, since there are billions and billions of fissions taking place each second in a typical reactor. About thirty million million (3×10 to the 13th) fissions are required to produce the release of approximately 1 Btu of heat. The energy equivalent available from the process of fission is astounding. The fissioning of one pound of U-235 produces the heat equivalent to that from the combustion of about 1400 tons of coal or 260,000 gallons of oil.

The average fission produces, among the particles released, about two and a half neutrons. The significance of this fact is that since it requires only one neutron to initiate the process of fission in a given nucleus, and since that process itself produces more than one neutron, a self-sustaining chain reaction can be initiated. Thus, a continuous evolution of heat is obtained because it is possible to maintain a self-sustaining chain reaction, given a sufficient quantity of the U-235 isotope.

The production of a controlled, self-sustaining chain reaction in uranium requires a proper balance between the neutrons produced during the process of fission and those absorbed by the uranium to produce another fission, those captured by other materials (including the U-238), and those neutrons lost by escape from the zone of the chain reaction (the reactor

---

[3] Harvey Graves, Jr., *Nuclear Fuel Management*, p. 140, John Wiley & Sons, New York, 1979.

core). When there is a balance between neutrons produced and lost to the reaction, the reactor is said to have reached criticality.

When this balance is achieved, the fractional difference between the production and loss rates of neutrons, called the *excess* reactivity, is zero. The achievement of this very precise balance proves to be a difficult engineering job in practice. Besides the usual requirements for materials, there is that of neutron economy, so there will be enough neutrons left over to make the process self-sustaining.

Simplified Description of a Nuclear Plant

The heat in a nuclear plant is generated in the nuclear core, which is in a thick-walled pressure vessel. Pumps circulate coolant water through the core. There is a pressurizer to prevent the water in the core from boiling. The heated coolant water flows through the steam generator transferring heat to water surrounding the steam generator tubes. The coolant water flows back to the core. The water on the secondary side of the steam generator boils. The steam flows to the turbine that drives the propeller shaft after the speed is reduced by reduction gears.

There are control rods to control the nuclear reaction. In addition to the main components there are a myriad of sensors, instruments, valves, etc.

Criticality[4]

It is necessary to have the proper amount of fissionable material in such a distribution that the reactor can be made critical. There also must be sufficient reactivity for proper control, not only at the start but throughout the useful life of the reactor. These problems can now be solved largely by theoretical calculations.

However, during the early phases of the project, it was not possible to calculate accurately the nuclear design, since the necessary theory and design constants were not sufficiently well known. For this reason, and to verify the calculations, a series of critical and exponential experiments was needed to determine whether the reactor design would be satisfactory from a criticality standpoint and to provide the basic data needed to permit the theoretical calculations to adequately determine flux distribution. These experiments would show where the theoretical approximations were in error and would permit corrections to be made so that other properties of the reactor could be predicted.

---

[4]R.J. Creagan and J.J. Taylor, Private Communication dated 1992.

The final critical experiment consisted of a nearly exact nuclear mock-up of the reactor and all of its controls. These experiments also helped determine the spatial distributions of neutrons for various control rod positions. In a sense, these experiments serve as the low-power proving grounds for reactor designs.

An exponential experiment consists of a lattice of assemblies and a source of neutrons. The neutron flux is measured from the source of neutrons as it falls off exponentially. By measuring the slope of the exponential curve, we can calculate the reactivity of the infinite lattice. The experiment determines the reactivity of the assembly under investigation, as if it exists in an infinite regime, the so-called infinite lattice.

In the case of the exponential experiments conducted at the Brookhaven National Laboratory (BNL), a graphite reactor provided the source of neutrons, and the finite lateral size of the experiment was dictated by sheets of cadmium surrounding the experimental lattice. In general, critical experiments are easier to operate and produce more accurate results, but exponential experiments can be run with much less fissionable material as subcritical assemblies.

One example of experiments with slightly enriched rods was run at Brookhaven by placing the rods adjacent to each other, so that there was practically no water (moderator) in the assembly. The significance of this experiment was that there had been a question of whether, in a natural uranium-water assembly, even if packed tightly together and of infinite size, the fast neutrons from fission of U-235 would cause enough fissions in U-238 to achieve criticality. Since U-238 only fissions with neutrons above a threshold energy, the less moderation the more fissions. Because such a lattice, even if infinite in size, might not be critical, only an exponential experiment could determine its reactivity.

Thus, while the Brookhaven exponential experiments confirmed many of the measurements made at Bettis in the critical experiments, additional data valuable to reactor physics experimental knowledge were determined. These incremental data included the lattices that would not go critical at Bettis, either because there was not enough moderation, or at the other extreme too much moderation, so that the uranium concentration was not enough to go critical.

The first criticals for the Mark I were done in the Building 316 at Argonne National Laboratory (ANL). Creagan left Bettis for ANL around Thanksgiving 1950 and found Building 316 just completed, with nothing inside. He and Adolph Toepfer, an experienced Westinghouse engineer, worked together getting the necessary apparatus assembled. Carl Benz was the ANL man in charge, but he was more than willing to be guided.

Sid Siegel, the Bettis physics department manager had taken Creagan to Sacandaga, where General Electric was doing its criticals for the intermediate breeder reactor, which eventually became the submarine intermediate reactor (SIR). Sid also took him to Oak Ridge National Laboratory (ORNL) to observe Dixon Callihan performing criticals for the gaseous diffusion plants, done in connection with its safety program.

Because of fatalities at Los Alamos, which had resulted from critical experiments, it was decided that remote control was mandatory, from the other side of a 4-foot-thick wall. The necessary equipment was assembled, which included a tank to contain the uranium assemblies, water, and control rods. Control instrumentation was specified and built by Tom Brill, manager of the ANL instrument department.

The fuel elements consisted of zirconium slats, to which had been glued friction-tape-like strips containing uranium oxide. The uranium powder had been mixed with plastic and extruded to form the strips. The zirconium slats were so brittle that they shattered like glass if dropped. Zirconium becomes brittle if it is not melted in a vacuum. Physics calculations and safeguard analyses were performed by the Bettis physics department, and were reviewed by the Advisory Committee for Reactor Safeguards.

A source of neutrons was necessary, because the assembly had a very low neutron background level. U-238 fissions spontaneously at about 15 fissions per second per kilogram. But in contrast, because odd nuclei do not fission spontaneously, U-235 does not emit much radiation. Beryllium bombarded by alpha particles emits neutrons, because the last neutron is held more loosely to beryllium than to any other isotope in the periodic chart.

A neutron source made of a mixture of beryllium and polonium was used. The scheduled first criticality was attended by Rickover and Zinn. Rickover had to fly to Oak Ridge before criticality was achieved, but he checked back.

Experiments were run to confirm calculations of critical mass, control rod worth, neutron distributions, transient response, and temperature coefficients of reactivity. Extra reactivity is provided for maximum xenon override. The control rods must also compensate for this extra reactivity when the xenon is not present. The experimental transient power ramps were measured with various amounts of excess reactivity. Control rod worths to start and stop transients were measured.

Neutron flux profiles were measured to confirm calculations of neutron diffusion, which would enable hot channel factors to be established, which in turn would determine power output capability.

Meanwhile, Bettis built a critical facility that consisted of uranyl nitrate solution stored in tall thin containers, to be pumped into the reactor cavity for criticality. This facility was designed to help develop reactor theory, and the uranium solution could conform more closely to a sphere or cylinder than mechanical fuel elements. Experimental results were well predicted by reactor theory, except at interfaces, such as between reactor and reflector, or at the surface of control rods. On the control rod surface, neutron levels should go to zero, except there was an extrapolation distance in the theoretical boundary condition. Extrapolation distance, as used in reactor physics calculations, refers to the incremental distance between the boundary where the neutron flux actually goes to zero and the boundary where theory assumes it goes to zero. Therefore the actual radius of the reactor is smaller than the dimensions calculated by the extrapolation distance.

Reactor theory in use at the time was two neutron group diffusion analysis, using constants obtained from the "barn book," so called because neutron cross sections were given in units of barns. A barn is 10^{-24} cm^2, so the term *barn* was scientific exaggeration.

The two neutron groups in the theory corresponded to fast neutrons from fission and a second group, which diffused while slowing down. This second group caused almost all fissions, thus producing fast neutrons. Cross sections used for the two neutron groups were later determined by computer averaging of cross sections over neutron energies involved in the two different groups. The breakpoint between the two groups is that energy above which scattering collisions cause neutrons to gain negligible energy.

Bettis later did some basic critical experiments that served the nuclear industry for many years. Uranium was procured in three low enrichments. Critical experiments were performed with these enrichments, at three metal-to-water ratios, fabricated into rods of three diameters. The three points for the three different variables determined curves that allowed interpolation for many possible experiments. These critical experiments were so important to providing the basis for nuclear design that six of them were operating at Bettis by the 1960s, including a high-temperature critical.

The data obtained from these experiments were so basic that it was verified at Brookhaven in exponential experiments on top of the large graphite reactor, as a source of neutrons. This series of experiments constituted the reference for reactor theory for many years.

An essential experimental input, at a more basic level, was the measurement of neutron cross sections of pertinent isotopes carried on by universities and national laboratories using accelerators. This was a little noticed, but highly vital, contribution to the development of nuclear power; it was carried out by extremely competent experimental nuclear physicists.

A special center was created at BNL to collect all cross-section data, check for consistency among different experimental groups, and provide an evaluated cross-section base reactor designers could use. The Evaluated Nuclear Data File (ENDF) center is still in existence and is depended on for both military and commercial reactor design work.

Bettis performed criticals on advanced submarine reactors, surface ship reactors and Shippingport. Because of the seed-blanket type of core for Shippingport, critical measurements were extremely important for reactivity, control, and heat-transfer performance.

The experiments, however, could not provide insight into the depletion of fuel and burnable poison and the buildup of heavy isotopes. That task had to be analytical, since there were no long-term core operations or subsequent destructive hot-cell examinations that could provide such experimental data.

Flux Distribution

Heat production in a reactor is proportional to the product of the neutron flux (density times speed) and the fission cross section of the fuel material. Where the fuel material is uniformly distributed, heat produced at any point is, therefore, proportional to the neutron flux at that point. The neutron flux has roughly the same distribution for the same control rod positions regardless of power level.

That distribution depends on control rod position, since control rods cause local depressions in the flux. If the fuel is uniformly distributed, the power produced by the reactor is proportional to the volume integral of the flux.

For a given power level, the location of thermal hot spots depends on the gross peak flux due to power level and local perturbations that might cause thermal peaks. Hot spots can cause increased local corrosion, unduly high thermal stress due to the magnitude and gradient of the temperature, or burnout of the fuel element because of an inability to carry away heat under the boiling condition.

Proper design of the reactor can do a lot to decrease the ratio of peak-to-average neutron flux, and thus decrease the peak-to-average temperature. This can be done by properly locating and sizing control rods, by modifying the coolant volume fraction, and by varying the distribution of the uranium fuel. Variable loading and coolant orificing are both largely dependent on having a predicted flux pattern. They are difficult to determine, if flux distribution is changed due to nonuniform motion or patterns of control rods.

In 1950 there were really no suitable digital computers for performing reactor physics calculations, so we had a room full of people operating mechanical calculators, later making some adjustments using analog computers. When digital computers were developed, we always had the biggest and most advanced IBM, Philco, Control Data, or Cray equipment.

Even the reactor physics theory that existed was not sufficient for our needs and had to be developed further. The same was true of many of the instruments required. Three very-low-power critical experiments were conducted at Bettis. These were nuclear reactors, with a sustained nuclear chain reaction, but at very low power. Results of theoretical calculations for very simple geometries were compared with the results of the critical experiments, and then the theory was modified. Slightly more complex geometries were then tried and the theory again revised.

Contributions to Reactor Physics[5]

The 1950s and 1960s were years of rapid progress in the field of reactor physics. Developments occurred in parallel at many laboratories and, since much of the work first appeared in classified reports or internal memos, trying to establish original authorship would be a formidable task. It is fair to say, however, that Bettis was a major contributor.

The person chiefly responsible for guiding the reactor physics effort at Bettis was Sid Krasik. His name appears only rarely as an author in publications. But it was he who thought through what was needed, hired qualified people to do the work, suggested approaches, recognized good ideas when they appeared, and then gave all the credit to those he had inspired. He had a wonderful technique: Upon listening to someone give a fuzzy, confused description of some technical point, he would say "Do you mean . . .?" and then state clearly the idea the person was fumbling with. The person would say yes, and think it was his idea, and Sid would always describe it that way.

Reactor physics work at Bettis had a major advantage: It was aimed at predicting correctly the neutronic behavior of reactors that were going to be built. The need to analyze reactors that were geometrically complex led to development of a basic strategy quite different from that used for the early reactors at Oak Ridge and Hanford—a strategy that has become standard practice for design of thermal power reactors.

The central idea of this method for predicting neutron behavior is the recognition that—in a reactor composed of large, fairly homogeneous material zones (for example, the core and reflector)—the energy distribution of the neutrons throughout most of each zone is independent of location

---

[5]Allan F. Henry, Massachusetts Institute of Technology, Private Communication dated March 16, 1992.

within the zone. Where this separation of spatial shape and "energy shape" is valid, a complete description of the space-energy behavior of neutrons can be calculated by finding the energy and spatial shapes separately.

At Bettis, it was Bob Hellens who was responsible for working out this strategy (a major reason for his later winning the Atomic Energy Commission's E.O. Lawrence Award). To obtain an equation for the energy behavior of neutrons in a homogeneous material, one takes the Fourier transform of the space-energy equation and then splits the energy part into many "energy groups."

Early on, a computer code, MUFT (Multigroup Fourier Transform), was programmed to solve the energy-group problem. MUFT and its descendants are still used by many utilities and vendors to design and analyze power reactors.

The original version of MUFT was programmed by Ben Mount for an IBM card program calculator (CPC). That machine required masterful machine language programming to accommodate, in an acceptable running time, the 54 energy groups Hellens thought appropriate for the physics of the problem. That choice of 54 energy groups for the first MUFT cross-section libraries was a lasting one. Many codes today still use the 54-group library structure.

The MUFT code determined the energy shape of the neutrons for a range extending from 0.625 eV—another choice made by Hellens that still survives—up to the highest energy at which neutrons appear from fission. For "thermal" neutrons in the energy range 0 to 0.625 eV, special considerations are necessary. Slow-moving neutrons can actually gain energy when they scatter. The quantum-mechanical calculations required to provide accurate "thermal scattering kernels" for describing thermal neutron behavior have now been performed. However, in the early days at Bettis, many simple prescriptions were explored. One of these, the "mixed number density model" devised by Bob Breen, is still used by a number of utilities in the United States.

To describe the *spatial distribution* of neutrons in a reactor made up of large, nearly homogeneous regions, a good approximation is to assume that neutrons diffuse from one location to another. When the "energy-shape" is independent of position within the reactor, i.e., when the space energy shapes are "separable," a single spatial shape is sufficient. However, near interfaces between regions, separability breaks down. For this reason, various pieces of the overall energy shape appropriate to a given homogeneous material are permitted to have different spatial shapes throughout a region containing the material. The approximation is called *few-group diffusion theory*. It is now used universally for design and analysis of thermal reactors.

Bettis was a pioneer in the development of numerical methods for solving few-group diffusion equations using digital computers. The procedure was to approximate few-group equations (second-order partial differential equations) by a set of algebraic equations, the unknowns of which were the values of the group fluxes at a lattice of mesh points throughout the reactor. These algebraic "difference equations" were specified by Ely Gelbard, and a sequence of computer programs was written by Hank Bohl, Orv Marlowe, and Bill Caldwell (often assisted by a team of machine-language programmers). These programs culminated in a three-dimensional few-group code called PDQ. Today almost all the U.S. utilities use a version of PDQ to analyze their pressurized water reactors.

Since some PDQ problems involve more than a million simultaneous algebraic equations, it is very important to create efficient ways to solve the equations, to prove that the solution obtained is unique, and to estimate how close the numerical solution is to that of the basic differential equations. At Bettis, fundamental work on these problems was carried out by Dick Varga and Garrett Birkhoff (a professor at Harvard and consultant at Bettis) and extended by Lou Hageman. The very first efforts to improve the iterative methods by which solutions were found led to a reduction in computation time by a factor of 10.

From a nuclear standpoint, the first submarine cores were fairly homogeneous. However the Shippingport reactor was just the opposite. The blanket contained large amounts of natural uranium in plates and rods, and the seed contained large, cross-shaped control rods as well as lumps of burnable poison dispersed throughout the fuel. Aided by theoretical work at a number of other laboratories and by experimental work on lattices at Bettis and Brookhaven, Stanley Stein, Ely Gelbard, and Bob Hellens developed ways of finding "equivalent" homogenized few-group cross sections for natural uranium fuel. Charlie Maynard and Mark Goldsmith did the same thing for control rods. Specifically, they applied a procedure called *blackness theory* to the control slab problem. The method they developed is still used today.

The need to use "equivalent" and "homogenized" few-group cross sections for Shippingport and the naval reactors designed after *Nautilus* led to some concern that design methods were becoming patchwork. There was thus strong motivation to create calculating tools that avoided the separability assumption and the approximate ways of dealing with transport problems associated with neutron absorption in fuel rods and control slabs.

To investigate the separability approximation, Ely Gelbard specified and Hank Bohl programmed PIMG—a one-dimensional, multigroup code for solving equations specifying the P1 approximation to transport equations.

(P3MG, which extended the code to deal with the more accurate P3 approximation to the transport equation, followed shortly afterward.) These codes were among the earliest of their kind. They helped confirm that the few-group diffusion theory model led to acceptably accurate results.

For situations involving small lumps of highly absorbing material (burnable poisons) distributed in complicated geometrical patterns, the Monte Carlo method was developed. It is a statistical scheme in which thousands of "case histories" of neutrons are followed from birth to death in order to determine what will happen to the "average" neutron. (With today's large, parallel machines, millions of case histories can be followed.) The results are equivalent to those obtained if the transport equation were solved for the problem at hand.

At Bettis, Jerry Spanier did some basic work in establishing a firm mathematical foundation for the Monte Carlo method. He also justified mathematically some of the "variance reduction techniques" used to increase statistical accuracy of the average results inferred from a given number of case histories. He and Ely Gelbard specified a number of codes that were then used to determine correction factors for improving the accuracy of few-group diffusion theory cross sections.

With these properly corrected diffusion theory cross sections, the predictions of PDQ could be quite accurate, and there was a strong motivation to apply it to three-dimensional situations. However, the computational expense became (and remains today) too high. For a detailed description of a large PWR, in excess of six million mesh points would be required. To circumvent the very long computation times required to obtain such a solution, an approximation scheme called *flux synthesis* was invented at Bettis.

A forerunner of the synthesis scheme was developed by John Meyer. It consisted of stitching together two-dimensional PDQ solutions appropriate to different axial segments of the reactor. The stitching was done by homogenizing the two-dimensional solutions over the various radial planes and then using the resultant homogenized group cross sections to solve a one-dimensional diffusion equation for the axial direction.

The synthesis scheme that grew out of this early approach was developed by Stan Kaplan, with help at a later stage from John Yasinsky. Synthesis represents the three-dimensional flux shape as a mixture of a number of radial shapes appropriate to different axial elevations. Equations for mixing coefficients were found by requiring the composite three-dimensional selection to obey the defining, three-dimensional group diffusion equations in an integral sense. Once two-dimensional solutions are found, the computation time required to find the mixing coefficients is trivially small.

Design limitations for any power reactor are determined by transient behavior. The plant must be able to operate in a routine fashion during all planned maneuvers and must be able to withstand the consequences of unexpected events. Finally, the public must be protected from the consequences of all catastrophic accidents that, although extremely unlikely, are deemed credible. To meet these design and safety requirements, it is necessary to be able to predict both the thermal-hydraulic and the nuclear behavior of the plant under a great variety of conditions.

Calculation methods for transient analysis were among the first to be developed at Bettis. The first of these involved a simulation of the entire plant and led to a set of some 40 coupled differential equations, which were solved in the early 1950s on a digital computer at the National Bureau of Standards. These calculations provided the first analysis of the now-standard list of reactor accidents: loss of flow, inlet temperature drop, and control rod ejection.

The analysis showed for the first time that light-water-moderated reactors could be designed with a high degree of inherent protection against the consequences of such accidents. Moreover, they showed that ordinary operation of the plant would be much simpler than had been thought. A demand for greater power at the turbine could be met automatically by the reactor without any need to move control rods.

Since digital computers were at the time primitive, further work at Bettis on transient plant analysis was carried out on analog machines. In fact, an analog plant simulator (another "first") was constructed showing a small-scale model of the plant complete with water flowing through glass tubes that glowed red as the "temperature" increased. One could turn knobs to change flow rates, turbine settings, or control rod positions and see the predicted consequences on various dials.

The 40-odd equation digital model and early analog models were somewhat qualitative. The favorable prediction they presented about kinetic behavior of the plant led to continued efforts to improve accuracy. In the case of the nuclear part of the model, these efforts had two facets: to put the kinetic model on a firm ground theoretically and to improve the detail with which kinetic behavior could be predicted.

In the early 1950s the behavior of the neutron population during a transient was almost always predicted by the point kinetics equations, the output of which was the total reactor power. This power behavior was determined by a quantity called *reactivity*. The best definition of reactivity (and several other terms appearing in the point equations) was a matter of some dispute.

Al Henry clarified this situation by showing that point equations could be derived in a formally exact manner directly from the space-time-energy–dependent transport equation. The "exact" definition of reactivity required knowing space-energy flux at the instant in question. In what is now known as the *point kinetics approximation*, that exact shape is approximated by a time-independent neutron distribution. The theory thus suggested a method now used to provide complete space-time analysis of neutronic behavior—namely, to update the shape from time to time as the transient proceeds.

Work at Bettis to improve the detail that neutronic space-time analysis could provide was at first limited by the speed and memory capabilities of digital computers. However, the laboratory was among the first to take advantage of the increase in machine capability. In the early 1960s, a one-dimensional, few-group, transient digital code was programmed, and a few years later it was extended to two dimensions. Then, when Stan Kaplan invented the synthesis method, that idea was extended to the time domain. Thus, as early as 1968, Bettis had the capability of predicting transient behavior at more than 500,000 mesh points through the reactor. Those calculations were performed on a Philco-2000 computer, a primitive machine by today's standards.

Nuclear Core Design[6]

Nunzio (Joe) Palladino, one of the Westinghouse engineers who had worked on the Daniels pile at Oak Ridge and later was part of the team at ANL, came to Bettis in 1950 and was in charge of thermal, hydraulic and much of the mechanical design.

Fuel Cladding Material

We believed despite the lack of knowledge about zirconium that it was the best bet as a cladding material. So early on, we mounted a major development effort in this area, which is covered in detail later in this chapter.

The maximum fuel element surface temperature that could be used was limited by several factors. It had to be below the boiling point of water for the system operating pressure to prevent bulk boiling. It also had to be sufficiently low to guarantee a satisfactory lifetime for the fuel elements, from a corrosion viewpoint. Obviously, it was desirable to have this metal surface temperature as high as possible, to attain the maximum thermal efficiency in the overall plant. A complicating factor was the variation in metal temperature due to power transients and the possible variation in system pressure, which could not be accurately known.

In determining average coolant temperature, it is necessary to strike a proper balance of weight, space, and cost of the primary coolant system

---

[6]N.J. Palladino, Private Communication dated December 17, 1991.

components—a nice example of necessary engineering compromise. For a particular steam generator and core design, increasing the average coolant temperature requires a larger and heavier core and reactor vessel. Increasing the average coolant temperature reduces the size, weight, and cost of the boiler.

Higher total primary system flow rates (lbs/hour) reduce the temperature differential across the reactor and hence allow use of less complex designs. Lower primary system flow rates allow use of smaller pipes and valves. Although higher fluid velocities (ft/sec) were considered to reduce component sizes, practical pressure drop, corrosion, and material considerations set an upper limit of about 25 ft/sec.

High flows permit cores that are less complex. Low flow rates permit smaller pipes and valves. In any event, there is a practical upper limit of about 40 ft/sec for coolant flow anywhere in the system.

Core Mechanical and Thermal Design

Having determined the fuel element material, it was still possible to design the core with an almost infinite variety of shapes and configurations. Fuel element geometry must be satisfactory from a nuclear viewpoint and also from the standpoints of heat transfer reliability, fabrication, and cost. There must be no excessively hot spots in the core occasioned by nonuniform distribution of fuel or coolant that would cause corrosion damage or excessive thermal stress.

Fuel element samples[7] were fabricated to simulate as closely as possible core-manufacturing procedures, and they were tested in U.S. government test reactors such as the Materials Testing Reactor (MTR), the Engineering Test Reactor (ETR) in Idaho, and at the Hanford Site as well as the Canadian NRX and NRU reactors at Chalk River. To properly test such samples, loops were designed and installed within these operating reactors to duplicate temperatures, pressures, water composition and flow rate, power output, and—to varying degrees—thermal and fast neutron fluxes.

This continuing in-pile test program before adequate operating experience could be obtained distinguished the naval reactor development from that of commercial reactors, which rely on examination of fuel elements and components withdrawn from the reactor cores for development and quality-evaluation information. In addition to such loop testing, samples of fuel elements, pressure vessels, cladding, etc., were exposed in capsules and tested after irradiation in hot laboratory facilities for dimensional and mechanical property changes.

---

[7]B. Lustman, Private Communication dated July 12, 1991.

Assembly must be accomplished so that failure to pass inspection, at any stage of manufacture, results in a minimum loss due to scrapping of material. The entire assembly, of course, must be able to withstand high-impact shock.

Fabricating methods and quality control of core manufacture entailed an extraordinarily large amount of development work. Fuel element samples were made and tested in existing reactors to verify that fuel elements could satisfactorily resist corrosion and radiation damage. Though the development proved that fuel element design was satisfactory, the materials and processes did not assure an acceptable fuel element, as the quality varied. Each ingot of the material, and then each fuel element, had to be corrosion tested.

Boiling Heat Transfer

Although the PWR seemed to be the approach most likely to succeed, it was not without its problems. The most severe of these was a shortage of knowledge about boiling heat transfer under various regimes of phase change and various levels of forced pump flow. Because of limitations that can be encountered, boiling heat transfer is more difficult to characterize than single-phase liquid flow.

The various regimes of boiling heat transfer are defined in the following list:

1. *Local boiling* is boiling in which the bulk liquid temperature is at or below saturation and in which bubbles collapse.
2. *Bulk boiling* is boiling in which the bulk liquid temperature is above the saturation temperature, and bubbles do not collapse in the fluid.
3. *Nucleate boiling* is boiling in which bubbles form at discrete points on the surface of the reactor. Nucleate boiling can take place with either local or bulk boiling.
4. *Film boiling* is boiling in which the heating surface is blanketed with vapor (as a result of very high heat throughput), which severely impedes heat transfer.
5. *Transition boiling* is boiling in the region between nucleate boiling and film boiling.

Nucleate boiling is an effective means of removing heat from hot surfaces. However, when the heat flux (Btu/hr ft^2) reaches a level where film or transition boiling into the film region is encountered, the heat removal effectiveness drops off significantly. This condition is known as *departure from nucleate boiling* (DNB), which is important because it can lead to melting of a fuel element in a reactor (often called *burnout*).

It was important to avoid DNB under all operating conditions, including transient operating conditions and accidents (such as loss of coolant flow).

Prior to the 1950s, there had been experiments on DNB using electrically heated wires in pools of stagnant water. However, there had been no significant work on determining DNB under various flow conditions and ambient pressures in flow configurations to be faced in the proposed core designs. Hence, an extensive program had to be undertaken with appropriately shaped fuel element mock-ups electrically heated in the presence of appropriately shaped flow channels with various flow rates at high pressure.

A special ac motor-driven dc generator was needed to provide electricity for these tests. Plans to obtain and install this equipment brought about considerable consternation among local AEC management, which hadn't been kept abreast of the DNB problems. The generator was literally hanging from the crane when a stop order came from the AEC local front office. After the situation was resolved the generator and other equipment were installed and the experimental program was begun under the direction of S.J. Green. Other major program contributors included R.A. Debartoli, L.S. Tong, and J.E. Zerbe.

This was one of the most valuable experimental efforts carried out at Bettis. Some of the early results are highlighted in the book *Boiling Heat Transfer and Two-Phase Flow* by L.S. Tong. The DNB correlation, which was used in analyses of early prototype Bettis fuel designs, was reported in the 1958 USAEC Report WAPD-188 entitled "Forced Convection Head Transfer Burnout Studies for Water in Rectangular Channels and Round Tubes at Pressures above 500 psia," authored by R.A. Debartoli, S.J. Green, B.W. LeTourneau, M. Troy, and A. Weiss.

To allow for both analytical and experimental uncertainties, the allowable value of DNB flux used in analyses of the reactor was only two-thirds of the DNB calculated from experimental work.

Design Criteria

Early in the design of nuclear reactors for the production of power, it became clear that the design process would involve compromises of many technological judgments. These judgments had to be made usually with little experience to draw on and often with incomplete engineering information on which to base them.

In starting a reactor design, one of the first questions that had to be faced was establishing criteria for design that would assure meeting overall design targets and plant safety under all operating conditions. The particular criteria to be used differ from one reactor to another; many times, they even differ from time to time during the development of a particular reactor. Today there exists (in PART 50, APP. A of Volume 10 of the *Code of Federal Regulations*) a well-integrated array of design criteria for commercial nuclear power plants.

One early design challenge was the establishment of criteria on which to base core features or operating characteristics. In general, each criterion used in design was, in essence, a statement about a design feature or a limiting value of a design variable that ensures fulfillment of a design objective or implies assurance of core or plant safety. For example, we had to explore the need to limit the maximum internal temperature of fuel elements to avoid a metallurgical phase change in the fuel or, in other cases, melting of the fuel. We also had to limit the maximum surface temperature of fuel cladding, to avoid boiling of coolant on the fuel element surfaces and to limit the coolant velocity passing fuel elements to avoid surface erosion.

Attention also had to be given to limiting temperature differences that produced untenable thermal stresses on fuel elements, fuel assembly structures, core structural supports, or pressure vessel walls, heads and pipe connections. Also, a reactor had to be capable of being shut down with one control rod stuck out of the core.

Other criteria were also developed as designs of naval reactors advanced and as the Shippingport design proceeded. Among the most important of those were the criteria associated with design of containment structures to avoid leakage of radioactive material in the event of a primary system pipe failure.

Hot-Channel and Hot-Spot Considerations

In evaluating whether fuel element temperature meets the design criteria, consideration must be given to variations in the heat generation rates in the core during operation. Inasmuch as radial power distribution is nonuniform, some of the channels are producing less power than the average channel and others are providing more. To determine the maximum value of the fuel element surface temperature, we must examine the fuel element producing the highest rate of heat in the core. In addition, we must recognize that the fuel element surface temperature depends not only on the heat generation pattern, but also on the rate of heat removal.

Thus, any effects that influence the rate of cooling will also affect the temperature in the average and hottest channels. To take these factors into account we used hot-channel factors. These are factors that relate hot-channel conditions and hot-spot conditions in the channel to the conditions in an ideal average channel. These factors are divided into nuclear and engineering hot-channel factors.

Nuclear Hot-Channel Factors. Nuclear factors arose out of the power density distribution pattern in the core as established from nuclear flux patterns determined by nuclear analyses and tests. Neutron flux patterns were difficult to predict until the core was designed because of their complex relationship to the material ratios and flux geometry in the core as well as the nature of the reflector and the number, size, shape, and location of control rods.

Preliminary analyses provided estimates of these factors as influenced by core materials, core geometry and control rod configurations on the neutron flux patterns. These analyses had to include not only gross neutron flux patterns but also local neutron flux peaking that, in the case of water reactors, arises from the channels of water between subassemblies and in channels vacated by control rods upon withdrawal during operation. Every reasonable effort was made to reduce flux peaks by avoiding such water holes, whenever other design features permitted.

Usually, initial analyses result in so-called axial peaking flux factors, radial peaking flux factors, and waterhole or other core peaking functions. An axial flux factor must have an axial flux shape associated with it; hence it is not an independent factor. A high axial peaking factor near the coolant inlet end of the core is not as bad since the temperature margins are greater at the inlet as the same peaking factor would be farther along the coolant flow path.

Engineering Hot-Channel Factors. Fuel element fabrication tolerances, when expressed as variations in the channel and fuel element dimensions, imply respective local variations in the heat-transfer characteristics of the core. In addition, there were operational tolerances arising out of fuel element geometries such as rods in cross flow.

Engineering factors were grouped in three categories:

1. A factor influencing the peak-to-average temperature rise in the coolant.

2. A factor influencing the peak-to-average temperature drop from the fuel element surface to the bulk coolant stream.

3. A factor influencing the peak-to-average heat flux in the core.

Flow Tests

Two sets of flow tests of particular interest were conducted in the development of the prototype core. One set included tests on a full-scale steel quadrant of the pressure vessel and internals of the core with one side of the quadrant made of glass to permit observation of flow patterns during operation. Tests on this device were concentrated primarily on assuring adequate cooling flow around thermal shields, particularly those at the entrance end of the core at the bottom of the pressure vessel.

Significant changes were found necessary in these flow passages. Means were developed for evaluating flow through these passages by treating the plates and support rings that formed the flow paths as an array of parallel-series paths controlled by orifices.

The other tests were air-flow tests on a full-scale wood and plastic mockup of all reactor vessel internals including the core, control rods, and core structural supports. By keeping the air flow in the incompressible regions, good correlation could be obtained between air flow and water flow.

The major result of these tests was a determination that two control rods, which because of space limitations could not be completely shrouded against cross flow, might fail to scram when the primary system pumps were at full output. The cross flow provided enough force to hold the control rods against the side of the channel.

This was confirmed early in the operation of the prototype core in Idaho by a special individual test on each of these two control rods with full flow. This problem was alleviated by limiting pump flow to slightly below the value that would cause the problem. This was possible because the pumps had excess capacity over that required by design. The design of these two shrouds was later modified to avoid this problem in the *Nautilus*.

Fasteners and Bolt Devices

Problems with fasteners and bolt-locking devices are faced in a wide range of materials, structures, and mechanical devices. To this day, problems arise in the use of bolts and locking devices in nuclear reactors despite the lessons that should have been learned from previous applications. Discussed next are three sets of fastener problems incurred in the design of the Mark I prototype.

Extension Bracket for Holding Fuel Element Assemblies

One challenging problem was the development of means for fastening fuel element assemblies to the upper structural supports of the reactor in such a way as to withstand naval shock requirements. The problem was inventing a scheme that would work with an already designed Zircaloy fuel element assembly and a stainless steel holding plate.

The solution turned out to be a stainless steel extension bracket made with a lip to slip into the top of the fuel element assembly where provisions were made for riveting the bracket to the sides of the upper part of the assembly. The outside of the bracket was the same size as the outside of the fuel element assembly. The inside of the bracket was shaped to provide a central path for cooling water to flow through. Corners were designed to receive bolts with locking devices to fasten the assembly to the flat support plate at the bottom of the upper support structure. A similar approach was used for holding the bottom of the reactor fuel assemblies.

The system, when analyzed, met the shock requirements of the Navy, assuming the rivets were properly made. For later designs the extension brackets were made of Zircaloy so they could be welded to the Zircaloy fuel element assembly.

In retrospect, the solution to this fastening problem seemed quite simple, but it took a team of nine engineers (including three loaned from ANL) to come up with a workable solution. The final solution came from observing designs used for fastening lavatory wall brackets to the floor. It took picturing the outside of these brackets as the inside of the desired fastener and the outside as the shape of the fuel assembly. The design worked very well.

Bolts

An application that ran into just about every type of bolt problem was the use of socket-head bolts to hold in place L-shaped brackets for limiting sideways motion of control rods during battle shock. The problems encountered, however, had nothing to do with battle-shock loading; rather they arose during normal operation of the Mark I core in Idaho. A number of the L-shaped brackets and bolts were broken from the core structure to which they were fastened and were transported to the inlet side of the steam generator, where they were found during a steam-generator inspection.

From an explanation of the reasons for this situation, the following important factors were found:

- The recess in the socket head was too deep, so the head could not withstand the load placed on it when tightened and exposed to the reactor environment.

- Inadequate attention was given to the differential expansion between the bolt and the bracket, so the bolt head was loaded much more than expected during reactor operation. Preoperational analyses of the design took into account that the bracket and bolt material were made of different steels with different coefficients of expansion; however, consideration was not given to the higher operating temperature of the bolts due to poorer heat transfer from the bolts.

- Poor thread form and burrs on many of these bolts, because of quality-control inadequacies, led to seizing of bolts in some cases and reduction of tightening load in others.
- Locking devices used in this application did not prevent the broken bolts from moving around in the pressure vessel and steam generator during operation of the plant. Consequences could have been worse than experienced.

Locking Devices

The performance of locking devices was considered so important that a special task force under B.F. Langer was organized to develop guidelines for design of acceptable locking devices for various circumstances. The results of this work are summarized in WAPD-CE-37, by B.F. Langer. Five classes of applications are described next.

Class A Locking Devices. Fastenings exposed to primary or secondary fluid and located where failure could result in a loose fragment being carried to a vital part of the system should be considered as completely inaccessible and must be designed for the highest degree of reliability. Some general rules should be followed.

The locking device must be positive. This means that it must not depend on friction. As a result of this requirement, two operations must be performed in securing the fastening. One is the tightening of the screw thread itself, and the other is a separate operation usually involving the plastic deformation or welding of a member to a position in which it interferes with the loosening of the main holding member. This rule eliminates the use of lock nuts, spring lock washers, and self-locking nuts.

The deflection used in setting the locking device must be large enough so success of the device does not depend on a high degree of skill on the part of the worker, great precision in the machining of the parts, or a high degree of dimensional stability. It must be obvious from cursory inspection whether or not the lock has been secured. This rule eliminates most peening and upsetting and all prick-punching operations.

The locking operation must not involve plastic strain sufficient to crack the material. Stops and guides should be provided, so that the prevention of cracks is not dependent on the skill of the worker. The sharpness of allowable bends must be consistent with the ductility of the material being used. Devices involving plastic deformation must not be used more than once. This rule eliminates most cotter pin designs and makes lock wiring undesirable, except as described below.

Locking members must not be so thin that slight corrosion can cause failure. Wires should be avoided in regions of high fluid velocity and should not be less than 0.050 in. in diameter in any case. Sheet members should not be less than 0.015 in. thick.

The design should be such that if the loaded part of the fastening (e.g., bolt shank) fails, its fragments will be held captive by other members or by the locking device itself.

Class B Locking Devices. Fastenings exposed to primary or secondary fluid, but located in regions of low velocity and in places where fragments could not be carried into the main fluid stream, are also to be considered as completely inaccessible and should meet Class A requirements.

Lock wiring 0.050 in. in diameter or larger is acceptable, provided the manufacturing drawing specifies size and material of wire, size and location of holes, and configuration of the wire. The configuration must be such that loosening of the fastening results in tightening of the wire. Fastenings must be wired in pairs, except that for an odd number of fastenings one group of three may be used. Lock wiring must be tight.

Class C Locking Devices. Fastenings not exposed to primary or secondary fluids, but which are located in positions that are inaccessible for inspection, tightening, or replacement during power operation, must be provided with locking devices meeting the three Class A requirements listed earlier.

Class D Locking Devices. Fastenings not exposed to primary or secondary fluids and which are accessible for frequent inspection, tightening, and replacement during power operation must be provided with locking devices that have proven satisfactory for operation in power plants of the type being designed. These include star washers, spring lock washers, cotter pins, lock wires, peened bolt ends, and certain proprietary self-locking nuts, as well as any device that meets the requirements of Classes A, B and C above. Snap rings are acceptable, except that for naval use they must be proven shock-proof by test. Set screws and lock nuts are not recommended.

General Requirements and Exceptions

When welding is used as a locking means, due consideration should be given to the weldability of the material, the effect of welding on the corrosion resistance of the assembly, on the cleanliness of the adjacent members, and the possibility of crack formation. Whenever used, size and location of the weld must be specified. A note calling for tack welding is not acceptable.

Locking devices are not required when heavy bolts or studs larger than 2-in. nominal size are tightened up to 75% or more of their maximum allowable capacity under carefully controlled conditions involving heating and measuring of bolt stretch.

Controlling the Reactor[8]

New frontiers were the order of the day in the design of the plant controls and instrumentation. Many unknowns had to be addressed to meet the stringent power and maneuvering requirements of the submarine. In addition, there was no precedent for steam-turbine propulsion in a submarine. A plant had to be designed that could be started rapidly, maneuvered at high rates of power change, and controlled under normal and upset conditions. All the operating modes of the submarine had to be considered. Safety and reliability were overriding objectives. A favorite Rickover question—Would you put your son on this boat?—made one think carefully about whether a design proposal was reliable enough to meet the Admiral's standards.

For the *Nautilus*, control system requirements included those needed for startup, power operation, and shutdown. Such operations required coordination of the functions of the primary and secondary systems while bringing the reactor to power from source level to the power range. The increase in reactor power had to be coordinated with that of the turbine.

The control systems development required the integration of the functions and the limitations of all parts of the plant. Such information was made available by the many disciplines at Bettis. Dick Cunningham, assisted by Walt Esselman, had the responsibility of coordinating this information and developing a control system that supplied these functions and recognized the limitations. Design concepts were reviewed with Bob Panoff who had Naval Reactors Branch (NRB) responsibility in this area.

Startup Range Control

A reactor is controlled by adjusting the balance between neutrons produced and neutrons absorbed. This is usually done by the insertion or withdrawal of control rods, containing materials that absorb neutrons. It is fortunate

---

[8]Much of this section is from W.H. Esselman and R.J. Creagan as a result of many communications, both written and oral.

that some of the neutrons emitted by the fission process are delayed by radioactive decay processes (delayed neutrons). Thus a control system that operates with time constants in the order of seconds is possible, rather than one operating in milliseconds. Such a fast system would be required if all the neutrons were emitted during the process of fission itself (prompt neutrons).

Because only even-numbered isotopes spontaneously fission, U-235 does not produce neutrons—thus the need for a neutron source for startup. We had to measure neutron level at the subcritical condition and compute accurately the neutron multiplication under all conditions up to power level. This system required improvement of the BF-3, a neutron detector using boron-coated electrode surfaces inside an ionization chamber.

With a neutron source in the reactor, nuclear instruments had to measure power over a range of 10 or 11 decades. Nuclear instruments required from subcriticality to power operation included boron trifluoride (BF-3) neutron proportional counters, fission counters, and ionization chambers. Issues related to the neutron instrumentation system included the number of instruments required for redundancy, the transient response requirements, and the range of measurements of each type of instrument.

Once criticality has been reached, the power level rises exponentially as a function of the amount of reactivity above critical. If withdrawal of control rods continues, excess reactivity—and thus the exponent—increases, and power rises at an ever-increasing rate.

If the power level is not known until it is about 1% of full power, it might exceed full power within a fraction of a second. To prevent such a rapid rise in power, it is measured at extremely low levels, and the rate of rise is adjusted to an acceptable level. This is done by use of a neutron detector.

This ability to measure such a small amount of power is essential in the control of nuclear reactors. However, the ability to measure almost zero radiation has haunted the nuclear industry. Thus, any release of radiation surrounding a nuclear plant is viewed with great alarm by the general public, though it might be even thousands of times smaller than that which people receive in their normal daily activities. Cosmic rays send approximately one charged particle per second through each square centimeter of a person's body.

Power Range Control

Since the fundamental kinetic characteristics of the reactor and plant cooling system were not well understood during the STR design, the nature of the control system required was a major uncertainty. Initial discussions centered on the need for feedback loops for the reactor, the primary system, and the interaction with the steam plant. The magnitude and importance of the feedback effect of the negative temperature coefficient of reactivity on transients were not clearly recognized. Strong feelings were expressed that the control system should be able to cope with a condition that exhibited very little effect of negative temperature coefficient feedback. As a result, a neutron feedback loop was developed that consisted of the output of the ionization chamber and a fast center rod drive that responded to a power level demand.

From 1950 the fundamental principle of reactor control design was to proceed on the basis that the temperature coefficient could be zero. This was always the backup position. Decisions were made that the reactor control system would be based on temperature control of the reactor coolant. The aim was to control the reactor without neutron instrument feedback. Early in the design process, a simplified linear systems Nyquist analysis made by Esselman and Orv Meyer became the basis of the control systems design.

By January 1951, Monte Schultz developed an early analog control model of the reactor plant that was quickly expanded to include the entire coolant system, steam plant, and turbine. With this tool the control system design was greatly simplified. As confidence grew in the validity of the analog model, it became the fundamental basis for Mark I control systems analysis. The reactor model was based on a number of horizontal zones. Besides the reactor, the plant model included the primary coolant system, steam generators, and steam system to the steam turbine. Nelson Grace made important contributions to the development of the analog computer techniques.

Transient responses of the reactor to turbine power increases and decreases were extensively studied. An understanding of the effects of potential plant malfunctions was developed from results of analog computer studies. Simulator studies of various upset conditions were used as inputs for determining the reactor shutdown (scram) settings.

As the reactor kinetic analysis and the control system design progressed, the benefit of the negative temperature coefficient became more evident. We began to appreciate and talk about the inherent stability characteristics of the system. It was exciting to begin to understand the effect of the temperature coefficient of reactivity. Changes in reactor power level that would tend to increase the temperature of the core were counteracted by the temperature coefficient. The rapidity of the response of the reactor depends on the heat capacity of the core and the temperature effect on the reactivity or neutron multiplication. However, sufficient uncertainty existed about the negative temperature coefficient of the reactor such that the control system was designed to function over a range of possible reactor characteristics.

The initial self-stabilization goal in reactor development was in the area of reactivity control: to use intrinsic design features, such as negative temperature, power, and fuel coefficients as "governors" to control rapid transients and power ramp rates in the reactor. The fuel coefficient is mainly the Doppler effect, but mechanical expansion can change the amount of water moderator and thus the reactivity. Such coefficients reduce the fission reaction rate instantaneously as temperatures or power increases, and they do so without the aid of external devices.

Fundamental issues related to control of the nuclear steam supply included how reactor power would be determined, how inlet and outlet temperatures would be controlled, and how the reactor coolant flow would behave. Control system design had to consider the various factors that will affect reactivity of the core and the power level of the reactor during power operation.

A knowledge of primary system characteristics was essential, and much work was required to develop an adequate understanding. A key aspect of this understanding was the thermal-hydraulic conditions in the core, including the transient effects of reactor power changes on the temperature and flow in the reactor and the design limits for materials. Control design had to consider hot-spot temperatures in the reactor. As discussed earlier hot-spot temperature is the function of the coolant flow, the fuel design, the general design of the reactor, and the primary coolant system design. Each of these systems and their effect on the reactor core had to be analyzed.

The effectiveness of the negative temperature coefficient in achieving such stability was identified by Bettis in 1951, for the first time by anybody. We performed transient calculations for the entire plant on the only general-purpose computer in the world that had such capability, the SEAC at the National Bureau of Standards (NBS).

Al Henry did the overall analysis of submarine control, which included all the coupled time-dependent differential equations involving reactor reactivity coefficients, coolant flow, pumps, heat transfer, hydraulics, steam generation, pressurizer, turbine—all the way out to the screws of the submarine. It was a wonderful, all-encompassing analysis, which established the overall stability of the nuclear power plant and served as a model for reactor dynamic analyses thereafter.

The control principles adopted on the *Nautilus* and its prototype are still being used today. Analog computer simulations were extensively used for plant system analysis for about two decades, but digital computer simulation techniques have now become the norm.

Notwithstanding the analyses, which had been made before power operation began, there was, quite naturally, considerable uncertainty and concern about the control of power level in the early operation of the prototype. After all, this was the first instance of significant nuclear power generation. The uncertainty was so great that a special, rapidly moving control rod was designed and installed in the prototype core. However, it was quickly determined through operation of that core, that the reactor was so stable such a rapid response control device was not needed. The special control rod was therefore not installed in the *Nautilus*. This simplified its control system, permitting increased core loading with resultant extended core life.

Confirmation of that calculation of reactor stability was a dominant goal in those early days of the naval reactor development program, culminating in its verification during initial operation of the *Nautilus* prototype.

In fact, the stability and reactor response, which result from strong negative reactivity coefficients in the naval plants, allow them to maneuver rapidly—surprisingly, more easily than is possible with oil-fired ships. This feature has been most useful, for example, in aircraft carrier operations; the ship's power plant can be more quickly brought to the high-power output needed to produce steam for high ship speed and catapults associated with flight operations.

In parallel, reactor self-regulation was also being sought in liquid metal-cooled fast reactor (LMR) development, because the feasibility of control of a fast-neutron cycle was in question. The Doppler coefficient in uranium oxide fuel was identified to be sufficiently negative that it would act as a governor, even in the fast-neutron cycle, and thus ensure safe control using conventional control rods, neutron flux detection, and control instrumentation. Analytical methods were developed, laboratory experiments carried out, and the SEFOR[9] test reactor was built. It successfully demonstrated the inherent stability of the LMR.[10]

Requirements for such intrinsic stabilization in reactor control were subsequently established and became codified in the standards and regulations of the nuclear power industry. Setting such standards for stability entailed significant effort, not only because quantitative accuracy needed to be established but also because it was an intrinsic feature. Some features that stabilize for normal power operation can be destabilizing in abnormal transients.

---

[9]L.D. Noble and C.D. Wilkinson, "Final Specifications for SEFOR Experimental Program," GEAP-SS76, General Electric, January 1968.

[10]Karl Cohen et al., "The Southwest Experimental Fast Oxide Reactor," APED-4281 VIII, Nuclear Congress, Rome, Italy, June 1963.

A case in point is the sudden introduction of cold water into the reactor to cause an intrinsic power surge and the need to set requirements, which would prevent significant cold water intrusions. The importance of intrinsic reactor control standards was highlighted at Chernobyl, where inadequate intrinsic reactor stability (i.e., a strong positive void coefficient) in that RBMK graphite-moderated, water-cooled channel reactor design was a key causative factor of the accident.

Power changes in submarine plants had to be made rapidly, without undesired transient effects. The reactor was controlled by electrically driven control rods. The primary coolant temperature responded to changes in the power demanded by the steam turbine. The control system was designed to maintain constant average coolant temperature into the reactor. The reactor temperature coefficient of reactivity led to an inherently stable power plant. Slow changes in control rod motions were needed to account for fuel burnup and changes in xenon poisoning. This approach led to a simple and reliable control system.

Of special interest to the control engineer was the accuracy to which the power level must be limited. The allowable transient during a power rise, for example, had to be limited so it was below the "scram" settings of the neutron instruments. Also core temperatures had to be maintained below tolerable levels.

To accomplish these various nuclear steam supply control functions, it was decided that power would be controlled to maintain reactor coolant temperatures and that the neutron instrument would not be needed for control during power operation. This approach allowed the reactor power level to be controlled independent of any vacuum tubes. Of course, neutron instrument signals were the prime indication of reactor power on the control panels.

Much discussion followed on whether the control system should be controlled from the hot leg temperature, i.e., temperature of water leaving the reactor, or the average temperature. Each alternative had both nuclear steam supply and control advantages. The T_c cold leg temperature control was selected.

Xenon and Other Fission Products

Another unusual reactor problem arises because of the existence of radioactive fission products, some of which are extremely strong absorbers of neutrons. The long-life poisons that were generated resulted in slow withdrawal of the rods from the reactor. Shorter lived poisons, such as xenon-135 and samarium-149, caused more serious problems in the reactor control design.

The reactor control rod drive had to be able to compensate for neutrons absorbed by the poisons generated during operation. As a reactor operated at power, the xenon and samarium poison effects built up and reached an equilibrium value. One of these in particular, xenon-135, produces an effect in reactors known as the *xenon override* problem. The fission product is one of a radioactive decay chain. Its mother nucleus and daughter nucleus are not strongly absorbing. In operating a reactor, a balance is achieved between the production of this xenon poison through radioactive decay from its mother and the loss of the xenon poison, both through radioactive decay and the process of absorbing neutrons.

When a reactor is shut down from high-power operation, the xenon poison at first continues to appear in the reactor at its preshutdown rate; but its loss rate is greatly decreased, since none of it is now lost through neutron absorption.

Therefore, once a reactor has been shut down from a high-power level an appreciable increase in the amount of xenon poisoning occurs. Extra reactivity is provided so the reactor can be started while the xenon is at a higher than normal operating level. The control rods must be capable of handling this extra reactivity when excess xenon is not present. If this extra reactivity is not available, it will not be possible to restart for many hours, i.e., until the process of normal radioactive decay has caused most of the xenon to disappear. A reactor that can be made critical when the xenon concentration reaches its peak is said to have *maximum xenon override capability*.

Not only does the magnitude of the xenon change but its distribution also changes. Because xenon is such a strong absorber of neutrons, a change in local xenon concentration can result in changes in power production shape. For a large reactor at high power, this situation can lead to a spatial oscillation in power with a period of about a day. When this occurs the reactor is said to be experiencing xenon oscillations, however, we found out that significant xenon oscillations did not occur in submarine reactors because they are quite small.

There was no way to verify these phenomena experimentally for the *Nautilus* core, except by running the power reactor itself. Critical experiments used to verify many other core physics characteristics were of no use, since they did not operate at high power. Verification had to await completion and testing of the reactor. Since the magnitude of the xenon effect was uncertain, the design was based on the highest calculated rate of xenon burnout.

As mentioned earlier, many fission fragments are radioactive. They continue to decay and produce heat for a long time after the reactor has been shut down. In normal operation, the cooling system prevents this afterheat from causing the nuclear fuel to overheat. If there is an accident, this function is performed by the emergency core cooling system.

Fuel Depletion

Another design consideration with respect to reactivity requirements of the control system is compensation for fuel depletion. This effect is not demanding on the speed of control rods, but it does establish the range of reactivity required in the control rod system.

One of the young scientists, Milton Danzker, came up with the following idea: If hafnium was added to the fuel elements, it would absorb neutrons and the absorption would cause depletion of the hafnium, which would slowly reduce the level of absorption. This would increase the reactivity, roughly in proportion to the loss in reactivity from depletion of the U-235. In effect, the "burnable poison" was the equivalent of a control rod that was being slowly withdrawn, but it did not cause distorted flux patterns.

The idea was developed and applied in highly sophisticated ways and it included the use of other materials, notably boron, that had "burnable poison" characteristics. Although Alvin Radkowsky may not have been the conceiver of the idea of using burnable poisons, he was largely responsible for its acceptance because he pushed very hard for its use.

The result has been the ability to achieve very long core lives, which would not have been possible using only conventional control rods. We were thrilled in those early days at the prospect that we might design a core that would last five years. Much progress has been made in this area since then. Currently the goal is to last the life of the submarine, with maximum xenon override. On a recent trip to one of the newest carriers, which had come to port for its first overhaul after five years of operation, John Taylor, after reminding the skipper that he did not have a security clearance, asked him when his ship would be coming in for a refueling. The answer was 17 years.

These physics core lifetime design advantages from burnable poisons are obtained at the expense of increased swelling and embrittlement of the fuel, because formation of helium during neutron absorption by boron adds to the swelling resulting from xenon and krypton formation during U-235 depletion.

Such a technique is not a safety requirement but a design option, which has proved extremely successful in the simplification of control systems, achieving greater reliability and improved economy. Fixed burnable poisons are used in most light water reactors in the world today. Development work is still under way to refine further the performance of fixed burnable poison/fuel combinations and to explore their potential to handle cold-to-hot reactivity swings.

Shutdown

Besides having sufficient fissionable material in the reactor to guarantee being able to attain criticality at all times, it is necessary for there to be a sufficient means of control to shut the reactor down, even in the most reactive condition. The difference between reactivity in the most reactive and least reactive conditions is caused by changes in temperature, presence of equilibrium and transient fission products, chemical additives in the water, and burnup of fissionable materials as power is produced.

Control also must be available to decrease the reactivity rapidly enough to prevent dangerous power overshoots caused by possible transient events.

Heat-transfer considerations and space considerations made it desirable to introduce the minimum acceptable number of control rods. This required extensive calculations and experiments to determine the amount suitable under all conditions. The control rods were made of the rare-earth metal, hafnium. Manufacture of these rods required extensive developments in the metallurgy of this unfamiliar metal.

One problem that has to be considered with respect to control rods is their nuclear depletion. Control rods perform their function by having an unusually high absorption cross section for neutrons. This very fact causes the atoms of the control rod material to change gradually to isotopes of higher mass. This means that the absorption cross section correspondingly changes, usually decreasing with the absorption of each neutron. The life characteristics of the rods, therefore, must be calculated carefully to determine that they will be adequate throughout the life of the reactor core.

Hafnium has many advantages as a control rod material: It is radiation and corrosion resistant and absorbs neutrons at intermediate (resonance) and thermal energies. It has many isotopes, so neutron absorption just changes the isotope to another that is also a high neutron absorber. Therefore, absorption does not decrease as it would, if hafnium atoms were destroyed.

Different plant shutdown conditions resulted in the need to consider a wide range of operating requirements. Startup had to be feasible from either a cold or hot condition. Alternative methods of heating the primary system were evaluated, such as heating by means of the main coolant pumps. Expected levels of xenon poisoning in the reactor core and the rate of burnup during restart were important requirements of control rod insertion rates. Safety shutdown systems were designed to consider all possible situations during normal and abnormal operating conditions.

A fundamental reactor safety system philosophy had to be established and a safety shutdown system designed. Each cause-and-effect candidate for inclusion in the safety shutdown system was subject to extensive review.

The constant problem was to find the proper signals to shut down the reactor without developing so sensitive a system that plant reliability would be affected. Allowable time between a shutdown signal and reactor rod insertion was a major design decision.

Control Rod Drives

Initial discussions about the design of control rod drives centered on whether the rods should be mechanically ganged or individually driven. The first design early in 1950 consisted of individual hydraulic motors for each rod. These motors were to be water actuated and controlled so that the rods could be accurately positioned. A serious problem with this design was getting the needed information on the position of the control rods.

Later the mechanically linked system, designed by ANL, was favored to assure that they would move in unison. The rods were to be coupled by a mechanical device called a wobble plate. In this manner, they would be mechanically tied together. This design also needed only one penetration through the reactor vessel wall.

Many tests of materials and hydraulic drive components were undertaken, all with poor results. The difficulty of finding materials compatible in high-temperature water under the rubbing conditions required by a hydraulic-mechanical drive was formidable. Hundreds of different materials were tested.

The problems encountered in these tests discouraged further development of this type of ganged system, and thoughts of electric drives increased. Dick Cunningham was the driving force in the selection and development of the control drive mechanisms. When it became obvious to us that a new system was required, Walt Esselman was probably the first to suggest electric motor drives. A small reluctance-type motor was used in the elevator division, where he had experience with them. The reluctance motor had advantages because its rotating parts needed no windings. As soon as we believed that the electric motor drive could be successfully developed, we recommended abandoning the reference design, a three-gang mechanical drive.

R. Robinson had responsibility for design of control rod drives and with W. Roman received a patent on the collapsible control rod concept. E. Frisch made many contributions to the design of these drive systems.

A decision was made to concentrate on canned-motor control drives. Questions arose concerning allowable steps in reactivity and the speed at which a reluctance motor could move the rods. The stepping motor did require substantial power and had limited torque. A key question was, could they be designed to move the rods rapidly enough to change reactivity in response to reactor control needs? For example, could the control rods move rapidly enough to take care of the xenon override situation? At that time, the calculations on the magnitude of xenon poisoning in the core were in doubt.

In a meeting with Sid Krasik, these various points were discussed, and it was decided control rods would be ganged in three groups, i.e., the electrical motors would be driven in unison in three groups. We were not certain if we would need a rapid change in reactivity to control the reactor properly, so a fast-acting linear motor-driven control rod was included in the center of the core. Esselman filed a patent disclosure on the center rod linear motor.

The prototype did not have the collapsible rotor drive used later. For shutdown purposes, the rods were injected partway by springs. The selection of the canned rotor stepping motor drives was a major decision in the *Nautilus* project. It received high-level attention.

Instrumentation

Nuclear Instrumentation

The key purposes of nuclear instrumentation were to monitor the startup of the reactor and to supply indications to the reactor shutdown system. Power level depends on the total number of fissions taking place per unit of time within the core. Neutron flux is, therefore, proportional to power. However, the proportionality constant varies depending on the peak-to-average ratio of flux. It is also a function of the attenuation of the flux from the core to the instrument that measures it. This constant is determined initially by calculation, and it is verified and calibrated by actual operation of the power plant. The variation in flux at the instrument, from startup to full power, presents a considerable instrumentation problem.

A key point to be highlighted is the major consideration given to the reliability of neutron instrument circuitry. Redundancy was an important issue. Major credit for design and development of this reliable circuitry must be given to Monte Schultz and Basil Lide. These electronic systems, which are very similar to those that continued to be used for many years, were largely developed as a part of this initial effort on the *Nautilus* project.

To assure that instruments were available, R. Bayard set up a laboratory at Bettis to make the necessary units for the prototype. These instruments later became a product line for Westinghouse's Elmira, New York, division. When the reactor was operating in the power range, the control system was independent of the nuclear instrumentation. However, the nuclear instrumentation system continued to supply key signals to the reactor shutdown system during power operation.

The decision was made very early that the power operation of the power plant would not depend on the use of vacuum tubes. Esselman's personal prejudice against vacuum tubes, on a reliability basis, had a great influence on the adoption of this principle. Then the only solid-state device available was a rectifier; therefore, we selected magnetic amplifiers for the main control function. We made this decision with no lengthy analysis and a minimum of discussion. It was a very important decision with broad consequences. It was proposed, and in a short time NRB accepted it.

By not depending on vacuum tubes, we had to use instrumentation sensors, which were designed with low-impedance electrical devices, to provide substantial current or voltage changes. For example, resistance thermometers were used instead of thermocouples. Similarly, magnetic indicators were used on control position indicators, pressure sensors, differential pressure cells, flow meters, and other instruments. Sensors of this type were more rugged and reliable than other types we could have used. This led to selection of pneumatic devices for measurement of the primary system parameters.

Primary System Instrumentation

At the time of the development of the *Nautilus* plant no instrumentation was available for application in the submarine and nuclear environments of the *Nautilus* plant. Major efforts were, therefore, undertaken to develop a series of pressure, temperature, and flow sensors to meet submarine requirements. In each case, the aim was to develop sensors with sufficient signal strength to avoid the need for significant amplification of the output signal.

Important primary system fluid measurements included the following:

- *Temperature:* These measurements had to be extremely accurate to determine reactor power so resistance thermometers were selected.
- *Pressure:* Bellows with a magnetic differential type pickup were selected for this application.
- *Differential pressure:* These sensors used a bellows arrangement similar to the pressure measuring device. Measurements of pressure differences were used for a variety of purposes. Pressure drop across a venturi was used to measure feedwater flow to the steam generators.

- *Pressurizer and steam-generator levels:* Measurements of pressurizer and steam generator levels were made by sensing the pressure difference caused by the weights of the liquid. Sometimes, the height of the liquid being measured was compared with a reference of known height.

The reliability of these instrumentation systems was of key importance. Each sensor design had to be closely integrated with the fluid systems. Particularly challenging was the development of level instrumentation for the pressurizer. The prime difficulty was associated with finding a reliable reference leg filled with a known amount of water. For reliability, the instrument systems selected for application were pneumatic. Pneumatic amplifiers were used for many plant control and computational purposes.

An important part of instrumentation development was the design of panel displays to serve as the interface with plant operators. Traditional instruments of that period were too large to consider for use on panels in the limited space allocated for the operators. A series of small linear gauges were developed to display the system temperatures, pressures, and other variables in an easy-to-read format. These instruments were the forerunners of much of the panel instrumentation used on current commercial nuclear power plants.

At the time of this development, no instrumentation was available for the pressures and temperatures that existed in a pressurized water reactor. Major developments included the design of instruments with magnetic differential transformers.

Primary System Valve and Equipment Control

Besides control of the reactor, a key part of the control system was the design of equipment and systems for the control of the primary system. This system design included control of the valves and fluid systems of the reactor plant. The main coolant pumps and the steam generators were key primary system components. Pump operation had to be coordinated with reactor power so fluid and reactor temperatures were maintained within the desired limits. Extensive analysis and simulation studies were made to determine transient flow and temperature conditions following a loss of pump power. These results, plus studies of other primary and secondary system initiating events, were used as inputs to the design of the safety shutdown system.

It was necessary to develop proper interlocks between the various plant systems to prevent undesirable conditions from occurring. For example, interlocks were necessary to prevent opening valves in cold loops, because this could cause drastic changes in reactivity. A decision we made early was that the valves should be of a sealed type, i.e., there would be no packing, with the potential for leaks. This decision was an important one and led to the design of valves, operated by magnetic pilot valves.

Secondary System Control

Very early in the design it became evident that the integration required between design of the nuclear steam supply system and the steam plant control was an important issue. Operations of the steam plant could cause substantial transients on the primary system and the reactor. It was therefore necessary to understand thoroughly the effect of the steam plant on the reactor system, particularly for the first-ever submarine application.

Much attention was given to the effect of thermal shocks on the primary system that could be caused by unusual operations of the steam plant. In addition, it was recognized that under conditions of rapid shutdown of the reactor and cool down of the plant, the temperature of the reactor could drop rapidly, thus introducing positive reactivity. We called this latter condition the *cold-water transient*. To assure that it was not necessary to shut down the plant rapidly, in case power had to be decreased, we added turbine bypass valves. Steam flow could be maintained by bypassing the propulsion turbine.

Integrated Plant Controls

The design of the control systems to integrate the many reactor primary system and steam plant systems into an operable power plant was assigned to Esselman. He was ably assisted by Al Sanderson and John Nolan who were responsible for plant controls from the primary system to the steam turbine.

Orv Meyer and Ray Mairson played key roles in the development of the reactor controls. Reactor shutdown and alarm systems required sensor inputs from all aspects of the STR prototype. The electrical systems, which included vital supplies for plant operation and the submarine hotel loads, were designed by Bill DiPietro. All these engineers went on to have important roles in future atomic power developments.

Operator surveillance and control of the reactor-related coolant systems, steam turbine systems, and all of the auxiliary systems were accomplished from a small area called the Maneuvering Room. This room, which was barely big enough to house four or five operators, contained all of the information needed to run the plant. Control equipment was located throughout the crowded hull. Special small instrumentation displays were developed to get all of the needed information on the available panel space. All the equipment had to be both reliable and pass the required shock resistance tests.

Since the control area allocated for the crew had to be kept small, a compact design of the control panels was essential. Since steam plant instruments had not been used in submarines, special small indicators had to be developed.

A great amount of thought was given to the layout of the panel in the Maneuvering Room. Each Westinghouse proposal was reviewed by Panoff and Jack Grigg of NRB. Finally a group of Human Factors engineers from the Naval Research Laboratories visited Bettis to review a mock-up of the key panels and required operator actions. Today we would say the objective was to make the controls user friendly.

Using the Heat

The amount of energy that may be released within a reactor, and hence the amount of power that may be produced, does not depend on the size of the reactor as determined by nuclear considerations. Any amount of power, from a few watts up to hundreds of thousands of kilowatts of heat, may be produced from any size reactor—provided this energy can be removed from it. In smaller reactors, the limitation on power production is not nuclear. It depends on the ability of the engineer to remove the desired amount of heat from the small volume of the reactor.

Application of the heat released in a reactor to the production of useful power is achieved through a conventional thermodynamic cycle. This does not mean that this part of the system is necessarily identical with a similar cycle used for conventional fuels. Indeed, the reverse is usually true because, when using conventional fuels, the engineer is given a starting temperature level of approximately 3000°F in the combustion gases. By proper design, the materials used to make boiler tubes or gas-turbine blades can be kept considerably cooler than combustion gases.

A conventional power plant usually operates at the highest practicable temperature. In a nuclear reactor, this is not as important. The materials in the core of the reactor proper must be the hottest in the entire system. These are the very materials that have the most severe requirements already imposed on them with regard to neutron absorption, radiation damage, strength, and corrosion. Where efficiency in a fossil-fueled plant normally means increased cruising range and decreased fuel consumption, in a nuclear plant that big incentive for increasing efficiency is not as important, because the cruising range is almost unlimited.

The weight of the nuclear fuel inventory itself is an inconsequential part of the overall weight of a power plant. Thus, improvements in efficiency do not mean that the overall weight is materially reduced. If thermodynamic efficiency is increased, the ship will not have significantly increased cruising range, as is true for conventional fuels. A fossil-fueled plant usually operates at the highest possible temperature, whereas the incentive in a nuclear system is to reduce total plant equipment space and weight and to increase reliability and ease of operation. Safety, reliability, and ease of operation were Rickover's prime goals.

The thermodynamic cycle chosen for the submarine thermal reactor (STR) permitted operation at just about the minimum material temperatures that would allow generation of useful power. The cycle selected uses ordinary water maintained under high pressure, so it does not boil, to transfer heat from the reactor into a steam generator. This generator is a shell-and-tube-type feedwater heater. Ordinary boiler water at low pressures on the other side of the tubes (from the primary coolant) boils and forms steam. That steam is used to drive steam turbines for propulsive power and to generate electricity to meet the submarine's needs.

The Power Plant

The equipment to be mounted in the hull of a nuclear-powered submarine is so functionally interrelated that it must be considered as a single power plant. However, it requires hull space of such size that it must be divided it into two physically separated compartments to provide adequate watertight integrity and proper balance, in case there is battle damage to the submarine hull.

The reactor compartment contains the nuclear reactor, all steam-generating equipment, and the auxiliary systems. The engine room contains propulsion equipment, all steam-driven items, and associated control panels and switchgear. It is the main control point for equipment in both compartments.

The reactor and the primary coolant system are contained in the lower reactor compartment. Energy from the reactor is transferred by high-pressure water to the primary side of the steam generators. Being at a higher temperature, this primary-side water supplies energy to produce secondary-side steam. Water from the steam generators then passes through the main coolant pumps to maintain the required flow back to the reactor.

The nuclear core is contained within a large carbon-steel pressure vessel, with the inner surface clad with stainless steel. There must be no leakage during operation, but it must be possible to remove the core for refueling.

Before Westinghouse entered the program, the Navy had given a contract to the Babcock & Wilcox Company (B&W) for development and manufacture of this vessel. The plan was to build a vessel with a hemispherical head, which would be about three inches thick, and to provide cutting machines and welding techniques for cutting the head open and rewelding it after refueling.

The difficulty came in the rewelding of this low-alloy (SA-212) steel head after refueling. Heavy welds in this material required post-weld heat treatment, but it was not feasible to heat treat this weld. B&W attempted unsuccessfully to develop a procedure for rewelding.

The final design was a two-inch-thick lip on the vessel and the head, which were welded together. This weld was leakproof but was not considered a strength weld. For strength, studs were employed that were prestressed to keep the weld always in compression. The vessel head kept the core in place, but the shrinkage of the weld had to be accommodated in the design. Westinghouse developed a method of using zirconium as a crushable material, which accommodated the shrinkage and kept the nuclear core in place.

Another technical uncertainty[11] faced in the early days of the naval program was the embrittlement of steel under heavy neutron bombardment. Radiation experiments carried out at ORNL had shown a tendency for radiation embrittlement of steel, but data were sparse at that time. There were very few reactors in operation and the Materials Testing Reactor (MTR) was not yet in operation. Because of this concern, a conservative decision was made to place a sheath of steel, called a *thermal shield,* between the nuclear core and the reactor vessel wall to reduce the fast neutron flux that would impinge on the reactor vessel. Since this sheath was not a structural part, embrittlement would not be a problem. Subsequent studies of the phenomenon have verified that this additional shielding was necessary.

Reactor Compartment

One of the key design decisions, early in the program, was to make the reactor plant a hermetically sealed system. This minimized the possibility of a primary coolant leak into the atmosphere of the submarine.

The decision had a dramatic effect on the design of all primary coolant system components with moving parts such as the main coolant pumps, valves, and control rod drive mechanisms. In a conventional plant, such components are driven by electric motors and employ rotating or linear seals to separate the driving motor from the fluid passing through the component being driven.

---

[11]J.J. Taylor, Private Communication dated September 9, 1991.

Several main coolant pumps were built and tested, and a pump with a thin can, which sealed the windings and the stator iron from the fluid, called a canned-motor pump, was adopted. For the control rod drives, a linear-canned-electrical motor was developed. Valves were designed to be hydraulically operated, using primary coolant as the operating fluid. The sealed system requirement dictated the use of welded pipe; any flanged joint had to be backed up by a seal weld.

The primary coolant pumps are in the lower part of the compartment. During the war, Ben Cametti, at Westinghouse Research Laboratories, developed a small pump for a uranium separation process; all electrical parts were sealed from the fluid being pumped. This first pump had a capacity of only 30 gallons per minute. Later, it was enlarged to 150 gallons per minute. Cametti was transferred to Bettis, as happened frequently when expertise was available within the corporation.

For the *Nautilus*, a 2000-gal/min pump was required. This was a major design challenge. These centrifugal pumps are driven by three-phase induction canned-motor units. The thin can, which sealed the windings and the stator iron from the fluid, was backed up by the stator itself to absorb full system pressure.

A major problem was the operation of bearings in low-viscosity, highly corrosive, high-temperature water. The complete pump unit has an electrical stator assembly, canned-motor assembly, and a pressure housing with integral cooler, bearings, pump impeller, and lower bolting ring. The drive unit is assembled into the volute casing that forms part of the primary loop.

Within the motor, cooled primary water is circulated by an auxiliary radial-vane impeller. From the impeller, water flows downward through the "air gap" to the lower radial and upper and lower thrust bearings—and into the cooling tubes that form part of the heat exchanger. The water is then brought to the top of the pump, where it enters the motor frame and the auxiliary impeller suction. Some water from the auxiliary impeller is bypassed from this circuit to circulate water to the upper radial bearing. The bypassed water flows through the bearing, then directly back to the auxiliary impeller suction.

Motor heat is dissipated to the primary water recirculated within the pump. A close-fitting labyrinth-type seal on the shaft below the thrust bearing and a double Belleville spring arrangement prevent the motor cooling water from circulating freely with the high-temperature primary coolant water. Heat from the end turns is transferred by conduction to the pressure shell.

The upper and lower journal bearings are graphitar, backed by stainless steel shells with spherical seats for self-alignment. These seats rest on cylindrical rings of Stellite. Shaft-journal surfaces are Malcomized stainless steel.

The upper thrust bearing, a tapered-land type, is made of graphitar with runners of stainless steel. The lower thrust bearing is a fully equalized pivoted shoe-type bearing.

Conductor slots are kept as narrow as possible to reduce the stress in the can where it bridges the slots. Beyond the end of the stator punchings, support is provided by the backup cylinders to which the can ends are welded.

Contracts for pump development were also given to Byron Jackson and Allis Chalmers, but in the end only Westinghouse pumps were used, because the others did not meet the requirements.

This original *Nautilus* pump design was so successful that it was used on all nuclear ships for many years. The manufacture of these pumps was transferred to the newly formed electro-mechanical division at Cheswick, Pennsylvania. This division extended its product line to include control rod drive mechanisms, valves, and pumps used in naval and central station nuclear plants.

Outboard of the pumps is the steam-generating equipment. The steam generator transfers heat energy from the radioactive primary water to nonradioactive steam. In operation, the primary water enters the bottom end of the steam generator and passes through the rolled and welded tubes. As the secondary water flows across the bundle, boiling takes place. In the upper part of the compartment above the steam-generating equipment, from which emerge the main steam headers that lead aft to the engine room, is the steam separator. The pressurizer unit is aft of the main pumps.

The pressurizing system is designed to maintain a controlled pressure in the primary coolant system and its auxiliary systems under all operating conditions; it also provides surge capacity. Pressurizing is obtained by providing sufficient electrical heat to build up a head of steam in a pressurizing tank. A spray is provided to reduce pressure by collapsing part of the steam bubble. The pressurizing tank is connected to the primary coolant system and transmits the developed pressure to the system.

The compartment also contains many auxiliary systems, which are needed for such functions as filling, draining, and chemical control of the primary water. The system must be filled with pure water (million-ohm) initially, and addition and purification of water must be possible at any time. There are many charging lines to the main loop, so a section isolated for servicing can be recharged by operating the isolating valves, without a shutdown or risk of sudden loss in pressure.

The coolant loop is provided with pilot-operated hydraulic stop valves to permit isolation of parts for maintenance purposes. From the separator, steam is carried through pipes that penetrate the bulkhead and diverge in the engine room to supply the main-propulsion turbine and the ship's service turbine-generator sets. Condensate is pumped back to the steam generators. The piping is designed to take up its own expansion, as well as that of the components to which it is connected.

Shielding the radioactive power plant components, without adding a prohibitive amount of weight, was a major arrangement problem. As finally resolved, the shielding is such that essential controls and equipment can be manned while the reactor is operating. Necessarily, piping and electric cables penetrate the shield. The penetrations must be watertight and airtight to prevent leaking of radioactive airborne particles, and they also must be arranged in a labyrinth fashion to attenuate properly any radiation leakage through them.

These components outline the main areas in the reactor compartment. The spaces between them are filled with secondary equipment and related piping and cables, electrical and pneumatic control panels, valves, motor-generator sets, and instruments.

Engine Room

The engine room is next to and aft of the reactor compartment. This space is divided into upper and lower levels. Each level is provided with a walkway running fore and aft along the horizontal centerline. Components also have to be located on intermediate levels because of their varying size. These intermediate levels are accessible from either the lower or upper level.

The primary purpose of the steam plant is to convert the heat energy developed by the reactor into useful shaft horsepower for ship propulsion. In addition, the steam plant also supplies electrical power to all auxiliaries for power plant operation and to all electrical equipment and lighting systems, including sufficient power for charging the battery.

The steam plant is patterned after a conventional surface ship installation insofar as the handling of the steam condensate and feedwater is concerned. The steam system is subdivided into identical port and starboard steam-turbine plants that can be operated alone or simultaneously, either cross-connected or isolated from each other.

The main engines are geared, marine-type steam turbines. There are auxiliary-power turbine generators with auxiliaries such as condensers, air ejectors, condensate pumps, and lubricating-oil systems.

The design of the condensers and the circulating-water piping and pumps for all units became special problems. The circulating-water system of a conventional plant, land based or shipboard, is designed for low ambient pressure. In the submarine, the equipment in the circulating-water system had to be designed for high pressure to contain the seawater for static conditions corresponding to maximum submergence.

Power from Turbine to Propeller

To convert shaft horsepower from the low- and high-pressure turbines into useful thrust, power passes through the reduction gears, a clutch, and finally through the propulsion motor shaft. Power is then applied directly to the propeller. The propulsion motor provides a small amount of power when steam turbine power is not available.

The propulsion system clutch enables the propulsion motor and propeller section to be disengaged from the turbine and main reduction-gear section. If a clutch were not used, the propulsion motor, when in operation on batteries, would have to overcome the inertia and friction losses of the turbine rotors and reduction gear, besides overcoming the normal propeller load. The clutch also permits some angular and parallel misalignment of the drive shaft. It incorporates a thrust link that serves to locate the reduction gears axially and transmit to the main thrust bearing any axial forces due to these gears.

Maneuverability

A submarine must be quite maneuverable. In terms of a nuclear-powered submarine, this means that the power plant must be capable of varying its output rapidly and changing quickly from a quiet operating condition to one of full power ahead. For more on how well this was accomplished, see the section on the Ireland committee in Chapter 3.

The power plant for the *Nautilus* was designed to function in two ways. Normal operation involves propulsion by the main geared turbines, with steam furnished by the reactor system; emergency operation uses electric motor drive, with power supplied by a battery or a diesel generator.

Special Requirements

A basic design problem for submarine equipment is compactness. The space for equipment is much less than would be available for a commercial design. For example, heat-exchanger designs had to be chosen not for optimum heat transfer, but for space reasons. Certain design features were employed that would not be commercially feasible in a conventional steam plant.

Equipment weight is important, because it cannot exceed the amount set by the displacement of the submarine. For stability reasons, the center of gravity is required to be low in the hull. In locating heavier plant components this factor had to be considered.

Normally lead ballast is used, and a ballast estimate is made at the time the overall weight is computed for the vessel. If the components weigh more than estimated, then the ballast is removed from the ship to maintain buoyancy. In doing this, so much ballast may be removed that a center-of-gravity problem is created. Such a situation would require that the ship size be increased to accommodate the needed ballast and maintain stability.

Submarine personnel must be protected from nuclear radiation. The shield reduces the radiation to a level such that, during a cruise lasting the life of the reactor, the average crew member will receive less radiation than during a lifetime exposure to cosmic rays and natural radioactivity in the sea, air, drinking water, and the ground; routine chest and dental x-rays; television screens; and luminescent instrument dials.

The radiation-monitoring system for the submarine, which ensures proper radiation levels for personnel protection, has air-particle detectors, gamma detectors, boiler-leak detectors and a discharge system-activity indicator. Air detectors sample the activity of the air in the shielded area and in adjacent compartments. Gamma detectors are installed in various sections of the ship. The boiler-leak detectors warn of a ruptured boiler tube in the steam generator that would allow radioactive coolant to enter the unshielded steam system. The discharge system activity indicators ensure that radioactive water is not discharged at a dock or elsewhere, which would produce a hazard.

One unusual and quite difficult problem was the design of relief valves that would work when the submarine was at maximum depth as well as on the surface. Reclosing after a safety valve tripped was the main difficulty.

To capitalize on the benefit of nuclear power, it was necessary to design the reactor plant for a long life, and the longest possible periods between refuelings. This required knowing the effect of radiation on the fuel and the effect of the buildup of the fission fragments on reactivity. Early objectives were relatively modest, but eventually the goal was that the fuel last for the expected life of the submarine.

Testing Philosophy

The principle of individually testing each item in a system's environment was followed to the maximum extent possible, so that the final assembly represented a test of the complete power plant and the interaction of the components, rather than a test of individual items. In addition, where secondary service systems were involved, the systems themselves were tested individually before inclusion in the final plant. Finally, the entire nuclear power plant was tested as a whole at the Mark I facility. This test program proved its worth in that the power plant installed in the *Nautilus* was similar to the prototype that had over a year's operating experience.

Steam Cycle

Having selected the basic reactor type, as described, it was then necessary to determine overall plant design parameters. A variety of steam cycles, including various throttle pressures and back pressures, were considered, and the most suitable one—from a weight and space viewpoint—was selected to be compatible with the temperatures available from the primary plant. The reactor heat load was determined, as well as primary system water temperature, pressure, and flow. Maximum fuel element surface temperature was selected by the study.

Radiation and Heat of Structures

The nuclear core is a source of intense radiation of gamma rays, neutrons, and alpha and beta particles. Absorption of this radiation creates heat, depending on the type and amount of radiation. Because of the amount of absorbed radiation in the material that supports the nuclear core, great care must be taken to have sufficient coolant flow pass these surface areas. Nonuniform heating can cause increased thermal stresses and lead to distortion of the structures.

Shielding[12]

One problem that permeates the entire field of nuclear engineering is that of shielding. With mobile plants, shielding is particularly difficult. For stationary power plants, where size and weight are not particularly important, ordinary materials such as concrete, iron, and water are used for shielding; economic considerations usually determine the materials.

In mobile reactors for ship and space applications, emphasis is placed on weight and space, so that unusual materials must be investigated. Some materials with the best shielding characteristics do not have good structural properties, so their possible utilization in reactor shielding is a matter of concern to the nuclear engineer. Besides this, the shield must be arranged for maximum accessibility to equipment for maintenance purposes.

It is first necessary to know the attenuation characteristics of the materials and combinations of materials. After this, the problem is still a difficult one, because radiation does not emanate from a single point or even from a simple distributed source such as the core itself. Rather, it emanates from the core and from radioactive corrosion products, as well as radioactive oxygen and nitrogen, in the coolant.

---

[12]Major contributions came from J.J. Taylor and R.J. Creagan.

In addition, these corrosion products, over time, distribute themselves throughout the reactor system, concentrating in certain places such as crevices. There is an additional complicating factor, namely, the self-shielding of the pipes, the core itself, the coolant and other pieces of equipment such as valves and pumps. In addition, the amount of radiation that can be allowed to emanate from the shielded areas depends on the access required by the crew.

In contrast to the reactor physics situation, there was no effective way to calculate the attenuation of radiation through the many orders of magnitude necessary to assure crew safety.

In reactor physics, it is primarily neutron diffusion, absorption, and isotropic scattering effects that determine neutron reaction rates and flux distributions. But shielding involves attenuation of a factor of many millions. Thus, the influence of multiple scattering and conversion of one type of radiation to another (i.e., neutron capture producing gammas within the shield material) dominates, and transport theory applied with the calculational capability then available could in no way predict the result accurately enough to determine the minimum amount of shielding.

Thus, an empirical approach was required. A major shielding experimental program was set up at ORNL to provide these empirical data, and John Taylor found himself there every other week, working with ORNL engineering (Blizzard and others) on design of experiments and the evaluation of data. Without ORNL, Bettis could never have produced an adequate basic shield design. A group under W.C. Redmond at ANL also participated in the shielding effort for Mark I.

Another problem was that we did not know the magnitude of nitrogen-16, the dominant source of radiation in the coolant system, outside the reactor shielding. Nitrogen-16 was produced by absorption of a neutron in oxygen-16, the dominant oxygen isotope in water. But nitrogen-16 has only a 7.5-second half-life, so the conventional method of measuring the neutron cross section with accelerators was ineffective.

To fill the gap, Taylor designed a rig to circulate water in a clean circular pattern. He installed the rig in an irradiation loop in the MTR in Idaho to generate nitrogen-16. Water was pumped rapidly enough through the loop such that there was significant detectable radiation in the portion of the loop outside the reactor. With appropriate corrections for time delays, an effective cross section was derived for O-16 absorption and the N-16 production to use in the shielding design. This was reported in *Physical Review*, since it was a basic physics measurement, though obtained through an engineering experiment.

A third problem seemed even more intractable. Before the naval reactor program got under way, "whiz kid" and major contributor to the H-bomb development, Fred de Hoffman, collected data on scattering of gamma rays from the reactor compartment into the water outside the submarine and back into the submarine. His calculations showed that the radiation the crew would receive, on the other side of the reactor compartment, would be harmful.

Shadow shielding was explored mathematically to the extent of many millions of dollars, not only because of the naval program but also because of its importance to aircraft and space reactors. The biggest uncertainty was neutron and gamma scattering data as a function of angle and energy. Blizzard of ORNL, with his shield test facility and later the reactor suspended in air by cables, was a major contributor to shielding development.

The highest level neutron sources available were installed in a large tank in the physics lab to try to measure backscatter. ORNL was also asked to devise experiments to measure the effect. Neither group was successful in taking measurements at a level sufficiently above background to be definitive, because they didn't have neutron and gamma sources intense enough or of the right energy spectrum to simulate the source from the *Nautilus* reactor. For some time, therefore, a major lead shield was designed to be placed on the hull to reduce this backscattered radiation to a safe level.

Taylor continued to remain suspicious about the need for that shielding because of our inability to get any indication from experiments. After further review of de Hoffman's paper, Taylor found an error in the calculations. We were then able to remove that lead shielding on the hull from the design.

Needless to say, we weren't satisfied until we had demonstrated the adequacy of this decision on the prototype. So the reactor section of the propulsion plant was installed in an enormous water tank to simulate the submarine in the ocean. We tested the whole shield design, also checking out the magnitude of the backscattering effects. Sure enough, they were inappreciable and the lead shielding was not needed. Later, we were told that the shipbuilder had added to the ballast the tons of lead we had removed from the hull, just in case we were wrong!

The *Life* magazine cover picture of the *Nautilus* prototype showed a disorderly bunch of cables running all over the top of the facility. These were the cables for our tests. Admiral Rickover told Taylor, during the design, that if the tests couldn't measure any radiation of consequence outside the shielding, he would be a failure. Of course, he would have been more of a failure if the radiation had been too high.

It came out just about right. Excess radiation, measured outside the shielding in Mark I, was reduced by providing a polyethylene collar around the reactor vessel. This was an important benefit of Mark I.

Another feature of nuclear power plants, associated with shielding, is the high degree of leakproofing required of the coolant systems, which must contain radioactive coolant. Leakage of a few cubic centimeters of most reactor coolants will release enough radioactivity to make the surrounding area uninhabitable. An ordinary steam power plant is subject to copious leakage when compared with this standard. Valve packing glands, pump-shaft seals, pipe flanges and turbine-shaft seals usually all leak to the extent that many gallons of makeup water per day must be added to the boiler.

Heat exchangers, piping, the reactor vessel, and other primary coolant system components, which make up this necessarily impervious envelope that contains the coolant, are of rugged construction and good engineering practice ensures great integrity. However, it is necessary to transmit certain motions and pressures, as well as intelligence and considerable amounts of energy, through this envelope. Therefore the points at which these pass through the envelope are points of weakness. A great amount of effort was applied to achieve the necessary soundness and reliability at these points.

Repairs

Another consideration associated with radioactivity and shielding is maintenance. In the development of shipboard nuclear power plants, we must ensure that one can do maintenance on all parts of the system if necessary. Sometimes this involves ingenious mechanical devices to perform remote operations. It always necessitates plans for repair as part of the design, because once any part is radioactive, taking it back to the machine shop for alterations is difficult. Maintenance requirements mandate a much greater than normal attention to details of design and installation.

Metallurgy Development[13]

Strange as it may seem, pure water under high temperature and pressure is a most corrosive liquid, as corrosive as some acids. We set up an extensive program to develop the alloys needed for the components—primary loop piping and pressure vessel cladding—that would be in contact with the coolant water. These were tested under simulated reactor conditions. We installed a few autoclaves at Bettis, but when the magnitude of the problem was realized, we knew many more would be required. We rented

---

[13]Major contributions to this section were made by B. Lustman, D.E. Thomas, R.H. Fillnow, and J. Theilacker on all discussions of materials.

space in an old distillery in Large, Pennsylvania, and installed an extensive autoclave facility, which was operated by the materials engineering department.

For application to components such as mechanisms, valves, and pumps, we had to find materials that would withstand the corrosive effects of high-temperature water and have good wear characteristics, when in rubbing contact with other pieces of metal, without any lubrication other than high-temperature water. Most of the materials that had the best corrosion-resistant properties displayed inherently poor wear-resistant properties, and vice versa.

So a thorough investigation program was required to determine the combination of metals that presented the best compromise for various applications. It was first necessary to perform many tests for short periods using simple shapes and later to make actual full-size components and test them under actual service conditions in high-temperature water.

In some applications, it was necessary to use a material compatible with either salt water or high-temperature, high-purity primary coolant water. Stainless steel, with poor corrosion-resistant properties in salt water, was not suitable. Other materials were identified as having suitable properties. This raised the difficult problem of making either a transition weld or a mechanical transition between these other materials and the stainless steel in the primary-loop piping.

Many materials or alloys were ruled out, because their wear particles would become activated by exposure to the neutron flux during passage through the core and consequently would adversely affect habitability. A prime example is cobalt, which is often used as an alloying element in non-nuclear wear applications.

Besides the applications mentioned, it was necessary to select materials suitable for such uses as springs. Springs must be made of a material that will withstand the corrosion of the coolant and withstand whatever radiation is involved at the point of application while not losing strength due to prolonged exposure to high-temperature water. Springs, when under stress, are also subject to stress corrosion.

Another little-known problem was that of contact corrosion—the tendency of two metals, when in contact or near contact, to form a corrosion film between them when they are together for long periods of time. The resultant corrosion products, in effect, cause a welding of the two pieces. This was of particular concern where parts, immersed in the coolant, were subjected to intermittent motion.

Materials used in the core needed to have a low absorption cross section for neutrons, in order not to increase the amount of fissionable material required. Our core mechanical and heat-transfer designers did not have the freedom to select materials from among those whose characteristics were well known. Nor could they choose materials on the basis of ease of manufacturing.

For the necessary heat-transfer surface to be obtained in a nuclear core, highly enriched uranium fuel must be alloyed with other elements and then clad. The cladding material on the fuel elements must not only prevent fission products from contaminating the primary coolant, but also must itself be corrosion resistant so that radioactive corrosion products do not enter the primary coolant in any appreciable amounts. It is also desirable to have fuel elements that can be fabricated as cheaply as possible.

The materials used in nuclear reactors are subjected to various degrees of damage from radiation. The subject of radiation damage of solids and fluids is another one of those interdisciplinary fields lying among physicists, metallurgists, mechanical engineers, and chemists.

Irradiation testing[14] was vital to reactor system materials qualifications and to fuel element development and design. We initially were working with no data or theory of any kind concerning radiation damage. The MTR was an important asset, as were the Oak Ridge and Canadian test reactors.

Radiation damage is proportional to the time integral of the flux and is also a function of flux spectrum, type of material, and temperature. Determination of its effects, therefore, requires extensive tests for long periods and under conditions simulating actual reactor operation.

Fuel elements and other materials subjected to irradiation in a nuclear testing reactor became very radioactive and required special equipment for handling and evaluation. Bettis was at the forefront of technology in this area.

By the mid-1960s, after a major body of data had been generated from irradiation testing, Bettis initiated a program to develop analytical methods using computers for predicting fuel element behavior. The methods were successful in providing for further refinements in fuel design and in reducing the amount of irradiation testing needed, although by no means eliminating such tests. The result has been that, in both the naval and the commercial program, nuclear fuels met or exceeded expectations. One would wish that the portion of the plant considered conventional technology had done as well.

The most important factor controlling core materials choice, fabrication method, and fuel element design was the requirement that core operation be unimpeded and core life unaffected by the presence of undetected fabrication flaws that exposed fuel to the coolant or by the formation of such

---

[14]Partial contributions from J.J. Taylor, Private Communication dated September 9, 1991.

flaws during operation. As a result of this requirement, fuel element testing in both out-of-pile and in-pile facilities was conducted using simulated flaws through the cladding, which exposed fuel alloy to the coolant.

Zirconium and Hafnium

As mentioned earlier, zirconium appeared to be ideally suited as a cladding material for nuclear fuel, but there was a major problem: There were only a few pounds of pure zirconium in the world. We also found that only the purest zirconium was corrosion resistant. We built a production line, that produced pure crystal bar zirconium in an incredibly short time. Much that came off the line was not pure enough, but we got a sufficient quantity for the prototype and the *Nautilus*. Later, we found that an alloy of Zr with Sn, Cr, Fe, and Ni was far superior to the crystal bar. We developed such an alloy at Bettis. Hafnium was an excellent control rod material, but again it existed only in laboratory quantities and production quantities had to be made.

In late 1947, the Daniels Pile project was being officially closed down, but Rickover had the group doing a conceptual design for a submarine propulsion plant using pressurized water. It was at this time that Pomerantz and Kaufman found that it was the impurity hafnium that caused zirconium to readily absorb neutrons. Moreover, hafnium appeared to be suitable for control rods. This was the concept, but bringing it to reality was a long process. It was not until 1952 that zirconium suitable for fuel cladding was available. Zirconium, from then on, became the fuel-cladding choice. The only other parameter established at that time was light water as a coolant.

There were small development programs on other cladding materials such as stainless steel, aluminum, niobium, and beryllium. These had major problems, making them poor second choices if zirconium could be developed.

By 1948, Dr. W.J. Kroll, developer of the process for titanium production for the U.S. Bureau of Mines, had developed a technique for producing zirconium feed stock for the Van Arkel-DeBoer ("crystal bar") process. This was the reference process for purifying zirconium and similar elements of this type.

This method takes advantage of the reversibility of the reaction between zirconium and iodine, based on temperature. At low temperatures zirconium combines with iodine to form zirconium tetra iodide (ZrI^4) gas. At elevated temperatures the zirconium tetra iodide dissociates to form Zr and free I^2. Van Arkel and DeBoer had found that the pure zirconium from the dissociation would deposit on a heated wire in a vacuum chamber. The freed I^2 then repeated the process of reacting at a lower temperature with the zirconium sponge, resulting in a cycle that produces zirconium of the required purity.

Unfortunately, even by 1950 the production process had progressed only to the point where small pieces of crystal bar could be produced, about the size of a pencil. About 30,000 pounds were required for the Mark I.

The New York operations office[15] of the AEC contracted with Foote Mineral Company in 1950 for 30,000 pounds of crystal bar. The first 9000 pounds had a price of $110, with later deliveries to be at $50 per pound. These prices were exclusive of the cost of zirconium tetrachloride feed material. Sponge feed material at this time cost about $45 per pound. Only three years earlier the price was $2000 per pound not including feed stock.

Samples of early crystal bar showed widely varying corrosion-resistant properties. Metallurgists believed that impurities, due to poor control of the atmosphere in the Van Arkel-DeBoer vessels, caused this. Nitrogen was identified as a major problem. There were also titanium impurities from inadvertent mixing of zirconium and titanium sponges. Aluminum impurities came from gaskets, and tungsten resulted from improper use of tungsten starting wires.

It was believed that impurities were the major problem and that if the process could be conducted under sufficiently controlled conditions, zirconium of adequate corrosion resistance could be produced. Fortunately, by this time Kroll had developed a suitable source of sponge. This did result in obtaining small quantities of acceptable material, but we needed large quantities.

Crystal Bar Zirconium Factory.[16] Dr. Zalman Shapiro, realizing there might be a need for quantities of crystal bar zirconium, had a small group working on the scale-up of the Van Arkel-DeBoer process. By July 1950, Shapiro had shown the feasibility of scale-up. Les Weber was an important member of the team. Now the problem was getting the quantities required and in the time required. Rickover asked Weaver if he could be in full-scale production in three months. Weaver was very doubtful but agreed to try.

Again the tremendous Westinghouse technical ability, and the willingness to use it, came to the rescue. Someone with experience in running a high-technology production operation was needed to set up and run the factory. Al Squire, in the materials engineering department at East Pittsburgh, had operated several pilot-plant-type facilities on a round-the-clock

---

[15]H. Rickover, "The Decision to Use Zirconium in Nuclear Reactors," William J. Kroll Medal presentation, March 21, 1975.

[16]A. Squire, Private Communication dated 1975.

basis, among them a highly sophisticated brazing development and heat-treating laboratory.

Completely green about the whole project, Squire reported for work at Bettis the following day. Everyone at Bettis was so intensely occupied with the program that Squire wasn't even asked if he wanted to join the team; it was just assumed that he would. The challenge was presented—take the equipment designs and processes developed by Shapiro, get equipment built and installed, hire and train operators, and be in production in three months.

The retorts, tanks two feet in diameter and five feet high, had to be made from Hastalloy C, one of the few materials resistant to iodine attack. There was one supplier, and few experienced fabricators. The units had to be vacuum tight, with provision for passage of a high current of electricity through the tank to electrodes to which wires could be attached; the only suitable gasket material was gold.

The electric power to the units had to be closely controlled and variable to compensate for the resistance of the "hot wire" as zirconium built up on it. Large variable-voltage transformers had to be designed from scratch and built. Vacuum systems were required, as were cooling systems to keep the zirconium sponge at the right temperature during the reaction cycle.

George Kepner, corporate director of production, was brought to Bettis to handle procurement. He had wide contacts throughout industry, made in breaking bottlenecks for Westinghouse during the war. He had a persuasive manner that few suppliers' top management people could resist.

There was still the problem of the factory itself, and the people to run it. On his first day, Squire was shown an empty wing of the new Bettis main office and laboratory building. It was really empty! Apart from a crane, there was nothing except the floor, and that had to be dug up to make a pit for the retorts.

As Shapiro did the equipment design and Kepner moved heaven and earth to get material and equipment ordered and delivered, Squire worked on the facility arrangement and with contractors, who were excavating, building, completing facilities, and installing equipment as fast as it was delivered.

Meanwhile, there were operators to be hired and trained. Employees working on the highly classified naval reactors program were required to have an AEC "Q" clearance. Squire himself didn't have one but was issued a "P" clearance, requiring an escort. Expediting his "Q," which was necessary for access to classified nuclear information, was difficult, because he was born in Scotland. The McCarthy era was starting, and several employees who had associated in their younger days with liberal organizations were regarded with suspicion and some were not cleared. One by one, a team of about 30 people was assembled to run the operation on a 24-hour, seven-day-a-week basis.

There were former air force pilots, automobile mechanics, a barber, several clerks, and a number of people fresh out of high school. They all had one attribute—they were eager to learn and to do something important.

The future operators worked under Shapiro's three highly competent technicians—Fred Schrag, John Eck, and Nick Pollack. By mid-October, the installation was completed and the first product was made, just 14 weeks after the go-ahead.

Rickover was kept informed of progress and problems between the time the job was authorized and the time operations began. There was no interference by him at the working level. On the contrary, all red tape was cut, at his direction, and unusual procurement actions were authorized. If they hadn't been, the job would have taken months instead of weeks.

The factory process was a success, and the line was expanded 50% a few months later. But the product was so pure that its corrosion resistance was highly sensitive to small quantities of some elements that otherwise would not have mattered. The factory product was carefully tagged according to raw material and heat number, and it was cut into bars and tested in hot, high-purity water in autoclaves at 680°F.

Sometimes the material passed the test, emerging with an adherent purplish iridescent film. It was promptly passed on to the metallurgists to be melted in vacuum-arc furnaces and used for fuel fabrication process development. Other times, the bars emerged with white spots, or in one case the precious crystal bar turned into a white powder, zirconium oxide, at the bottom of the autoclave.

At one point during this period, Bill Johnson pointed his finger at Thomas and said, "If the Mark I fuel elements corrode it's your fault." Thomas was adequately impressed.

While we could get enough good material for Mark I, this situation was unacceptable for a major naval submarine program. Bettis metallurgists, led by D.E. Thomas, dug in hard. They found that the material was too pure and that small quantities of iron and nickel enhanced its corrosion resistance and ductility. To correct for this, they added small wires of iron and nickel to the zirconium sponge feed. The corrosion resistance and ductility of the crystal bar improved. Only an occasional few heats were bad.

We had believed that the purity of crystal bar was what gave it its corrosion-resisting properties, but we found that if there were even small amounts of the wrong impurities, it corroded quickly. The answer was a zirconium alloy. With the right alloying elements and the right percentage, small accidental impurities did not cause zirconium to corrode.

Of the 40,000 pounds of crystal bar required for Mark I, Bettis produced about 85%. In the period from November 1950 to June 1952 Bettis produced some 80,000 pounds. As difficult as it was, a sufficient quantity of satisfactory material for the Mark I was produced. That led some to want to let well enough alone, but that opinion did not prevail.

As one can see, at every turn in the development of nuclear power a failure could have occurred that would have changed the basic reference design in a major way. For example, if zirconium could not have been used, we would have essentially been back at square one.

Frequent mention is made of the short time allowed for reaching a certain goal. Of course missing one or two by a few days would not have been a major setback, but there were thousands of them. Taking each goal as sacred was an important factor in the success of the program.

Zirconium Alloy Development. It was recognized that unalloyed iodide crystal bar zirconium had disadvantages that precluded its continued use. Its corrosion resistance was more variable than desired and it was excessively sensitive to contamination by nitrogen during processing. Its mechanical strength was low. Its cost was high, and there was doubt that, even at its best, it would have corrosion resistance adequate for the longer lived reactor cores projected for the future.

Early in the development of zirconium, the AEC had established a broad research and development program in extractive metallurgy, hafnium separation, melting, fabrication, physical and mechanical metallurgy, analytical chemistry, and corrosion. Without this background fund of knowledge successful alloy development would have taken longer.

The onset of corrosion failure was evidenced by a rapid change from a protective to a nonprotective mode, marked by loss of corrosion product oxide by flaking and spalling. Termed "breakaway," the onset of failure was highly variable both from batch to batch and from within a given crystal bar. The geometry of the crystal growth process with time variation in the iodide atmosphere caused compositional variation across the diameter of the crystal bars that, after corrosion testing, showed up like growth rings in a tree. Some early clues concerning harmful impurities such as N, Al, and Ti were obtained by selectively analyzing these rings. Such efforts were hampered at first because of the uncertainties in analytical procedures and the lack of adequate analytical standards, not to mention inconsistencies in the corrosion-testing procedures; therefore, early spectrographic results were only semiquantitative. However, the technology moved forward rapidly.

In the early development period, major reliance was placed on zirconium-tin binary alloys. A parallel process development, by the Albany,

Oregon, station of the U.S. Bureau of Mines, was by this time beginning to produce a purer sponge product that showed promise of serving directly as a low-cost base for a corrosion-resistant alloy. It was also recognized that exceptional mechanical strength at temperatures in the 500 to 600°F range was not likely to be achieved in combination with good corrosion resistance with crystal bar zirconium.

A zirconium alloy corrosion committee was formed and was active in the 1951 to 1953 period. Dr. D.E. Thomas headed the committee, which consisted of C.R. Breden of ANL; H.A Pray and R.S. Peoples of Battelle Memorial Institute; T.T. Magel and D.S. Kneppel of Nuclear Metals, Inc.; E.T. Hayes of U.S. Bureau of Mines, Albany, Oregon; K.M. Goldman of Bettis; and F. Kerze Jr. of the NRB.

It was discovered by Battelle and the Ames Laboratory that tin additions counteracted the harmful effects of the impurities on corrosion resistance. By this time, it was recognized that nitrogen was the principal impurity with which we had to be concerned. We didn't realize that the corrosion rate increases with an increasing tin content, above the optimum level, at low nitrogen content. This lead was promptly followed up with additional work and a decision was made to adopt an arc-melted sponge base alloy containing 2.5% Sn as the reference cladding for the *Nautilus* core.

While this alloy proved to have relatively good corrosion resistance in the pre-breakaway period, after breakaway, rather than sloughing off as in the case with pure Zr, the ZrO_2 corrosion product formed a thick, adherent surface layer. Because of the formation of such a thermal barrier, fuel element center temperatures would be elevated leading to a reduced fuel element life. Thus, an immediate effort was directed toward development of a more suitable alloy.

Zircaloy-2. The overriding consideration in the development that led to Zircaloy-2 was the improvement of corrosion resistance. We had just seen that at tin levels of about 1 to 2.5% the further addition of iron produced a further improvement in corrosion resistance, with a maximum effect at about 0.25% addition. A similar effect was found if chromium was added instead of iron. Here the maximum effect also occurred at about 0.25% addition. Similarly nickel added to the Zr-Sn binary showed behavior similar to that of iron or chromium, with the maximum effect occurring at the same level. In addition nickel had been observed to enhance corrosion resistance in higher temperature steam tests.

It was postulated that the synergistic effects noted above would also hold in a more complex alloy containing tin, iron, chromium, and nickel, that is, there would be a minimum in the corrosion rate curve versus the combined iron, chromium, and nickel contents Σ(Fe, Cr, Ni) at a given tin content, given the metallurgical similarity of the transition metals. It was clear that the tin content should be lowered below the 2.5% of Zircaloy-1, because adding tin beyond the small amount necessary to counteract the impurities only served to increase the corrosion rate, and because fabrication difficulties had been experienced at the 2.5% tin level.

The iron content was set at a nominal level of 0.12%, primarily due to the above, and the chromium level was set at 0.10% for the same reason, while the nickel content was set at 0.05% because of its effect on high-temperature corrosion resistance. At the same time it was recognized that these three transition elements normally occurred in the zirconium sponge as variable tramp elements, and that it would be advantageous to increase their levels by further addition of those alloying constituents, which are more controllable, so as to produce a more consistent product. These selections produced a Σ(Fe, Ni, Cr) value of 0.27%.

Lowering the tin content would result in lower tensile properties, which would be less desirable to the mechanical designers. However, the addition of the transition elements would provide mechanical properties comparable to those of the earlier alloy with 2.5% tin. Similar reasoning applied to fabricability.

At a meeting in my office on August 28, 1952, this selection was approved by Rickover, without the alloy ever having been melted and fabricated, much less tested. However, Bettis had melted, fabricated, and tested several sets of similar ternary alloys. From a metallurgical point of view, it was practically a zero extrapolation to the Zircaloy-2 composition. Patent disclosures were being prepared on these alloys at the time. Besides the subsequent imposition of compositional ranges based primarily on fabrication feasibility, the composition determined then has persisted to the present. It is unlikely that—within the family of such alloy additives as tin, iron, chromium, and nickel—a significantly different composition would be selected today.

Zircaloy-2 and all of the Zircaloys referred to are covered by patent 2,772,964 issued to D.E. Thomas, K.M. Goldman, R.B. Gordon, and W.A. Johnson, all of the Bettis laboratory. In 1972, the American Society for Metals awarded its Engineering Achievement Award to the inventors, Bettis and Westinghouse. In 1985 the American Nuclear Society's Materials Science and Technology Division gave Thomas its outstanding achievement award for "pioneering work in the development of Zircaloy."

It is ironic to reflect that, of the three Group IV-A elements—titanium, zirconium, and hafnium—only zirconium shows this exaggerated sensitivity of hot-water corrosion resistance to purity and structure. In other respects, of course, the selection of zirconium turned out to pay unexpected dividends. The many slip and twinning modes available lead to tough, ductile behavior under most direct or cyclical mechanical stressing conditions, and they make zirconium relatively amenable to most metal-fabrication methods. The high solubility for gaseous contaminants such as oxygen and nitrogen facilitates welding or solid-state bonding processes, and it yields microstructures singularly clean of the inclusions that plague most other commercial metals. The low elastic modulus and low thermal expansion coefficient limit residual internal stresses and thermal stresses, although the poor thermal conductivity is unfavorable.

On the other hand, the anisotropic mechanical and physical properties, because of zirconium's hexagonal close-packed crystal structure, lead to dimensional instabilities. Strength properties at even moderately elevated temperatures are not notable. Hot water or steam corrosion resistance can be maintained only by strict control of chemistry, processing, and environmental conditions—and even then to only modestly elevated temperatures.

Despite the initial bitter complaints of manufacturing people—who predictably denounced the alloy as unmeltable, unworkable, and unusable—Zircaloy-2 was adopted as the reference core structural material for the *Nautilus*. The alloy was chosen, but the development of the fabrication process still had to be completed and time was getting short. Not the least of its attributes has been the reproducibility of its properties and its insensitivity to normal variations and accidents of quantity production.

Zircaloy-3. From a reactor physics point of view, cores with longer life were possible. There was a fear that though Zircaloy-2 had proved satisfactory for current reactors, another alloy was required that would permit these longer lifetime cores. Evidence from tests in the Chalk River reactor in Canada and other tests uncovered the possibility of plant corrosion products depositing on the fuel elements and causing increases in metal surface temperature. This also called for a better alloy.

Experiments showed that a content of 0.25% tin and 0.25% iron would provide corrosion resistance for the nitrogen levels expected. This alloy was optimized for best high-temperature water corrosion resistance. This alloy was superseded shortly by Zircaloy-4.

Corrosion tests on this material showed white corrosion indications, which were determined to be associated with the formation of stringers. Stringers consisted of intermetallic compounds of Zr, Fe, Sn, and impurity elements associated with bubbles of He and Ar incorporated into arc-melted ingots during melting and strung out during fabrication. Although changes in the fabrication process could correct this problem, the lower mechanical properties compared with Zircaloy-2 caused abandonment of this alloy, except for the one Mark I replacement core.

Zircaloy-4. In the course of making measurements on the notch impact toughness of samples of Zircaloy-2, it was noted that those samples slowly cooled from temperatures far below any phase transformation temperature and had much lower toughness than those cooled rapidly from those same temperatures. Examination of the microstructure revealed that the brittle samples showed the presence of a plate-like phase that was absent in the rapidly cooled, tougher samples. Fortuitously, experimenters at Battelle Memorial Institute who were investigating the Ti-H phase diagram (the Zr-H system was unknown) identified a similar plate-like phase as a hydride.

Further investigation showed that the source of hydrogen was atomic hydrogen released during the corrosion reaction, about 50% of that released being incorporated into the samples. Fortunately the molecular hydrogen in the coolant used to suppress CRUD[17] formation is shielded from entry into the metal by the adherent ZrO_2 corrosion film. In the submarine application the amount of corrosion and consequent absorption of hydrogen were sufficiently low over the design life of the core that the absorbed hydrogen level was deemed tolerable.

However, in the course of testing the Zircaloy-2 tubing/UO_2 pellet fuel element being developed for the Shippingport reactor, several chilling facts were uncovered. In the presence of a thermal gradient, such as would exist in an operating fuel element, hydrogen was found to migrate to cooler areas, such as the cladding surface, where it precipitated in amounts far in excess of acceptable limits. In isolated cases, the level of hydride precipitation was sufficient to result in severe local embrittlement. To reduce the level of hydrogen absorbed by the cladding, Ni, which was known to enhance hydrogen absorption, was removed from the Zircaloy-2 composition. The reduction in steam corrosion resistance was compensated by an increase in Fe content. The resulting zirconium alloy, containing 1.5% tin, 0.20% iron, and 0.10% chromium, was christened Zircaloy-4 and has become the workhorse of the PWR industry worldwide.

An arbitrary limit of 250 ppm hydrogen was imposed as an allowable level to accommodate localized hydrogen absorption and thermal gradient concentration. These changes in alloy composition and hydrogen limits were, to a large extent, negotiated over the phone with NRB personnel. The urgency of the schedule dictated that in many cases engineering judgment had to supersede conventional design procedures.

It is interesting that the USSR—lacking feed material of sufficient purity to be used directly, even as an alloy—has continued to use a secondary refining step. They have used niobium, which is less efficient than tin in combating the corrosion deterioration of nitrogen contamination.

---

[17]CRUD is discussed later in this chapter.

Creep Under Irradiation

The phenomenon[18] of creep acceleration due to irradiation and of irradiation growth resulting from the anisotropic properties of the Zr-based cladding were unknown at this stage of reactor development. However, information on irradiation growth of noncubic structure uranium and dimensional changes in graphite blocks in the Hanford production reactors was available and warranted caution. Thus, even though postirradiation mechanical property tests showed irradiation hardening, mechanical designs were based on use of unirradiated properties. It is probably just as well that we were ignorant of in-pile creep acceleration. I doubt that the structural design community at that time had the analytical tools to handle these effects.

Since zirconium[19] is a close-packed, hexagonal material with a high degree of ductility, its behavior in creep under radiation was considered to be a potential problem. R.H. Fillnow[20] did the early investigation of the irradiation-induced changes in the mechanical properties of zirconium. By late 1950, he had designed a capsule that would permit the placement of a variable tensile force on a specimen that, when inserted into the Hanford production reactors, could determine the time-dependent behavior of zirconium at stresses considerably above those of a normal creep test.

If the test could have been carried to failure, the proper test definition would have been a creep rupture test. The first irradiation results were difficult to accept in that there was an almost instantaneous effect on the porportional limit of zirconium. The results were repeatable with successive reactor shutdowns and startups and were reproduced in succeeding specimen capsules of this type. Although much data have been gathered over the years since those early results and scientists have now proven the effect of a neutron flux rate dependency, the initial data showing an almost immediate effect of reactor irradiation on the proportional limit of zirconium still stand in the literature.

Although later experimental work revealed that at low creep rates the creep of zirconium was increased by irradiation exposure, the early creep rupture tests provided necessary information to reactor designers to demonstrate that the mechanical properties would not deteriorate drastically during exposure in the reactor. Later experimental work on the postirradiation mechanical properties of zirconium showed increased tensile strength as a result of irradiation hardening, similar to the early creep rupture tests.

---

[18]B. Lustman, Private Communication dated July 12, 1991.
[19]R.H. Fillnow, Private Communication dated April 1991.
[20]R.H. Fillnow, Private Communication dated September 29, 1991.

In-pile creep tests under the temperature and stress conditions required for water reactors are extremely difficult to perform; no successful creep tests were performed in the NRB program at the time Fillnow left Bettis in early 1973. Stress relaxation tests and in-pile creep tests at Chalk River, together with irradiation growth measurements, have revealed an in-pile acceleration of creep.

Arc Melting

For the Mark I reactor prototype, crystal bar zirconium was chopped up and arc-melted in argon/helium atmosphere using tungsten-tipped, nonconsumable electrodes and water-cooled copper crucibles producing ingots about 4 inches in diameter and 12 inches long. This melting was done entirely at Bettis.

With the advent of Zircaloy-2 and the use of the Kroll sponge as the base melting stock, Bettis used an adaptation of a method that had been employed with crystal bar stock for melting alloys. That is, initial ingots were prepared by nonconsumable electrode melting of mixtures of alloying ingredients and zirconium sponge. The initial ingot was worked down to sheet or rod, chopped, blended, and remelted in a nonconsumable tungsten electrode furnace to produce a second ingot whose homogeneity was acceptable.

At this time, in October 1951, the Bureau of Mines, Albany, Oregon, began melting Zircaloy on a larger scale. Starting ingots were prepared by consumable arc-melting using 2-in. × 2-in. × 2-ft-long electrodes fabricated by briquetting sponge and alloying additions. In early experiments, the initial ingots were then either worked down to sheet; chopped, blended, and nonconsumably remelted; or fabricated into 2-in. × 2-in. bar electrodes and consumably remelted. The former procedure was employed when the soundness and homogeneity of the initial ingot were so poor that remelting with a consumable electrode would be judged inadequate to yield acceptable second-melt ingots.

The fabricated electrode technique turned out to be the most practical method and resulted in adequate ingot quality. This procedure consisted of an initial melt of the briquetted sponge and alloying elements to an 8-in.-diameter ingot, followed by fabrication to 2-in. × 2-in. bar and a final consumable remelt to an 8-in. ingot. About three tons of alloy a day were produced by this procedure.

Meanwhile, experiments were being done with various means of securing adequate homogeneity in single melts by measures such as magnetic stirring of the molten pool; none of these measures yielded ingots of acceptable soundness or homogeneity. There was a strong suspicion, both at

Bettis and at the Bureau of Mines, that the intermediate fabrication of the electrode for the second melt contributed importantly to achievement of an acceptable homogeneity level in the second ingot. It became apparent at Bettis that the Bureau of Mines' production rate of one ingot per day was inadequate to meet the requirements of the naval program. Also, both the AEC and the bureau charters discouraged buildup of activities at government laboratories that could be done by private industry.

Hurford and Gordon conceived the possibility of remelting the second ingot, using the first ingot as a consumable electrode with no intermediate fabrication. Thus, pursuant to a purchase order issued by Bettis, a series of second-melt ingots 12 inches in diameter were remelted by Allegheny-Ludlum by the end of July 1952, using the Bureau of Mines first-melt 8-in.-diameter ingots, with no intermediate fabrication.

The results were completely successful, resulting in almost doubling the yield of acceptable material as compared with the Bureau process. Apparently, contrary to normal expectation, as ingot size increased homogeneity improved, probably because of an increase in the size of the molten pool. At almost the same time, the Bureau of Mines also consumably remelted 6-in.-diameter first-melt units into 10-in.-diameter ingots with similar spectacular results. This double-consumable remelting practice became the standard, not only for zirconium, but also for the titanium industry.

The final step toward today's melting practice was the introduction of vacuum instead of inert-atmosphere melting. As double-consumable electrode arc-melting became increasingly routine, melting rates were increased, resulting in entrapment of bubbles of the inert atmosphere in the ingot. These were troublesome, not only because of formation of stringers, but because of formation of blisters during annealing of fabricated components. Vacuum melting also was desirable because of better removal of volatile contaminants from the melt.

Zirconium Handling

Zirconium is classified as a reactive metal. This was demonstrated in a spectacular way at Bettis in 1953. Since zirconium was scarce, all machining chips and other unused pieces were stored in outdoor concrete block bins until procedures for reuse were developed. While storing a quantity of newly produced machining chips, the exhaust of a forklift truck ignited the zirconium stored in a nearby bin. Adjacent bins also ignited, and the entire Bettis laboratory complex was covered by a glow like thousands of flashbulbs going off together. Fine white particles drifted down over the entire complex.

Dr. Lustman tells the story of his interviewing a prospective employee during the incident. As they were leaving the lab, after seeing the flash of light and with the white powder falling on his head, the prospective employee said he could see that Bettis was an exciting place to work, but he desired a more settled environment. This incident and another at Oak Ridge caused the AEC to set up rigid handling practice guidelines.

This capability of zirconium to ignite, when finely divided and exposed to water, has had to be considered in safeguard studies. The possibility of a violent explosion occurring in a reactor accident has been explored extensively. This was first studied in connection with Shippingport.

The question was raised whether ignition of zirconium, even if locally initiated, could spread in an uncontrollable manner to a major fraction of the zirconium contained in a core. The accident condition under which such a reaction was considered most likely to occur was a loss-of-coolant accident in which a coolant pipe would suffer a major break.

If it could be shown that Zircaloy-4 reacted with water in a predictable manner at all temperatures, up through its melting point, this should assuage fears of unpredictable and uncontrollable reactions. These data were required in less than three weeks. An experiment was devised by Bill Bostrum in which samples of zirconium tubing similar to those used in Shippingport were heated inductively while immersed in water; the released volume of hydrogen was collected and used to monitor the rate of the reaction. Barricades were set up around the water container and a system of mirrors was devised so engineers could control metal temperature during the reaction. Another engineer observed the change in hydrogen volume as it bubbled up through the water column. Temperatures were gradually increased in consecutive experiments until, as the grand finale, zirconium was melted under water and the amount of reaction monitored during melting.

These experiments, completed during the required time period, demonstrated that the occurrence of high-temperature reaction rates could be extrapolated from those measured at lower temperatures, and that no new or unexpected phenomena intervened that would endanger reactor plant safety. Crude as these initial experiments were, the kinetic data derived from them and the conclusions drawn have been supported by subsequent experiments and analyses. Obviously, adverse data could have had serious consequences.

Corrosion Testing

Corrosion testing was relied on to determine the suitability of each piece. It was argued by some that product could be accepted based on composition alone. However, the occurrence of stringer corrosion illustrated the error of this approach.

On the other hand, there were those who felt we were using up part of the life in making tests. Because with longer life cores the testing time is such a small fraction of useful life, one could argue that such a short test does not give assurance of suitable material. However, experience has shown that material produced and corrosion tested under carefully controlled conditions, even for a small fraction of the expected core life, is satisfactory.

Zirconium and Zirconium Alloy Fabrication

In parallel with the development of the crystal bar process, fabrication methods for zirconium were also being developed. Because zirconium was highly reactive to normal air atmosphere, rolling processes had to be completely performed in an inert atmosphere. Pieces of zirconium were initially enclosed in sheet material that was stripped off following the rolling or fabrication process. Bob Gordon and Walt Hurford were most instrumental in development of the fabrication processes—to the point where ultimately full-size ingots could be fabricated into shapes for follow-on fuel element processing.

In early zirconium manufacture, when material of any size became available, it was subjected to a wide variety of mechanical and corrosion tests to establish a basis for use in core design. This characterization of zirconium was almost entirely a Bettis-Westinghouse development.

Hafnium

Hafnium[21] is normally found in zirconium ores. The chemistry of the two elements is nearly identical, so the technology of hafnium has developed along with that of zirconium. Many early processes for separating the two elements were developed on a laboratory basis. However, when it was recognized by the AEC that pure zirconium actually had a low neutron capture cross section, an effort was mounted to develop a separation process on an industrial scale.

By 1951, when the need for hafnium metal became apparent, the Union Carbide and Carbon Chemical Division at its Y-12 plant in Oak Ridge, Tennessee, had developed a liquid-liquid extraction process, which was scaled up first at the Albany, Oregon, station of the Bureau of Mines, later

---

[21]D.E. Thomas and E.T. Hayes, Eds., *The Metallurgy of Hafnium*, USAEC Naval Reactors Division of Reactor Development, 1960.

to be taken over by the Wah Chang Corporation, and then by the Carborundum Corporation and Mallory Sharon Corporation. This process was called the hexone-thiocyanate separation process (MIBK-HCNS), and has proved to be the most successful one.

An alternative process, also a liquid-liquid extraction one, called tributyl phosphate-nitric acid process (TBP-HNO$_3$), was developed by the Iowa State College. This process was then taken over by the Columbia National Company.

Both separation processes yield the oxide form of the metal, which can be processed to sponge and crystal bar in much the same way zirconium oxide is processed.

As a by-product of zirconium production, quantities of hafnium salts of low zirconium content were obtained. This material had only about a 5% zirconium content. It was not known whether the zirconium reduction process could be adapted for hafnium. It was understood that hafnium had a high capture cross section for neutrons, but exact information on this and other properties was not known. The reference control rod material was a silver-cadmium alloy sheathed in stainless steel, but Radkowsky of NRB and W.A. Johnson and Sid Krasik decided to start an investigative program on an informal basis.

Bettis had grave doubts about the ANL silver-cadmium rods. At the time, I wondered how much of this was the NIH factor. Several of these rods were being tested at Bettis and were removed from the test autoclaves in time for a meeting with NRB. The rods showed evidence of the problem that was worrying us. The stainless steel sheath was not uniformly bonded to the silver-cadmium in spite of the use of an intermediate brazing alloy to facilitate bonding. The areas where the cladding was not bonded could be determined by tapping the surface of the rod. If there was a hollow sound, the bonding of the cladding to the silver-cadmium was considered to be inadequate. This was a pretty crude method, but it was validated often by cutting up the rods for a first-hand inspection. Later the control rods were coated with wax; when the rods were heated, the wax melted only where there was good bonding. This lack of bonding would have caused hot spots and further deterioration and perhaps binding of the rods. Failure of the rods to move could be disastrous. Rickover, in a characteristic action, hit the samples with a hammer. The cladding split off, and that was the end of silver-cadmium.

The zirconium crystal bar pilot plant was converted to the production of hafnium in mid-1952, when Zircaloy became available. The control rods were needed in a very short time, if we were to meet the Mark I schedule.

Test results showed that hafnium did not have as much corrosion-resistance deterioration with impurities as did zirconium and absolute corrosion rates were much less. The neutron capture cross sections, as a function of neutron energy and lifetime, proved suitable. The daughter products also were acceptable, actually even more so than silver-cadmium. Hafnium was still scarce—so scarce that all souvenir pieces were collected and made a significant contribution.

The initial scheme for control rod manipulation involved use of a complicated in-vessel mechanical device being developed in concert with ANL. Extensive tests were performed to duplicate the types of motion and wear that would be encountered in the final mechanism; many alloy compositions were tested using autoclaves to duplicate temperatures and wear geometries such as sliding, rotating, etc. Difficulties in performing these small-scale tests due to binding of moving parts illustrated problems that might arise from operation of the large in-reactor mechanism and were a factor in the decision to switch to individually driven control rods.

Fuel Development and Manufacture

The development program for establishing materials and fabrication processes for the STR fuel elements was initiated in 1949. A fuel element group was set up under the direction of Bob Gordon, who reported to William Johnson, manager of the materials department.

Early fuel element development efforts were fourfold:

- Interface with ANL, Battelle Memorial Institute, MIT, and Brookhaven National Laboratory, which had fuel element development contracts directly with either NRB or the AEC.
- Initiate a close in-house relationship with personnel from metallurgy, reactor design, physics, and chemistry—thus providing a forum for establishing basic requirements.
- Set up facilities and install equipment for fuel element development.
- Carry out development efforts aimed at establishing processes for fabricating fuel and structural components.

Perhaps the most significant of the problems, in the early days, was the complete lack of metallurgical processing information for zirconium as discussed previously. Not only was the concept of the utilization of zirconium for a nuclear application born with this program, but also the need to develop its complete physical and processing metallurgy in a time span of no more than two years.

A second problem was the lack of zirconium material with which to work. Previous sections have detailed the history of producing, initially, crystal bar zirconium and later alloys of zirconium. While progress in this feed material program was outstanding, it nevertheless provided only limited quantities to support the fuel element and structural component development efforts. Thus fuel element scientists lived on a hand-to-mouth basis with respect to availability of zirconium. This necessitated early development being performed on subsize components.

Another noteworthy problem had to do with the recruiting effort necessary to bring the urgently needed experienced technical personnel aboard. Much credit must be given to Westinghouse in that it was the sister divisions of Bettis, mainly the materials engineering department at East Pittsburgh, and the central research laboratories that gave many of their most promising engineers and scientists to the fuel development effort.

Mark I Prototype Fuel Development

This program, carried out under the direction of Bob Gordon, comprised the following major efforts: zirconium ingot melting and fabrication, under Walt Hurford; fuel filler process development under Lou Carapella; fuel components (elements, subassemblies, and clusters) and structures development, under John Theilacker; inspection practices and quality assurance, under Lloyd Kramer.

Fuel Filler Development. The basic building block of the core fuel component is the fuel element. In its design, fuel is totally encased in a sheath of zirconium (or zirconium alloy). Techniques utilized in producing a compositionally uniform alloy of uranium and Zircaloy to form a "fuel filler" for the element have been termed the *fuel filler* process.

Many problems of either a technical or nontechnical nature were encountered in this phase of the program, in which processes for fabricating uranium were put in place. It is necessary, however, to refrain from discussing those of a technical nature for security classification reasons. Practically all information in this phase of the effort was classified in the early days and remains so today.

Nontechnical problems requiring solutions included accountability for uranium, criticality control, and protection of workers and the environment from the dangers of radioactive contamination. All work with uranium, either natural or enriched, was carried out in separate isolated rooms in order to control contamination, and to provide a level of accountability that minimized uranium material loss. Strict accountability for uranium was practiced primarily with the introduction of enriched uranium. From that point on, every gram of material had to be accounted for, requiring daily physical inventories and a system of bookkeeping that not only showed the location of all enriched uranium but also specific quantities.

In addition, a detailed separate physical inventory was independently conducted each month by the Health Physics Department. During this inventory, production was shut down and resumed only when health physics was satisfied that complete accountability was assured. Accountability was required not only because of the value of the enriched uranium, but also as a nuclear weapons proliferation safeguard.

The physics department aided in handling the criticality problem by laying out critical zones in the facilities and specifying the amount of enriched uranium that could reside in each zone. The criticality zones were separated by heavy wire fences. Material could be moved from one zone to another only when authorized by the criticality control supervisor. If too much enriched uranium was in one place, a self-sustaining chain reaction might have taken place, causing both physical and radiation damage.

We were especially mindful of radiation and contamination hazards when handling uranium in any form. Not only were the zones physically isolated but each had its own air-handling system. Health physics personnel continuously monitored these areas. Personnel were required to wear special clothing with monitoring devices (dosimeters) attached to each employee to measure levels of radiation received. Also analyses were conducted at regular intervals of the urine of each employee.

Many factors have contributed to the success of Westinghouse in the nuclear business, but one that must be recognized is its ability to handle uranium and overcome the complexities associated with working this material in its various forms—and also similar fuels such as thorium and plutonium. Westinghouse has never, in any of its nuclear fuel endeavors, experienced any adverse publicity through fuel-related catastrophes, i.e., significant unaccountable losses of fuel materials, criticality incidents and excessive contamination exposure to workers, the environment, or the public.

This was due to the development of a proper attitude toward the handling of uranium. This attitude prevailed at Bettis from top management to the technicians in the laboratories. It demanded and received, from those who handled the material, a discipline that precluded sloppy practices, deviation from written procedures, risk taking, etc. It is this attitude that has been carried forward to subsequent fuel programs in both the naval nuclear and commercial nuclear programs. This was a factor in Westinghouse's success in being selected by the government to operate many of its laboratories and plants, such as West Valley, Idaho Chemical, Fernald Feed Material, Hanford Engineering and Development Laboratory, and Savannah River.

Fuel Components and Structures. While reactor design had considered many configurations—rods, plates, spheres—for the reference fuel element design, the selection of plate was made early, permitting development to proceed without dilution of work effort on alternative shapes.

Probably the most difficult of the many problems encountered was the requirement that the fuel filler had to be completely sheathed (surfaces, ends, and edges) with zirconium. In addition, all interfaces between the constituents (fuel filler and cladding) of the element had to be metallurgically bonded. Contributing to the solution of this problem were efforts carried out in the supporting laboratories, ANL, Battelle Memorial Institute, and MIT. But the most significant gains and final breakthrough were achieved at Bettis.

Numerous problems also had to be resolved in the development of processes for the fabrication of the subassembly and cluster components, i.e., joining a number of fuel elements into a subassembly and then a number of subassemblies into a cluster. One of the basic problems was establishing dimensional tolerances, and standards and specifications whereby these components could be manufactured and inspected. These parameters were finally established, but only after developmental fabrication of numerous components and analysis by the design group of information gained from their inspection.

Development of welding parameters for fabricating subassemblies required exhaustive analyses in which the resulting quality of welds and dimensions achieved were compared with parameters utilized. The weld parameters finally selected were optimized to give controlled shrinkage so that further machining would be eliminated and minimum distortion would occur. This development also included the design, manufacture, and placing into operation of large, atmosphere-controlled weld boxes, along with the associated jigs, fixtures, vacuum pumps, and power supplies.

Without question, the most critical fabrication steps in the entire process were the final two applied to the cluster, specifically—the acid pickle and the corrosion test. Failure to properly execute these operations could result in rejection of this large component—and with this the tremendous cost of materials and about two and a half months of processing time. The pickling operation was carried out in a large bath of hydrofluoric acid, nitric acid, and water. This was followed by a rinse in distilled water and finally the corrosion test. In this final operation, the cluster was loaded into a large autoclave, where it was exposed to 680°F water at 2700 psi pressure for three days. Personnel monitored this corrosion operation around the clock. Following the test, the cluster was removed from the autoclave for final inspection according to specifications provided by reactor design.

Inspection and Quality Control

Bob Gordon established an organization, independent of production, to develop the extensive and special inspection equipment and techniques required and to carry out the inspection and quality control function. Many of the engineers assigned to this group had special nondestructive test qualifications. In addition, considerable assistance in the development of equipment was provided by personnel from instrumentation and control, chemistry, and physics.

Special items of equipment developed to provide critical data included the following: a scintillation counter for determining fuel homogeneity in the fuel component; a fiber optics probe for visual inspection of weld penetration; a corrosion product in the coolant channels of the subassemblies and clusters; and a coolant channel inspection gage, which was multiprobe and gave the thickness measurement of coolant channels of subassemblies along their entire length.

Also developed was a system of release points in the fabrication of components. Processing of materials and components could not proceed beyond these release points until all inspections up to that point showed that intermediate requirements had been met.

Manufacture of the Prototype Fuel Components

While development of fuel components was proceeding at an accelerating pace during 1951, the decision was made to establish a separate facility dedicated solely to manufacture of fuel components. Since the Mark I core was needed at the construction site in Idaho by late 1952 and would take five months to manufacture, a date of June 1952 was set for operation of the new facility.

Thus, beginning about July 1951, the technical personnel of Bob Gordon's fuel element development section were devoting efforts, not only to materials and development efforts, but also to working with Lou Mechling's works engineers on a design for a building providing floor layouts, meeting service requirements, etc. In addition, corporate personnel were again brought to Bettis to handle procurement of equipment and other hardware. Many major items of equipment—such as forge presses, rolling mills, furnaces, machine tools, shears, and roller levelers—were expeditiously ordered, and promises of equipment delivery were extracted from vendors that were in most cases one-fourth of the normal delivery time.

Groundbreaking of the 19,000-ft^2 manufacturing facility, the "F" building, took place in November 1951, and the building was completed in May 1952. Equipment installation was initiated while the building was still under construction. Occupancy by fuel element personnel took place in late

May, and actual fuel fabrication with enriched uranium was initiated in July.

Many other efforts were carried out in the early to middle months of 1952 in preparation for manufacture of the prototype fuel components. Personnel were hired and when security clearances were obtained, indoctrination and training programs were begun. A system of paperwork was implemented to control movement of material and also to provide detailed procedures for carrying out each step of the manufacturing and inspection program. Because the quantity of enriched uranium in work at any time was substantial, a system of criticality control also was established in the fuel element, subassembly, and cluster areas. Bettis had by far the largest quantity of enriched uranium anywhere outside the weapons program.

In November 1952, just barely in time to meet the construction schedule, the last cluster was successfully completed.

Development and Manufacture of the Prototype Control Rods

Material selection for control rods was covered previously. The late decision to use hafnium for control rods, after failure of the silver-cadmium development, posed a major problem. Personnel had to be reassigned from fuel component development to the hafnium program. Later, a silver-indium-cadmium alloy for control rods was developed.

Completion of the hafnium development program in time to meet the construction schedule required that we solve a number of problems, many not even identified when development began. Again, the material, hafnium, was in exceedingly short supply. Complete success was achieved in the development of processes to fabricate the cruciform-shaped rods. Almost every applicable machine tool at Bettis was made available for the manufacture of hafnium control rods. These rods were delivered to the construction site in February 1953, about one month before criticality.

Fuel Component Development and Manufacture

Following completion of the fuel and control rod components for the prototype, development began on the *Nautilus* components. Many changes resulting from information gained in the prototype program were incorporated in the design, fabrication, and testing of essentially all components.

As detailed previously, a major change was using Zircaloy-2 in place of crystal bar zirconium. This material required numerous changes in parameters for processing fuel components, but because of its superior corrosion resistance it represented a significant step forward in manufacturability and inspectability of product.

Manufacture of the full complement of fuel components and control rods for the *Nautilus* was completed without difficulty and on schedule—as was the replacement core, the manufacture of which followed immediately.

Development continued in the fuel laboratories at Bettis on fuel for subsequent naval programs. Because of limited manufacturing capacity, only initial or prototypical nuclear cores were manufactured at Bettis. All follow-on fabrication was done in the vendor facilities Bettis had brought into the program. This included the Westinghouse fuel facility at Cheswick, Pennsylvania.

In early 1956, in response to the need for additional manufacturing facilities for the ever-increasing number of nuclear cores for propulsion of naval ships, a vendor development and qualification program was initiated. Contractors interested in participating were given the opportunity of qualifying. For the vendor this required establishing complete facilities, hiring and training employees, setting up the necessary quality control, security, and criticality systems, and finally fabricating a representative number of samples submitted to Bettis for acceptance evaluation. In this manner Babcock & Wilcox, and Combustion Engineering, Metals & Controls, Nuclear Materials, Olin Mathieson also became part of the naval program.

Bettis' responsibility in this vendor program was to provide close surveillance and audit of manufacturing operations carried out in these vendors' shops and also to monitor quality assurance to ensure product acceptability.

Water Chemistry Development

Water chemistry problems have plagued the nuclear power field from the early days. The problem is not so much corrosion causing structural failure of the material. It is rather corrosion products getting into the primary coolant water, causing seizing of moving parts due to contact corrosion or passing through the reactor core, and becoming radioactive. Radioactive material in the water then flows out of the reactor core and through the remainder of the system. This radioactive material, distributed throughout the primary coolant system and its auxiliary systems, increases the shielding problem and also makes maintenance more difficult. There is also the possibility that corrosion products may become deposited on the heat-transfer surfaces of the core itself or the steam generators and thus reduce heat-transfer efficiency, causing higher fuel element temperatures and increased possibility of corrosion.

This preferential precipitation[22] of corrosion products on heat-transfer surfaces was in fact noticed during operation of a test fuel assembly in a

---

[22]B. Lustman, Private Communication dated July 12, 1991.

loop in the Canadian NRX reactor in February 1952. A thick black film of iron oxide was heavily and preferentially deposited on fuel element surfaces. The test loop had been constructed of stainless steel, which was supposed to be corrosion resistant under the existing conditions. However, the deposits were determined by analysis to be corrosion products of construction materials in the loop. They were given the name CRUD (Chalk River Unexpected Deposits), a name that has survived to this day in water reactor engineering and mythology. If this contamination could not be prevented, reactor cores would have to be decontaminated about once a month, clearly an impossible situation.

Examination showed that there was more contamination on the sample from a radioactive environment than on a similar sample tested out of reactor. There was concern that the reactor environment might be the major factor. It was later shown that, with the proper water chemistry, this CRUD buildup could be avoided. Again, what might have been an insurmountable obstacle was overcome. Water chemistry work at Bettis, and later in the Westinghouse commercial activity, has been outstanding. Many of the world's reactor problems would have been avoided with proper attention to water chemistry—notably, steam generator tube failures.

Fortunately, water chemists, under the leadership of Paul Cohen, determined that the amount of corrosion products and their deposition could be controlled by maintaining an overpressure of H_2 in the coolant and by operating with water pH levels of 10. Proper water chemistry is essential for continued satisfactory operation of PWRs. Failure to provide it was the main contributor to steam generator problems.

There are strict standards for the purity of primary coolant water. Suitable analytical techniques have been set up to monitor the amount of impurities present at all times. The sampling system provided permits samples to be obtained from any part of the plant. Sufficient chemical analytical apparatus is provided on the submarine to make the necessary tests.

Due to the work at Bettis, the CRUD level was reduced to parts per billion. Of particular concern was the question of nucleate boiling on core surfaces and any deposition of supersaturated chemicals added to the primary water. Although naval reactor personnel were deeply involved with decisions relating to primary coolant water chemistry development, the technical bases for the chemistry that evolved were developed solely by Bettis personnel. The Navy now has the most advanced water chemistry in the world, developed by Westinghouse personnel.

A successful decontamination program was developed. Early cores had considerably more corrosion products than desired; corrosion added to the radiation level in the reactor compartment. The first primary system decontamination was carried out on the Mark I plant in the late 1950s. This procedure was developed to a degree that, by the mid-1960s, cores could be routinely decontaminated without damage to the core or pipe loops and hundreds of curies of activity could be removed. An example of how the work being done at Bettis led the field came to light when the need arose in the late 1960s for a Canadian reactor plant to be decontaminated. Their method of decontamination was at least 10 years behind that at Bettis.

Secondary plant water chemistry was also a significant development. Thus, it is possible with the proper water chemistry to reduce the problem of steam generator tube leaks. The proper water chemistry for the primary and secondary are different. It is a significant advantage of PWRs over BWRs that the appropriate water chemistry for both the primary and secondary systems can be used. Further, there is no problem of additives coming out in a PWR and getting into the secondary system as may be the case in the BWR. Essentially all of the chemistry developments were achieved under the direction of Paul Cohen.

Development of Standards and Procedures

Refueling Methods

Bettis-developed off-the-head refueling achieved a high degree of success with the S5W design. NRB's insistence, however, on going to a module-type through-the-head refueling scheme, from the mistaken belief that modules could be replaced at intermediate times of core life, did not result in a truly viable means of refueling. NRB could not be convinced, until a great deal of effort and money had been expended, that modules could not be replaced beyond 25% to 30% of core life. This was due to the power perturbation incurred, necessitating a severe downgrading of power rating for the remaining lifetime of the core.

As a result of NRB's insistence, two long-lived cores, one for a submarine and one for a surface ship, were designed with this capability. By the time the first refueling of the surface-ship long-lived core prototype took place, Bettis was able to prove to NRB the distinct advantage of off-the-head refueling, and this core was then refueled by a novel technique developed by Bettis. The head array, which supported the modules, was disconnected and lifted off as a unit, with subsequent refueling being carried out by the off-the-head method. George Hardigg and John Yingling played major roles in this development. Finally, with the development of an advanced surface ship core for destroyer application, Bettis was able to convince NRB that off-the head refueling was faster and cheaper than modular-type refueling. It reduced the refueling time by more than half of that required for the module refueling method.

As could be expected, the Advisory Committee on Reactor Safeguards (ACRS) was rather surprised to find out that the Navy was refueling ships in harbors in populated areas without containment similar to that employed in commercial nuclear plants. In spite of several reviews by ACRS and actual viewing of refueling operations, no action was taken to change the refueling method for naval plants. The ACRS, in its review, was quite impressed with the control exercised in the Bettis-developed procedures and the highly qualified Bettis personnel involved in all critical refueling operations.

Welding Standards

Bettis interaction with the shipyards increased in 1961, when the Navy welding standard that was written at Bettis under the direction of R.H. Fillnow was introduced. Until that time, and including seal welding efforts, the standard approach to solving a problem was all too often a repetitive welding of mock-ups until a solution was developed. That is the approach that was frequently seen at various shipyards and, often, to the severe detriment of the program.

With the introduction of the Navy welding standard into shipyards in early 1962, rather chaotic conditions developed. With the new standards, shipyard rejection rates climbed and were sometimes as high as 90% in carbon-steel systems. At this point, Naval Reactors asked Bettis for assistance. Fillnow formed a group under Paul Seran, who developed lectures and lecture aides and then went to shipyards and taught people on the welding floor how to weld. It was distressing to find out that up to that time, welders had their work rejected but were not told for what reason or how to eliminate a particular defect in future welding. The group traveled from shipyard to shipyard with the major thrust at the Portsmouth Naval Shipyard, where they experienced the greatest reluctance to change the way of doing business.

With the sinking of the *Thresher* in 1963 and establishment by the Navy of a strong position on improvement of quality, cooperation received in the shipyards improved tremendously. Paul Seran, Joe Gipson, and Lou Sorg literally spent months teaching people in the shipyards how to weld, how to perform radiography, and how to accomplish an adequate magnetic particle test. With this continuous effort, quality of workmanship began to climb and reject rates began to decline. It took about three years, but one shipyard that previously had a 90% reject rate on carbon-steel pipe decreased it to 10%.

Although not as reluctant as shipyards, commercial organizations were hesitant to accept the requirements of the new Navy welding standard. One of the most forward-looking commercial organizations was Babcock & Wilcox. Within a short time, they appreciated the fact that people qualified to the new standard produced a much higher quality with a much lower rejection rate; they understood it was to their advantage financially to keep the rejection rate down. B&W was the first to take formal action to have all welders for both naval and civilian applications qualified to the Navy standard.

As part of the introduction of the welding standard, a system had to be developed by which inspectors were qualified. Bettis developed the complete program for qualification of test examiners, who in turn were empowered to qualify the necessary inspectors in their organization. Bettis was responsible for setting up training schools, developing examinations, and finally acting in the role of the Navy by granting qualifications to test examiners. It is interesting that many recognized "experts" in the field could not initially pass the qualification tests. One person was a national officer in the Non-Destructive Testing Society and recognized worldwide as a radiographic expert; he took the test examination three times before finally passing.

Paul Seran played a major role in development of the training program and qualification of test examiners. The Navy had such respect for Seran that he was permitted to act as final arbitrator of disputes on interpretation of either the standard or the results of radiography. He was a most important cog in institution of the standard—both in shipyards and in vendors' plants.

Bettis also played a very important part in development of seal welding procedures for welds to be made around penetrations in the reactor vessel head. Until 1963, development of procedures was made within projects without too much support from the materials organization. At this time, while completing the refueling of a plant at NRTS that was intended to serve as a proof-test for the forthcoming *Enterprise* refueling, we observed unusually severe cracking in many of the seal welds. These welds were finally closed with great difficulty using a manual procedure. Clearly, the problem encountered was metallurgical in nature and required an in-depth investigation of materials properties and solidification phenomena. Since the refueling was to take place in six months, Fillnow took personal responsibility for heading a task force to develop a solution. For the first time ever, the problem was attacked from a theoretical viewpoint. Investigated were solidification kinetics, arc physics, and basic physical and mechanical metallurgical phenomena.

Within three months, the welding machine was completely rebuilt, a procedure was developed and qualified, and welding materials—including mock-ups—were in readiness for refueling the *Enterprise*. Whereas initially it took more than eight months to make all necessary seal welds on the first fueling, but they were done in five weeks at the refueling. This success, together with the Bettis effort to teach shipyard welders, earned the Navy's deep respect.

Nondestructive Testing

It was necessary to upgrade the capabilities of shipyards and suppliers on radiography and ultrasonic testing. Paul Seran and Joe Gipson helped develop the necessary techniques and obtain acceptance by industry. Much of the basic development of ultrasonic techniques was performed at Bettis. Introduction of Inconel Alloy 600 in 1964, as a major material of construction in pressurizers and steam generators, caused a variety of welding and inspection problems.

We played a singular role in solving two problems associated with heater well tubes on pressurizers and the welding of steam generator tubing into the tubesheet. With the pressurizer, phases of weakness were developed in the heavy walled tubing due to the fabrication process. A critical amount of cold work, followed by a reheat treatment, caused grain boundary diffusion and alignment of carbide particles to form a plane of weakness. This type of tubing was highly susceptible to cracking and potential rupture. Bettis developed a discriminating ultrasonic technique and led a team into shipyards and vendors' plants, with the result that many heater well tubes were replaced. Random testing proved they were defective in every case.

Steam generator weld fissures were observed in the tube-to-tubesheet weld and threatened to delay delivery of many steam generators to the fleet. Bettis developed a discriminating ultrasonic test method for critically examining and repairing these welds, with the assurance that the repair was high quality. With the development of modified weld procedures, International Nickel Co. also developed improved Inconel filler-rod material. These procedures and inspection methods are still employed in the manufacture of naval steam generators.

Fracture Mechanics

In 1961, we knew that some method of defining brittle fracture behavior of reactor vessels, other than use of Charpy data with nil-ductility transition temperature, had to be developed. Mike Manjoine at the Westinghouse central research laboratories had created a small specimen, called the wedge-open-loading (WOL) specimen. Its applicability to reactor vessels and, in particular, to irradiation damage of reactor vessels had not been investigated. Bettis had little capability for development or analyses of fracture mechanics as a design tool.

In early 1962, John Chirigos was selected by Fillnow to set up an organization including outstanding people such as Bob Cloud and Fred Nichols. Within a year, they were clearly established in the forefront of fracture mechanics knowledge in the United States. People working with Manjoine characterized the WOL-type specimens and their relationship to Charpy data, and with the help of Paul Daly's group they developed much more discriminating analytical techniques by which the brittle behavior of reactor vessels subjected to irradiation could be determined. This contribution culminated in a contract with ASME in 1972 to write a section of the new ASME code on fracture mechanics.

Design Basis and Stress Criteria for the Pressure System

The degree of integrity and the need for safety in any nuclear plant far exceed those in any fossil-fueled power plant or most chemical processes. A catastrophic accident in the prototype could well stop the whole nuclear power program. Boilers and pressure vessels had been designed and used for many years under the rules of the ASME Boiler Code, which had proven adequate for conventional fossil-fueled systems and chemical plants.

In 1950, the ASME code provided simplified procedures that were adequate for conventional pressure vessels. In 1950, it did not call for a detailed stress analysis, but merely set the wall thickness necessary to keep basic hoop stress below the tabulated allowable stress. It did not require a detailed evaluation of the higher, more localized stresses known to exist, but instead it allowed for these by the safety factor and a set of design rules. Thermal stresses are typically far less demanding in conventional steam power plants.

In the early 1950s, B.F. Langer at Bettis, W.E. Cooper at KAPL, and J.L. Mershon at the Bureau of Ships realized that nuclear power plants required a much more sophisticated procedure for stress analysis. They developed a new set of design criteria that shifted emphasis away from use of standard configurations and toward detailed analysis of stresses. Setting allowable stress values required dividing stresses into categories and assigning different values to those categories.

The result was "Tentative Structural Design Basis for Reactor Pressure Vessels and Directly Associated Components (Pressurized Water-Cooled Systems)," first issued in December 1958. In 1962, it was revised again for Navy use and was given security classification. However, the unclassified version of this and the Navy welding standard were used by ASME committees as the basis for writing Section III of the Boiler Code, which is now the requirement for all primary vessels and many of the secondary systems used in nuclear power plants in all states and is widely used in Europe also. New rules require:

- Detailed calculation and classification of stresses and application of different stress limits to different classes of stresses whereas the old rules gave formulas for minimum allowable wall thickness.
- Calculation of thermal stresses and a provision for allowable values for them.
- Maximum shear stress theory of failure instead of the maximum stress theory.
- Consideration of the possibility of fatigue failure and rules for its prevention.

The work of Langer, Cooper, and Mershon formed the basis for design criteria now used for all nuclear power plants in the United States and most of the plants built in other parts of the world.

Funding Nuclear Projects

The relative magnitudes of problems overcome can be understood by looking at the general breakdown of funds spent in nuclear engineering over many years and covering several projects including the STR (Table 12.1).

Roughly, 28.1% has gone into mechanical engineering including heat transfer; 36.8% metallurgy and metallurgical engineering work; 11.2% theoretical and experimental physics work; 11.3% electric and electronic engineering development; 6.9% chemistry and chemical engineering; and 5.7% operational engineering and testing. The main effort is clearly in fields of specialization other than physics. This lesson learned on the submarine project stood Westinghouse in good stead, as we prepared for the nuclear rocket contract competition in later years.

Even if one considers that since that time inflation has increased costs by a factor of 10, this was a remarkably small price for a development that revolutionized naval warfare.

Table 12.1
Costs of Nautilus and Its Land-based Prototype*

Prototype	(in $ Millions)
AEC research and development through start of prototype	$ 57.4
AEC cost for construction of prototype	27.3
Navy research and development through delivery of ship	18.6
TOTAL	$103.3
AEC cost for shipboard nuclear reactor plant	16.3
Navy cost for construction of ship	58.2
TOTAL	$ 74.5

*R.G. Hewlett and Francis Duncan, *Nuclear Navy*, University of Chicago Press, Chicago, 1974.

CHAPTER 13

The Shippingport Technical Story

Plant Description[1]

The basic schematic diagram of the PWR reactor and steam plant is shown in Figure 13.1. The principal elements of the reactor plant are the reactor vessel, containing the nuclear core, and four main coolant loops that circulate reactor coolant water between the core and the steam generators, with a pressurizer to prevent boiling of the water coolant. With three loops operating, the first core of the reactor produced approximately 225 MW of heat, which was converted to 60-MW net electrical output.

The reactor core, the heat-producing unit was composed of two types of fuel elements assembled into a right-circular cylinder. The active portion of the core, that part containing fissionable material, was about six feet high and about six feet seven inches in diameter. The core was of the *seed-and-blanket* design. Highly enriched uranium, which forms the "seed," was contained in 1914 zirconium-clad plates, the active portions of which were 2.05 inches wide by 70.75 inches long. The "blanket" contained natural uranium, in the form of uranium oxide, in about 95,000 fuel elements; each element was a Zircaloy-2 tube, 0.411 inches in diameter and 10.25 inches long, filled with natural uranium-oxide pellets. The fuel element tubes were so short because at that time it was difficult to obtain longer tubes without defects.

Design steam pressure was set at 600 psi, the highest reasonable figure for existing reactor technology at that time. The primary system coolant

---

[1]Based on data from Naval Reactors Branch of the AEC, *The Shippingport Pressurized Water Reactor*, Addison-Wesley Publishing Company, Reading, MA, 1958.

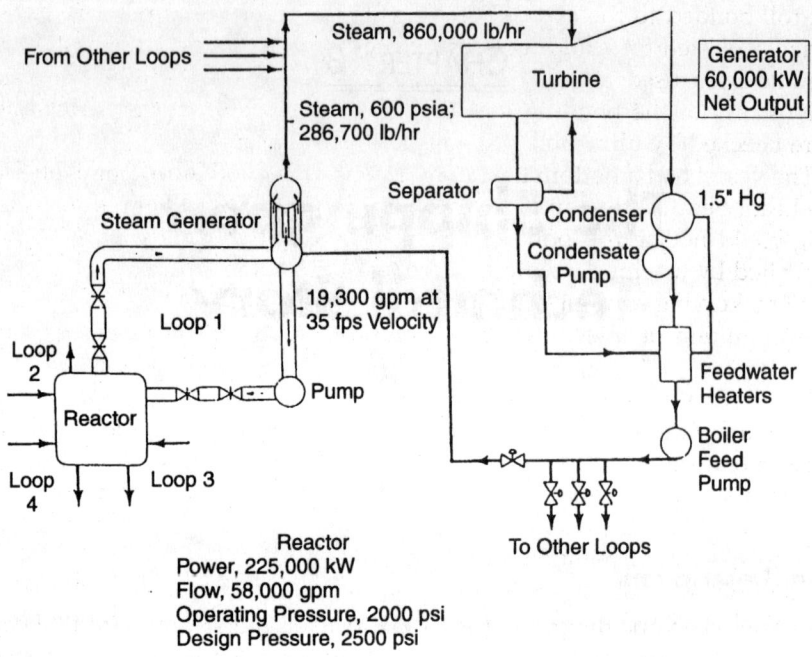

Figure 13.1 Basic schematic diagram of the Shippingport plant.

pressure was set at 2000 psia. The saturation temperature at this pressure is 636°F, which was considered satisfactory on the basis of then-current data on Zircaloy corrosion.

Reactor Vessel[2]

The core and its 32 control rods were housed in a reactor vessel that measured 109 inches in internal diameter and about 32 feet in inside height, 8-3/8 inches thick, and weighing 264 tons, which was about the maximum size industry could manufacture. The limitation was basically a function of the steel making and rolling capacity of the steel industry.

Direct contact of the carbon steel with the high-purity primary coolant was prevented by cladding the vessel interior with stainless steel that was

---

[2]Based on data from Naval Reactors Branch of the AEC, *The Shippingport Pressurized Water Reactor*, Addison-Wesley Publishing Company, Reading, MA, 1958.

hot-roll bonded to the carbon steel during the slab rolling process. No experience existed for cladding heavy plates by this process, since roll bonding had been used for only light plate and sheets on a production basis. The quality of the bonding and the lack of inclusions in the carbon steel were checked by ultrasonic tests of the finished plates.

The vessel had a bottom hemispherical head and a flanged hemispherical top head. Because the top head was 10 inches thick, the stainless steel cladding could not be roll bonded with existing equipment. Its cladding was deposited by machine welding. There had been some doubt about the ability to make a big enough vessel; however, these doubts were set aside based on recognition of relevant Babcock and Wilcox experience during World War II.

However, it was recognized that improvements in industrial capacity for heavier and larger pressure vessels would be required if significantly larger nuclear reactors were to be built. This might well be the limiting factor in the size reactor a country could build. This was one of the reasons for trying to determine, at the Atoms for Peace Conference in Geneva in 1955, the Soviet capacity for manufacture of large vessels.

Solutions of problems in design and development of the reactor vessel and head involved not only proof-tests to confirm design calculations but also extensive quality control tests—particularly on material—to assure a reliable product. Typical tests of both types are highlighted in this section, beginning with some of the production control tests on materials and extending to proof-tests on the closure seals and vessel head stresses.

Material Quality

The basic material selected for the PWR reactor vessel and head was Type SA-264, which is a modified ASTM-A-302, Grade B, carbon steel containing manganese and molybdenum. This material was selected to obtain high strength with good ductility and fracture toughness. It was clad on the interior with Type 304 stainless steel to obtain good corrosion resistance.

During the production of plates for the reactor vessel, tests were required for chemical composition, physical properties, and Charpy V-notch impact characteristics, the latter to assure good fracture toughness and ductility and give an indication of susceptibility to fracture in a brittle manner. The problem of brittle fracture had to be investigated in order to assure maximum reliability of the vessel during operation and the required life expectancy under neutron bombardment.

Although the Charpy V-notch impact test does not give direct information about the brittle fracture tendency, test data from the Naval Research Laboratory indicated that the nil-ductility transition temperature (NDTT) can be determined as that temperature at which Charpy V-notch specimens break at 30 foot-pounds impact. Because the temperature at which the material will fail in a brittle manner will vary between heats, this temperature had to be determined for all heats of vessel material; steel mills were requested to aim for a minimum of 15 foot-pounds at 0°F for the Charpy test.

Assuming that the vessel—in spite of all the precautions taken during its manufacture and handling—had a small imperfection, possibly in a weld, that could trigger a brittle fracture, a minimum temperature was established at which a fracture started in a plastically stressed region would be arrested in the elastic region. This temperature is identified as the fracture tension elastic (FTE) temperature and is approximately 50°F above the NDTT for this material. Minimum pressurizing temperature for the Shippingport vessel was set at 110°F on this basis.

The effect of radiation on material properties was investigated. Welding rods were copper plated to prevent rusting. This resulted in some copper in welds of thick plates that made them more sensitive to neutron flux. Results from in-pile tests of the vessel material indicated that the effect was only a ductility reduction of less than 10% under an integrated flux of 10^{18} neutrons/cm^2 cold. Effects of radiation on the material during operation were less, because intensity of radiation on the vessel was considerably below that used for the test; therefore, no corrective action was required to prevent radiation damage for the 20-year life of the plant. For a seed-and-blanket core, the neutron flux decreases in intensity from the seed exponentially through the subcritical blanket to the reactor vessel wall.

Closure Seals

To prevent leakage from the reactor vessel, a flexible type of seal (generally called an omega seal) was developed for welding the two mechanically joined components of the vessel and head. This seal consisted of a stainless steel tube, with a longitudinal slit in it, formed into a torus with the internal diameter of the torus equal to the vessel closure diameter. It was welded in turn to the vessel and the head. Integrally machined torus configurations were used for the smaller penetrations for control rod mechanisms and refueling ports in the vessel head.

The adequacy of such a sealing arrangement could be demonstrated only by proof-testing. While calculations for the vessel seal ring indicated acceptable steady-state stresses, the maximum "apparent" localized stresses during transient conditions were calculated to be about twice the 17,500-psi yield point for the material at 600°F. As a result, tests were made to establish endurance characteristics of the seal. These included room-temperature tests to destruction as well as tests at operating temperature and pressure on full-scale seal prototypes. This work was done by Combustion Engineering, as part of its design and manufacture of the reactor vessel and head.

In room-temperature tests, the vessel seal endured more than 3300 cycles of larger than expected rotations and displacements before failure. The operational tests were performed on a foreshortened full-diameter prototype vessel and head. The vessel and head were given a 3750-psi hydrostatic test and then thermally cycled 19 times between room and operating temperature at varying heating rates between 10°F/hr and 250°F/hr. The seal was then cut open and rewelded, and pressure cycling tests were begun with a target of 20,000 cycles with pressure variations between 1000 and 2500 psi. The latter test provided confirmatory information about the use of the seal after rewelding. The seal, however, was so designed that a new seal ring could be welded in place when necessary.

Cycling tests used for proof-testing the closure seal also permitted study of stresses in the prototype reactor vessel and head to establish acceptable heating and cooling rates.

Photoelastic Stress Tests

The Shippingport reactor vessel and head contained a total of 46 closely spaced penetrations. Analytical solutions for stress distribution in the head were difficult to obtain due to lack of symmetry, and verification tests were consequently performed. The tests were twofold and consisted of photoelastic tests for early design needs and full-scale tests on a steel model fitted with strain gauges for final stress verification. During the early stages of design, a need existed for analyzing several head penetration patterns, thus requiring the casting of several head models made from epoxy resin. This was a costly and time-consuming process, in part because of the curing necessary to eliminate undesirable casting defects in the heavy plastic sections.

As a result of early work with three-dimensional photoelastic models, it was found by personnel at Brown University working on the problem that two simplifications could be made. The first step was to eliminate the heavy model sections by analyzing only a portion of the vessel head containing the desired penetration pattern for influence of thickness on stress patterns. The absence of influence from the head flange moments permitted simulation of pure membrane stress at the edge of the test specimen. Small leakages were planned at the test plate edges to avoid influence from edge moments caused by contact. After loading the sector in this manner, stress patterns were frozen in the material, which was then sliced for analysis.

The investigation showed that stress concentration for an early penetration pattern involving a 32% ligament was 28,000 psi for a 10-inch-thick head.

As the second step, by comparison of three-dimensional tests on a sector with tests on two-dimensional specimens in tension, it was found that for a given head thickness a simple two-dimensional photoelastic model of a local area could be used to show changes in stresses caused by minor changes in penetration patterns. When the finally selected penetration pattern was thus compared with the early one, it was found that stress distributions were quite similar and that a 10-inch-thick head would be adequate under the prescribed design pressure.

Maximum calculated stress existing at the flange-to-head junction was found to be 31,500 psi. Of this, 5800 psi was a membrane stress and 25,700 psi was due to bending and included stress risers. Three-dimensional photoelastic tests on the complete model showed general agreement with the calculated stress value.

Specifications and Standards

The work on low-cycle fatigue by B.F. Langer[3] described earlier, which involved large thermal stresses imposed on nuclear reactor pressure vessel walls during their lifetime, was used in design of the reactor vessel. Early experiments and theoretical work done by Langer indicated that normal pressure vessel design methods prescribed by Section I of the ASME Unfired Pressure Vessel Code[4] needed revision. This work and supporting work done by pressure vessel manufacturers led to the development of a new section (Section III) in the ASME Unfired Pressure Vessel Code.

Specifications and standards similar to those for the naval reactors were developed by Bettis for Shippingport. These were designated as materials, equipment, and processes (MEP) standards. With only slight modifications these frequently became the basis for development of Reactor Development and Technology (RDT) Division specifications employed in the liquid metal fast breeder reactor (LMFBR) program. Similarly, many of these requirements have been included in newer versions of commercial specifications and codes. The welding standard for PWR Core-2 was directly applied to the final draft and subsequent release of Section III of the Code.

---

[3]B.F. Langer, "Working Stress Criteria for Nuclear Power Plants," presented at joint session of ASME Power and Metals Engineering Semi-Annual Meeting, Pittsburgh, Pennsylvania, June 20–24, 1954.

[4]Now called the ASME Boiler and Pressure Vessel Code.

Instrumentation

Core Instrumentation[5]

The Shippingport core contained flow- and temperature-sensing instrumentation for determining absolute and relative power production of selected fuel assemblies in the seed and in each of the four orificed flow regions of the blanket. Two of these were inside and two outside the seed annulus. The power production in the blanket varied with time due to the build up of plutonium.

Thermocouples were provided to measure axial temperature distributions in the fuel plates of selected seed fuel assemblies. In addition, there was a system for obtaining a sample of effluent coolant from each of the 113 blanket assemblies, delivering each sample to a monitor external to the pressure vessel for the detection of delayed neutrons, which would be indicative of a failure of a fuel rod cladding. Because of the compactness of the PWR core, little room existed for instrumentation components and leads.

Problems encountered in developing these systems can be categorized as follows:

- It was difficult to accommodate instrumentation components and leads within the restricted space available in a compact core and in a manner compatible with the scheme of remote refueling. This overall problem caused additional difficulties in accomplishing conventional functions, such as the joining and sealing of groups of sampling and sensing tubes, and the use of a flow-measuring meter under less-than-ideal flow conditions.
- Development of new devices to perform new functions for which there was little precedence, such as developing a device to obtain a representative mixed sample of effluent flow from a fuel assembly.

Assurance of satisfactory solutions to the problems in the first category was obtained through use of mock-ups and proof-tests. Solution of problems in the second category required extensive developmental testing. The extent to which these approaches were employed is illustrated in the following discussion.

---

[5] The source for much of the following material is N.J. Palladino, S. Cerni, L. Porse, and J. Sherman, "Engineering Development of the PWR Core and Pressure Vessel," WAPD-T-711, Westinghouse Electric Corporation.

Flow Measurement Instrumentation

The flow measurement system consisted of modified venturis and flow nozzles mounted below individual fuel assemblies as an integral part of the normal supporting structure. Tubing welded into these devices transmits the pressure signals from below the core through the reactor vessel head to differential-pressure cells mounted on the head. To simplify the method of remote refueling and to expedite the sampling system described later, the leads of this system were confined to the space available below and around the periphery of the core. Further, in order to facilitate assembly, the tubing was divided into three lengths: a lower section extending radially outward from the fuel assembly and vertically upward along the core structural supports; an upper section to the vessel head; and an external section of high-pressure piping to the differential pressure cells.

To provide adequate structure and containment for cooling flow, the 72 tubes comprising the lower section were constructed as 12 groups of six tubes, with each group contained within a 1-1/8-in.-o.d. tube. The upper sections were composed of three groups of 24 tubes, each group being contained in a 2-1/4-in.-o.d. pipe. Both the upper and lower sections terminated in tube sheets into which the pressure tubes were Nicro-brazed®.

Again, because of space limitations, a special seal was devised to join the sections. This consisted of short flexible lengths of thin-walled tubing connecting a pressure tube in one section to that in another section. Providing the seal was a threaded connection and a small-diameter gold ring gasket for each tube in each tube sheet. Design of the seal was perfected by a program of testing possible designs under simulated operating conditions of pressure differential, fluid viscosity, and temperature cycling. Each actual seal was pressure tested during core assembly.

To assure reliable and compact features, a full-scale wooden mock-up of the core structure was constructed on which the problems of tubing arrangement, grouping, and support were resolved. This device also assisted in determining tooling and template requirements and served to train shop personnel for assembly of the actual units. Additional fabrication development was provided by the construction of typical prototype units from each system to resolve the details of assembly, bending, Nicro-brazing, welding, cleanliness, and alignment.

A condition of changing flow geometry exists upstream from the flow meters in the core and because these meters were located in close proximity to the reactor vessel inlet pipes. Therefore, it was necessary to verify that the flow pattern through the meter was relatively insensitive to flow conditions that might exist below the support plate to which they were attached. Such verification was obtained by hydraulic tests of prototype meters mounted on a simulated section of the support plate, below which flow was introduced at various angles and rates. These tests also served to calibrate the flow meters over a wide range of flow rates. This system has operated in a highly satisfactory manner.

Temperature Measurement Instrumentation

Major design efforts associated with installation of exit-water temperature and seed-metal-temperature thermocouples were concerned with providing suitable structures for positioning and supporting the thermocouples, as well as making available the tooling required to insert and remove such assemblies. The thermocouples were chromel-alumel wires separated and insulated by zirconium-oxide contained within a stainless-steel sheath. The primary system seal was accomplished by a Nicro-brazed® joint between the sheath and a stainless steel plug, which in turn was structurally welded into the pressure-vessel head. Developmental Nicro-brazed® joints were completed for each geometry employed.

Above this joint, in the actual installation, a mechanical seal for each chromel and alumel wire was employed to provide a secondary seal. Between the primary and secondary seal, a joint was made between thermocouple leads and larger diameter wires that proceed to cabling and readout instrumentation. Development tests were required to perfect the mechanical seal for full-system pressure. During actual construction, each such seal was helium leak tested at 1000 psi.

During operation, the following deficiencies were discovered and isolated:

- The seed metal thermocouples were not operating satisfactorily. It was later found that an aging process, employed to stabilize thermocouple calibration, introduced local areas of brittleness in the sheathing of the 0.040-in.-diameter thermocouples, used to measure the seed metal temperature, by the precipitation of carbon along the grain boundaries. This brittleness, in combination with mechanical handling required during installation, caused minute local fissures and the subsequent introduction of system water and thermocouple failure. To correct the deficiencies, some replacement thermocouples were procured for the measurement of seed metal temperatures. These substituted Type 347 stainless steel for the Type 304 sheath. Precipitation of carbon along grain boundaries is inhibited in 347 stainless steel because of the addition of a stabilizing agent to Type 304 stainless steel, thus tying up the carbon.

- Some of the exit-water thermocouples gave sporadic readings. The wire juncture occurring between the primary and secondary seals employed a silver braze done in the field. This juncture occurred in a region of temperature gradient. Introduction of a third metal in a somewhat unpredictable manner caused an extraneous emf. All silver-braze junctures were replaced by a fusion weld. As a result, the thermocouples installed for the determination of water temperature operated satisfactorily.

These difficulties were encountered in systems for which major comprehensive tests were not conducted. Further, the difficulties were sufficiently pronounced that had a proof-test been conducted, to simulate an actual installation, it is quite probable they would have been discovered.

Failed Element Detection and Location

Arrangement of the 113 tubes needed to conduct samples of blanket fuel assembly effluent flow for failed element detection and location was worked out in conjunction with the flow measurement system as described earlier. Similar detailed attention was given to layout and fabrication of each assembly, and typical prototype units were constructed to develop processes.

Additional components in this system that required development tests include sampling devices mounted in the exit flow region of each blanket assembly, which would obtain sufficient samples throughout the cross section of flow to constitute a representative mixture. These sampling devices were evaluated in a cold-water loop by injecting dye as a point source upstream of the device under test. Water samples were collected from several portions of the sampling device, and the efficiency of the test piece was determined by comparison with a completely mixed sample.

A multiport valve mounted on the pressure vessel head, in an automatic consecutive manner, continuously passes samples to two monitoring stations. Experiments to determine torque requirements, position, accuracy, and wear were performed during development of the multiport valve and during proof-testing of the final design.

Monitoring stations were compact chambers constructed as helically coiled piping wherein a sample can be accumulated while continuously in transit through the piping system. The coil was surrounded by water and monitored by BF_3 counters.

Core and Core Structure Tests

Hundreds of tests were performed in the development of components and characteristics of equipment for each of the early reactor plants designed and built by Bettis. This section is not concerned with all of these tests or even most of them. Many of the tests on specific components, such as primary coolant pumps, are discussed in other sections. In this section, attention is given to some of the major tests on core and core structures, intended to fulfill one or more of the following purposes:

- Tests to develop new basic information
- Proof-tests and mock-ups to establish the adequacy of design features
- Production control tests to assure an adequate product
- Flow tests for the core
- Series-parallel orifices tests
- Determination of power output ratio (seed power/blanket power) change with time.

Inasmuch as Shippingport had a seed-and-blanket core, there were major differences in the amount of heat generated in various core regions. As a result, to make economical use of cooling flow, certain assemblies had to receive more coolant flow than others.

The PWR core was divided into five regions for orificing purposes—a seed region operating without flow restriction and four orificed blanket regions. Design of orifices for the blanket regions was beset with a twofold problem. The first was obtaining large pressure drops in a small space and the second was avoiding high velocities with possible attendant erosion.

The most promising method for overcoming these problems appeared to be the use of orifice assemblies consisting of several multihole plates arranged in series with the holes on successive plates offset from each other. The appropriate pressure drop and water velocity could thereby be provided for each region by proper selection of number of plates, the hole diameters in each plate, and plate spacing. Testing was required to determine the effect of these variables on pressure drop. Evaluation of maximum velocities in orifice assemblies was made by analytical means; velocity effects were confirmed by examination of the plates after tests.

Flow Distribution Around Control Rod Shrouds

Control rod shrouds were tubular members extending from the top of the seed clusters to the reactor vessel head to provide linear bearings for control rod drive shafts and to protect control rods, when withdrawn from the core, against hydraulic forces due to cross flow. Cross flow was produced by flow of water from the interior regions of the core to the periphery on its way out of the vessel. Should hydraulic forces from this cross flow produce excessive deflection of the shrouds, jamming of control rods might result.

The magnitude of these forces was dependent on flow-distribution patterns in the plenum chamber above the core.

Lacking information regarding flow distribution patterns in a geometry such as this we had to perform tests. Included as part of these tests was an investigation of the hold-down barrel designed to help distribute the flow and reduce the magnitude of accompanying forces. For this purpose, a quarter-scale model of the vessel and core components influencing flow was constructed and tested at Battelle Memorial Institute by use of air in the noncompressible range (Mach number less than one). General directions of air flow were observed using titanium tetrachloride smoke as a tracer. Estimates of hydraulic forces on the shrouds were obtained from the same model, and for maximum system flow a uniform load of about 32 pounds on the shroud was predicted.

Labyrinth Seals[6]

The thermal shield surrounded the core and was larger in diameter than the opening in the vessel for inserting the core. To permit removal of the PWR core cage without removal of the thermal shields, these components cannot be mechanically joined. Furthermore, clearance between them should be made as large as practicable to facilitate insertion and removal. Nevertheless, a restriction of flow must be provided in this clearance because water enters the reactor vessel at the bottom and exits at the top, different from the current practice of having both entry and exit at the top. Otherwise, the pressure drop across the core causes excessive flow to bypass the core through this space. Consideration was given to two means for restricting this flow, the use of piston rings and the use of a labyrinth seal.

Design, manufacture, and establishment of reliability of piston rings during installation, operation, and removal of the core appeared to have many problems that previous experience indicated might be difficult to solve. In addition, review of the literature disclosed no information that would provide a basis for evaluating the efficacy of a labyrinth seal for water. Hence, a series of tests was performed on a model wherein various configurations of labyrinth seals could be tested using cold water. It was on the data from these tests that the labyrinth seal design was selected for PWR core I.

---

[6]N.J. Palladino, S. Cerni, L. Porse, and J. Sherman, "Engineering Development of the PWR Core and Pressure Vessel," WAPD-T-711, Westinghouse Electric Corporation.

The test[7] consisted of a 2.100-in.-diameter flow channel into which were placed plugs 2.000 in. in diameter. Plug geometry was varied such that a range of zero to seven was obtained in the ratio of groove space to ring thickness. Three different lengths for each configuration were tested to obtain a pressure loss coefficient on a unit-length basis very similar to the friction factor for pipe flow. The data were correlated as pressure loss coefficients defined by the following equation:

$$f = \frac{\Delta P D_e 2g}{L V^2}$$

where f = dimensionless pressure loss coefficient, P = static pressure drop across seal in feet, D_e = equivalent diameter for flow based on minimum clearance in the labyrinth, L = length of seal, V = fluid velocity in the minimum clearance, and g = acceleration of gravity.

The study showed that the highest resistance to flow per unit length was obtained with a geometry having a ratio of groove space to ring thickness of about 4.5.

Fretting Tests[8]

The 145 fuel assemblies that form the Shippingport PWR core were designed to be individually removed and replaced by remote means. As a result, each of the fuel assemblies in the core contained a mechanical latch to hold it in place in the core cage. This latch must operate reliably after remaining immobile for as long as three years in high-temperature water. This latch was made of Type 304 stainless steel and was fastened to the Zircaloy-2 fuel clusters by mechanical means, because stainless steel cannot be welded to Zircaloy. Hence, the latch assembly was susceptible to fretting-type failure if a source of vibration existed.

Forces that would produce the small motions necessary to cause fretting could not be analytically identified. Calculations relating coolant velocity and impingement area to hydraulic thrust on latch components indicated that movement due to these forces is improbable because the weight of the members exceeds the probable thrust by a factor of 3. Furthermore, no mechanism could be identified by which the presence of either von Karmen vortices or a pulsating venturi effect would exist.

---

[7]N.J. Palladino, S. Cerni, L. Porse, and J. Sherman, "Engineering Development of the PWR Core and Pressure Vessel," WAPD-T-711, Westinghouse Electric Corporation.
[8]N.J. Palladino, S. Cerni, L. Porse, and J. Sherman, "Engineering Development of the PWR Core and Pressure Vessel," WAPD-T-711, Westinghouse Electric Corporation.

However, it was well known that analytical evaluation of the vibration phenomenon in complex flow patterns was not a satisfactory basis for identifying sources of fretting. Therefore, it was decided to proof-test the blanket and seed assemblies in a high-pressure, high-temperature test loop. Because of the differences in design features and flow velocities between the seed-and-blanket assemblies, the loop was designed with two separate test sections so prototypes of each could be simultaneously tested under operating flow and temperature conditions. Flow rates were comparable to those that would be expected during reactor operation.

The blanket assembly was tested for a period of eight weeks and no fretting was observed. In the case of the seed assembly, however, after approximately six weeks of testing, pronounced fretting failure was observed on the Zircaloy-2 surfaces in contact with the stainless-steel cylinders of the top extension bracket. Specifically, the Zircaloy-2 surfaces had been worn down by as much as 1/8 in. The seed assembly design was therefore modified by incorporating a spring washer into the joint between the Zircaloy and the stainless steel to prevent relative motion through the expected operating temperature range. The modified assembly was then tested for six weeks without any trouble.

Top Grid Tests

The problem of evaluating thermal stresses in combination with mechanical stresses in a complex structure was well typified by the stress problem in the top grid of the Shippingport core cage. This grid, because of its proximity to the core, was subjected to uneven heating by gamma rays emanating from the core; yet it must withstand the total hydraulic load placed on the fuel assemblies during operation without deflections that would impair control rod alignment. It was also desired that the grid be designed with small web thickness, so the space between assemblies dictated by this web thickness would not introduce prohibitive neutron flux peaks in the water occupying that space below the grid.

The question arose as to whether or not a grid of square modular construction, surrounded by a solid ring, could be adequately treated by simple modification of solid plate theory to obtain maximum stresses and deflections. By the method proposed, maximum stresses and deflections were calculated for a solid plane of the same total thickness as the grid divided by the ligament efficiency factor, which is the ratio of web thickness to modular pitch. Due to the modules being located on a square pitch and because of the low value of ligament efficiency, Poisson's ratio was taken to equal zero. To take into account the edge fixity due to the ring around the grid, compatibility equations were solved between the ring and the grid.

Confirmation of such an approach would simplify design methods without continuing dependence on complex numerical methods. Although such numerical methods were used on the Shippingport grid, test work showed that mechanical loads can be adequately treated by use of the simpler analysis. Stress and deflection calculations on the Core 1 top grid were checked by a test on a one-quarter-scale carbon-steel model of the grid using 60-strain gauges and six dial indicators. Loads were imposed by hydraulic pressure on a diaphragm that transmitted forces through pins to simulate action of fuel assembly latches.

Results of this test showed good correlation with the maximum stresses and deflections computed by linear-efficiency modification of solid plate theory. Measured stresses were within 5% and measured deflections were approximately 20% less than calculated. Further confirmations should be made before extending this approach to grids of significantly different configuration or dimensions.

The development of simple and accurate means for treating thermal stresses was not as successful. However, it was found that, if thermal stresses were small compared with mechanical stresses, a suitable approximation can be made by use of a one-dimensional treatment of temperature distribution through a solid plate. On the PWR grid, this method yielded results that were conservative by 45% when compared with the results of more complex numerical calculations of thermal stresses involving three-dimensional temperature distributions.

Control Rod Alignment

Because control rods, their drive mechanisms, and connecting shafts had to operate reliably in hot water without the benefit of normal lubrication, it was important to minimize bearing loads brought about by misalignment between these components. The problem was particularly important in Shippingport, because components were not accessible for measurement and alignment after installation in the vessel. Furthermore, in order to accomplish remote refueling of individual fuel assemblies, it was decided there would be no requirement for measurement and alignment at each refueling of a seed unit. Consequently, alignment had to be assured by a system of fits and keys involving the seed cluster, the cage, the thermal shields, the reactor vessel, the head, the control rod drive mechanisms and the shroud tubes containing bearings for the control rod shafts.

With so many components involved, either provisions had to be made to accommodate very large misalignments in the rod system or very tight tolerances had to be maintained in various fits and keys. Since the rod drive system involves translation of a control rod by a mechanism through a shaft, large misalignments could not be accommodated without large wear-producing lateral bearing loads. Accordingly, establishment of a balance between allowable misalignment and uneconomic or prohibitively tight clearances and tolerances was a difficult one. Such tolerance requirements involve cumulative tolerances that are very difficult to specify within allowable measurements, i.e., vessel, fuel, slot and blade dimensions. For these reasons finger type control rods were used for later commercial reactors.

This balance could not be established without experimental confirmation of the wear characteristics of components under various combinations of misalignments allowed by design clearances and tolerances. Three basic materials were involved. The control rod drive shaft was originally constructed entirely of precipitation-hardened (PH) Type 17–4 stainless steel. The lead screw of the shaft was actuated by drive mechanism rollers made of Stellite® 19. Guidance and support for the shaft was provided by linear bearings of Stellite® 6 material contained in the shroud tubes. Later, Rickover had all 17–4 PH SS removed due to the susceptibility of it cracking.

The problem with 17–4 PH SS is that if not heat treated correctly, it is likely to crack. Vendors frequently do not heat treat properly. This caused expensive problems for General Electric's (GE's) commercial nuclear plants and for Detroit Edison's LMFBR reactor.

Two types of tests were performed. First, a prototype control rod system was tested to determine the effect of various types and amounts of misalignment on the lateral forces imposed on components. These tests were used in conjunction with available wear data to select tentative design clearances and tolerances in the control rod system.

On the basis of these tests, for example, the maximum permissible misalignment of a control rod drive mechanism housing and the appropriate module of the top grid was selected as 1/8 in. Using this value, in combination with the other limiting conditions—such as clearance in the shroud bearings—the control rod system, including the drive mechanism, was operated in a loop under reactor conditions of water temperature and pressure. The mechanism and rod were cycled approximately 300 times. Maximum lateral force on any of the linear bearings was about 40 lbs during these tests. The components showed no adverse wear after tests.

Other Major Components

Pumps

Canned motor pumps used on the *Nautilus*, and described in Chapter 2, were many times larger than those developed during World War II. Those developed by Westinghouse for Shippingport were again many times larger than those for the *Nautilus*, having a capacity of 20,000 gallons per minute. One of these vertical single-stage centrifugal pumps was located in each of the four loops.

Steam Generators

Steam generators seem like a simple component; they are just a shell-and-tube type heat exchanger, in many respects like thousands that have been built. The difference is presence of corrosive high-temperature water on one side and a combination of steam and water on the other, with a high-velocity water flow and the attendant stresses this flow causes and because of the zero leak objective to prevent radioactive particles entering the secondary system. Deposits of corrosion products from the entire system get into small spaces and cause corrosion of the steam generator tubes. This has continued to be probably the major problem for PWRs because they are expensive to replace.

The decision was made to develop two types of steam generators—one with horizontal straight tubes and the other with vertical U-bend tubes—in order to determine which would be a better choice for the future. Thus, two manufacturers were to be involved, with the added cost involved. The straight-tube design was developed by Foster Wheeler and the U-shell design by Babcock & Wilcox. Three of the four steam generators could produce enough steam to generate 60 MW(e) and four could match the turbine 100-MW(e) capacity.

Other problems involved the method of securing tubes to tube sheets, stress relieving U-bend tubes, and supporting of the tubes subjected to the force of water flow. The problems of plugging leaking tubes and replacing tubes did not surface until much later.

Turbine Generator

The turbine generator was part of the balance of the Shippingport plant supplied by Duquesne Light. Westinghouse believed it was commercially necessary to supply this turbine generator rather than have a GE unit at this plant in the Pittsburgh area. Consequently, Westinghouse gave Duquesne Light an offer it couldn't refuse.

This turbine generator was an 1800 rpm machine rated at 100 MW(e). The steam at 600 psi was very low pressure by then-current utility standards.

State of the Art

Reactor Physics[9]

There were two basic core designs to choose from, the seed-and-blanket or the more or less uniform, slightly enriched uranium core. The seed-and-blanket concept consisted of a highly enriched uranium central ring, surrounded internally and externally by a natural uranium blanket. While this was debated, Bettis favored the slightly enriched concept, since the more complicated design was not necessary to prove the fundamental development of a significant nuclear electric power producer. We also felt it was economically more desirable.

The seed and blanket were based on using proven naval reactor fuel elements for the enriched seed and uranium dioxide pellets in Zircaloy tubes for the blanket. This concept seemed to require less enriched uranium, since it produces more power from natural uranium than does a uniformly enriched core. However, the significant comparison is the number of grams of uranium plus the separative work units (SWUs) required to provide the fuel inventory for the seed-and-blanket versus the slightly enriched core, including refueling. Economic objectives favored the slightly enriched core, not only from a SWU basis, but also from the greater power density available throughout the slightly enriched core. The seed-and-blanket core not only has a poorer peak to average power ratio, but it changes as plutonium builds up in the blanket. This requires greater coolant flow later in life, so fixed orificing cannot be as effective.

Because it was thought that enriched uranium was in short supply, the requirement for fewer grams of enriched uranium was considered an advantage, but SWU requirements meant more diffusion plant capacity was used for the seed-and-blanket concept. In fact, enriched uranium did not turn out to be in short supply. The seed is small and has a large leakage of neutrons to the blanket. The seed-and-blanket has a large negative temperature coefficient of reactivity and reduced requirements for mechanical control until the plutonium built up in the blanket.

Actually, there was not enough known about either concept to make a really informed decision; it had to be a command decision, and either would have worked. Subsequent knowledge and experience have caused the slightly enriched concept to be used throughout the world.

The seed-and-blanket approach was destined from the beginning to win, because the Naval Reactors Branch (NRB) considered it best; it had been

---

[9]Based on contributions from R.J. Creagan and A.F. Henry.

proposed by Alvin Radkowsky of NRB, who thought this could be a step toward a light water breeder reactor.

Even in the exploratory stage, with the complexity of the seed-and-blanket core, it was impractical to compute the necessary multiregion problems with hand calculators, even on a one-dimensional, two-group basis. Reactor-simulating analog computers were used, and when they were available, digital computers. The needs of PWR physics helped to furnish the impetus to develop two-dimensional and transport theory codes, which have been of great use to all reactor physics work. Compared with today, the ability to calculate reactors was very primitive. With neither an adequate theory nor computer capability, it was necessary to validate and correct calculations by means of critical experiments, i.e., almost zero-power reactors in relatively simple geometries and then more complicated ones.

The seed-and-blanket design took a lot more analysis and experimentation to prove, but it was completed successfully and in time to meet the Shippingport schedule. This added development contributed to the progress made in reactor theory.

The seed material was based on that of the submarine reactors—uranium enriched to about 93% in the uranium-235 isotope. However, blanket material consisted of natural uranium oxide pellets in Zircaloy tubes. Because at that time it was difficult to produce long defect-free Zircaloy tubes, the Shippingport core used tubes of only a little more than 10 in. These are arranged in an 11 × 11 lattice with one corner tube omitted for instrumentation. Seven such bundles are stacked for a core about 70 in. high.

Near interfaces between these two materials, both the energy and the spatial distribution of the neutrons changed drastically. It was necessary to partition the neutron population into four energy groups to account correctly for this behavior.

To predict the very heterogeneous flux shape (and hence power shape) throughout the reactor, a technique called *spatial flux synthesis* was developed. It consisted of blending together two-dimensional (PDQ) flux shapes (appropriate to the materials present at various elevations of the core) into a three-dimensional composite. In this way, neutron densities in the various energy groups could be predicted for more than a million spatial mesh cubes.

To verify the predictions of neutron behavior, two types of experiments were performed: clean lattice criticals and exponential experiments. The clean critical experiments, performed at Bettis (along with exponential experiments performed at Brookhaven National Laboratory) were aimed at understanding the behavior of neutrons in the blanket. When uranium-238, which constitutes 99.27% of natural uranium, absorbs a neutron, two electrons (beta-rays) are subsequently released, and the uranium-239 nucleus becomes plutonium-239.

This latter material, like U-235, can be fissioned by slow-moving (thermal) neutrons. Thus, as the seed depletes, increasing amounts of Pu-239 build up in the blanket regions, and power shifts from the seed to the blanket. The lattice experiments were used to verify and/or correct the theoretical predictions of various neutron reaction rates in the blanket. They are still used today to test the accuracy of new cross-section values or new theoretical lattice analysis codes.

The mock-up experiment was just what the name implies. It led to small adjustments of the theory, which then predicted the physics behavior of the actual reactor throughout life with remarkable accuracy.

The present, slightly enriched cores are less expensive than the seed-and-blanket configuration and have another feature that was not thought about in the early days, i.e., accountability for highly enriched fissionable material, to prevent proliferation of nuclear weapons. Slightly enriched uranium cannot be used to make a nuclear weapon, but highly enriched uranium can. There have been questions about whether the plutonium, which builds up in commercial reactor fuel as a by-product of the fission process, can be effectively diverted to weapon use. This is a far less significant issue in slightly enriched cores than would be the case with the highly enriched seed of the seed-and-blanket cores.

To prevent certain spots in the core from getting too hot, it was necessary to have a detailed three-dimensional map of the neutron flux. With the degree of precision that could be attained at the time, large margins of safety were needed, thus giving poor economic performance. This is an area in which great strides have since been made, both in computer capability and reactor theory—much of it pioneered by Bettis.

An interesting fact[10] is that, almost unique to any engineering discipline, reactor theory was formulated in an essentially complete form almost from the beginning of the program. The seminal article on neutron transport theory in the *Reviews of Modern Physics*[11] served as a bible for those in the Bettis physics department and served as a basis for a continuing series of seminars to assure a full understanding of the theory and its applications.

However, the fact that you could write equations for all these nuclear processes in an accurate form did not mean you could carry out computations with these equations, taking into account the entire energy spectrum and the three-dimensional geometry that a complete computational descrip-

---

[10] J.J. Taylor, Private Communication dated September 11, 1991.
[11] R.E. Marshak, *Reviews of Modern Physics*, Vol 19, p. 235, 1947.

tion would require. Here is where the computer came into play—by making it practical to carry out such complex computations. The theory of numerical solutions of these equations was advanced with increased computer power. Bettis hired Ph.D. mathematicians from the best schools who had a background in finite-difference calculations to help program numerical solutions to these equations on the computers available. As computers advanced they could handle increased geometric detail, more detailed energy spectral treatment, and improved treatment of transient behavior.

In parallel with increased computing power, mathematicians were able to devise new numerical algorithms for solution of equations, which resulted in major efficiencies in the number of calculations needed and, therefore, the time required to achieve a solution. Even the earliest computers were thousands of times faster than the previously used hand calculators. Many decades of improvement in both the speed of computers and the efficiency of the numerical process were achieved over the years as we moved from those desk calculators, with which the Mark I core was designed, to the supercomputers of the 1960s and 1970s.

Bettis was always vying to get the best new large-scale, general-purpose digital computer available. Only the weapons labs were able to put such new capability on line faster, because they had a similar technical motivation but a lot more money. In that time frame, nuclear development was the driving force for general-purpose computer development. Our first computer, the Circle computer, was inoperable much of the time and was replaced by the vastly superior IBM-650. Table 13.1 shows the growth of Bettis computer capability from the 1950s to the 1970s.

Fuel Elements

Highly enriched uranium seed fuel elements had been extensively investigated previously, and the technology was in good shape. The major part of fuel element development was involved with the natural uranium blanket, i.e., that part of the core around and inside the seed.

Table 13.1
Computer Capability at Bettis

Period	Computer	Relative Capacity
1956–60	IBM-650	1
1956–60	IBM-704	10
1960–68	Philco-2000	200
1966	CDC-6600	600
1970	CDC-7600	5000

There were two basic candidates for the blanket fuel: Zircaloy-clad UO_2 and co-extruded Zircaloy-clad gamma phase alloy (uranium molybdenum alloy). Most of the experience in those early days had been with metal alloy fuels; the fuel for naval reactors was a metal alloy, and NRB was pushing the co-extruded Zircaloy-clad fuel elements.

Of special concern was the behavior of UO_2 during reactor radiation, particularly in regard to the following:

- The heat-transfer characteristics of fuel elements operating both undefected and defected, and measurements of the heat fluxes at which center melting occurred, were considered basic.
- The high thermal expansivity, low thermal conductivity, and poor tensile strength of UO_2 indicated that it might fracture under the thermal gradients attained during fuel element operation. The extent of such fracture and determination of conditions under which possible progressive fracture might cause the cladding to deform or fail as a result of a "ratcheting" process were tested both in the reactor and out.
- Similarly, it was visualized that during reactor down periods, water might enter an unbonded UO_2 fuel element through a small cladding defect and subsequently flash to steam when the reactor was brought to power—leading to gross failure and blockage of coolant-water channels. This was called *water logging*.
- It was necessary to establish experimentally that failure of one fuel element did not cause progressive failure.
- Because the UO_2 was not bonded to the Zircaloy cladding, the release of fission products into the coolant from a defect in the cladding would be much greater than in the case of bonded corrosion-resistant metallic fuel. The nature and amounts of fission products released and their effects on coolant contamination were prime concerns.
- The effects of radiation on both the crystal structure and the metallographic structure of UO_2 were also considered to have a vital bearing on the performance of fuel material.

When the issue of fuel choice for Shippingport arose in the context of higher burnup requirements, Bill Johnson, manager of the materials department, introduced the possibility of going to oxide fuel on the basis that at higher burnups there was a strong potential for metal alloys to swell. On the other hand, there was concern that if a leak developed in the cladding, water could enter the fuel rod and react with the oxide in the phenomenon called water logging, as mentioned earlier.

There were lengthy discussions as to which was better—or perhaps, which was worse—and there were insufficient facts to make a clear choice. While the French had very early employed UO_2 in a reactor, there was little technical data on that available. We had virtually no information concerning the corrosion resistance or ability to withstand irradiation of either fuel candidate. We also had almost no knowledge concerning manufacture of either one.

However, some work done under Ben Lustman had shown the superior long time irradiation properties of ceramic UO_2 pellets. Johnson was convinced intuitively that the oxide fuel was preferable. He probably turned the tide one day in a major presentation, which included NRB people, by showing slides of uranium alloy fuel samples that had been irradiated to the point of major swelling. A picture of those failed fuel samples swayed the decision to oxide.

The latter approach was dropped on May 9, 1955. The effort on fuel elements with slightly enriched gamma phase alloys was then carried forward on the A1W aircraft carrier project in plate geometry.

In the end, the blanket consisted of clusters of fuel rods. These rods were Zircaloy tubes, into which UO_2 pellets were inserted with plugs welded into each end.

No one had ever manufactured uranium-oxide pellets. To do this, we had to develop the process and then construct the manufacturing facility. Little was known about uranium oxide, e.g., the melting point given in the literature varied by as much as 400°C. The heat conductivity was also unknown. These two parameters must be identified in order to determine that the center of the fuel element will not melt under operating conditions. Development facilities had to be obtained and the development, design, and manufacture completed in less than three years to meet our schedule.

Safety

Philosophy

In the design, construction, and operation of Shippingport, every necessary precaution was taken to guard against hazards to operating personnel or the surrounding area. The principal safety problem was to prevent, in the event of any conceivable accident, release of harmful quantities of fission products from the reactor core.

Thus, the fission products were enclosed within several independent barriers. The first for the blanket was uranium dioxide pellets, the second the Zircaloy cladding of the fuel pellets, the third was the thick-walled pipes and components that contain the coolant water, and the fourth was the containment building shell. If fission products leaked into the coolant through flaws that developed in the fuel element cladding, the coolant pressure boundary would contain them or ultimately the containment walls. For the seed the fission products were first contained within the zirconium uranium-235 filler. There were provisions for locating and removing failed fuel elements from the reactor.

We decided that the plant should have safety containment, i.e., the entire reactor primary system would be enclosed in such a manner that the containment could withstand pressure built up by expanding steam in the event of a rupture of the primary coolant system. This fourth barrier consisted of a series of four large interconnected, vapor-tight steel pressure vessels housing all parts of the plant that contain coolant at high temperature and pressure.

This decision established the policy that nuclear plants in the United States would have containment. The Soviets, in the same time frame and with much the same facts, decided against containment. For many years, Soviet reactors were built without containment until other countries in the Soviet Bloc, to whom they wished to sell reactors, insisted on it. The extensive radioactive contamination that occurred from the Chernobyl accident was the result of the earlier poor Soviet decision.

Most of the early reactors that had been built did not include containment. These reactors had been either material testing or low-power experimental reactors, with little buildup of radioactive fission products, or they had been production reactors at remote sites such as at Hanford, Washington. The first reactor with containment was the GE built submarine prototype at West Milton, New York. This was done so that it would not have to be built at the National Reactor Testing Station in Idaho.

Since the Shippingport plant was to be partially underground, containment was designed so the reactor vessel was in a sphere and each of two loops in a cylinder, with an auxiliary cylinder to give sufficient volume to withstand the pressure if there was a rupture of the primary-coolant piping and one steam generator.

Container design pressure was 52.8 psi at 280°F as well as for a 3-psi vacuum, with a test pressure of 70.0 psi. The diameter of the cylindrical section was that required to contain the equipment. Thickness was set at the maximum that could be welded without a stress relief step. There was concrete shielding between the two loops in each cylinder. With this configuration, some maintenance could be performed while the nuclear plant was operating. This was a realistic design for an experimental plant, but much less expensive aboveground spheres or domed steel or reinforced concrete cylinders are now used.

The design-basis accident was a brittle fracture of the reactor vessel, either from an inherent brittle flaw or from radiation-induced embrittlement. After study of all available data, it was determined that the reactor vessel could be manufactured so as not to be inherently brittle and that neutron impact radiation damage occurring to the vessel would be sufficiently annealed at operating temperature that brittle fracture was not a credible accident during plant lifetime.

The design-basis accident was considered to be a double-ended guillotine break of a main coolant pipe. Later commercial plants replaced the simplistic Shippingport assumption—which considered only a sudden guillotine break of the largest primary pipe—by a much more detailed consideration of various leak rates, which resulted in many changes in the emergency cooling systems and containment for later plants. This contributed to the regulatory instability that characterizes the U.S. nuclear program. The primary reason for that instability has been the nagging question of "If you design a plant with containment, you must believe there is a possibility of needing it—so does the containment cover every eventuality?"

Another contributing factor in the Chernobyl accident was the lack of a negative temperature coefficient of reactivity. This is covered elsewhere.

Safety Analysis

The safety analysis is much too complicated to cover in detail here. Though the analysis seems primitive compared with what is required today, it was far more sophisticated than that for the *Nautilus* and the prototype. This report was concurred with by the Advisory Committee for Reactor Safety (ACRS), chaired by Dr. Edward Teller. ACRS was, by definition and legally, an advisory body, not a regulatory one. The ability to make the comprehensive analysis required today simply did not exist. Had it been required, the Shippingport plant would not have been built.

To acquire the information or the capability to make today's required safety analysis almost mandated that a reactor be completely designed, if not built and operated. It is inconceivable that any development program so comprehensive could have been funded—a real Catch 22. Nobody stopped us, so Shippingport was built, operated successfully and safely and is now decommissioned. Other nuclear plants have been built, much has been learned, a sophisticated safety analysis can be made, and current reactors are now even more assuredly safe.

CHAPTER 14

The Astronuclear Technical Story

The NERVA Engine

The initial performance objectives of the NERVA engine development program were to demonstrate an engine with a minimum specific impulse of 800 seconds,[1] 55,000 pounds thrust, 10 minutes duration at full thrust, and programmed closed-loop control with ability to start up with no external energy sources. This is called bootstrapping.

From a tank above the reactor about 70 pounds per second of liquid hydrogen is pumped into the nozzle to regeneratively cool it as shown in Figure 14.1. While the hydrogen is liquid as it enters the nozzle, it is gaseous at about 90 K/162°R when it enters the beryllium reflector in the annulus between the core and the reactor vessel. The flow then proceeds through the reflector and on to a shield. After passing through the shield the hydrogen enters the core.

The purpose of the shield was to protect the engine components, vehicle, and payload from excessive radiation doses and heating. The heating of the hydrogen in the propellant tank had to be reduced to levels that would not produce pump cavitation.

The hydrogen-cooled epithermal reactor consisted of a right circular cylindrical graphite core surrounded by a reflector consisting of 12 beryllium sectors, each containing a control drum. Boral plates in these drums supplied the poison needed for reactor control.

---

[1]Defined as pounds of thrust per pound of propellant used per second.

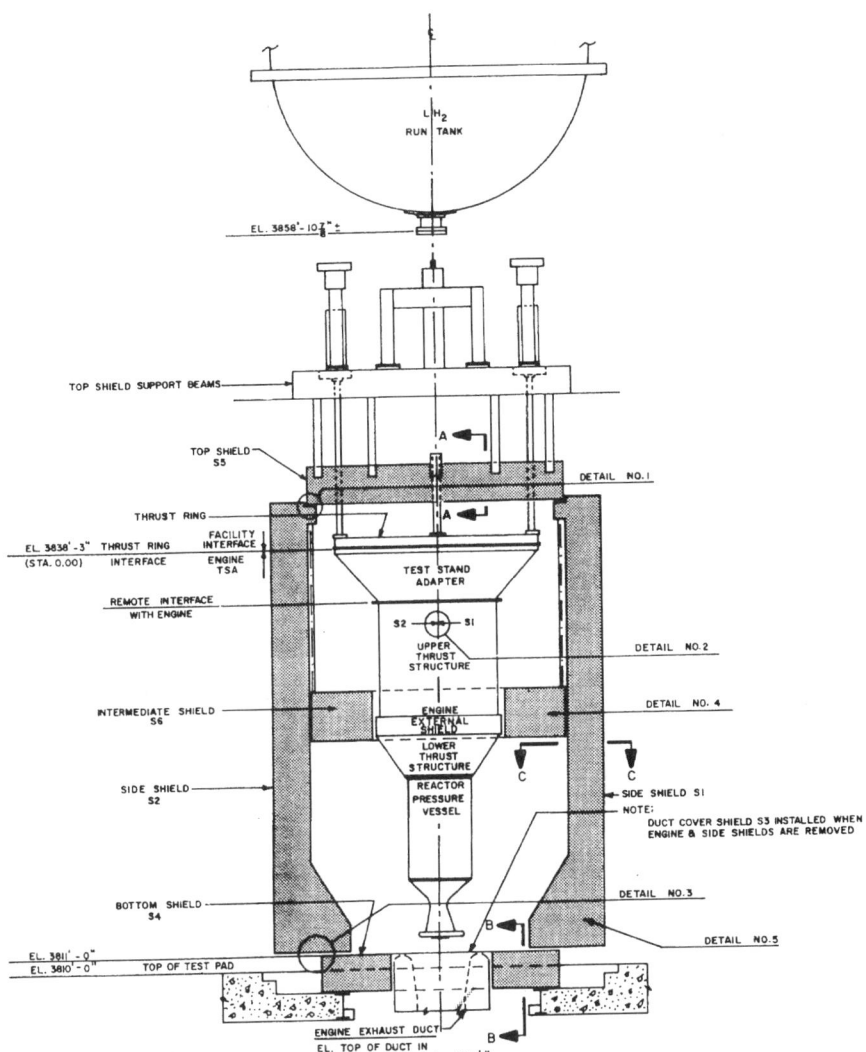

Figure 14.1 Engine test compartment radiation shields.

The reactor core was composed of clusters of hexagonal graphite fuel elements, the majority of which contained six fueled and one unfueled element as shown in Figure 14.2. Each fuel element contained 19 coolant channels so that there were approximately 32,000 coolant channels. The elements were fueled with pyrographite-coated beads of uranium dicarbide and were coated with niobium carbide for corrosion protection.

Each cluster was supported by an Inconel tie rod, which passed through the center of the unfueled element. A lateral support and seal had to be provided to prevent hydrogen from bypassing the core.

Maximum material temperatures within the core reached 2828 K/5090°R when calculated on a statistical hot channel basis. Core exit temperature and pressure were 2772 K/4090°R and 560 psia. Reactor power densities were about 2 kW/cm^3 and heat fluxes more than 10 times that experienced in water reactors. A hot bleed provided hydrogen for the turbine-driven pump. When expanded through the nozzle this configuration could supply 56,000 pounds of thrust at a specific impulse of about 800 seconds.

The NRX reactors were designed to meet flight requirements, such as the environmental loads that would be encountered during boost and flight

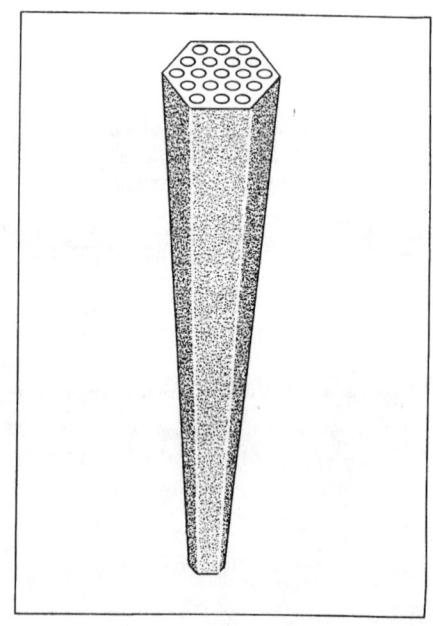

Figure 14.2 Hexagonal graphite fuel elements.

- SUSTAIN CONTROLLED NUCLEAR HEAT GENERATION

    PYROCARBON COATED UC$_2$ FUEL BEADS DISPERSED THROUGH AXM GRAPHITE MATRIX (630 MG/CC MAXIMUM FUEL LOADING)

- LIMIT TOTAL REACTIVITY LOSS TO $1.00 AT END OF LIFE

    CARBIDE COATING OF FLOW CHANNELS

- AFFORD HEAT TRANSFER FROM FUEL ELEMENT TO H$_2$ PROPELLANT

    19 FLOW CHANNELS IN EACH 3/4 INCH HEX, x 52 INCH LONG FUEL ELEMENT

operations. The reactor was supported by a titanium ring contained within the flanges of the pressure vessel and attached to the reflector. The top support plate was mounted on a titanium cone that rested against the reflector. Structural support of the NRX-A differed from that of the KIWI-B4A, because of boost and flight requirements. An NRX reactor being prepared for shipment to the Nevada test site is shown in Figure 14.3.

Most of the chemical engines had a specific impulse of no more than 400 seconds compared to the 825 seconds for the nuclear rocket engine. This higher performance is made possible because specific impulse is proportional to the square root of T/M where T is the nozzle gas temperature and M is the molecular weight of the exhaust gas. Since hydrogen has a molecular weight of two, a nuclear engine can achieve much higher specific impulses at the same operating temperature than a chemical engine. Nuclear engines with solid cores, such as NERVA, have the potential to achieve specific impulses of 900 seconds or more.

Flight Engine

During the NRX/XE test series, SNPO mission studies showed the merits of a 75,000-lb thrust engine for a Mars mission. The engine, which was to power a third stage of the Saturn vehicle, was designated the flight version. The design of this engine was completed but did not proceed to fabrication or to test. This engine incorporated the lessons learned from the NRX/XE test series. The distance from the bottom of the propellant tank to the nozzle exhaust was 20 ft. Key features that were added included the following:

- Approximately 200 additional fuel elements were added to accommodate the added power requirements. The design requirement was to produce a specific impulse of 825 seconds. To achieve this performance level, great care was taken with the thermal design so that the peak-to-average hot channel factor was reduced relative to the NRX/XE design.
- The tie-rod cluster support was replaced with a counterflow tie tube design that eliminated the core bypass flow. This concept was first tested on the Phoebus tests.
- An internal shield composed of boron carbide-aluminum-titanium hydride (BATH) composite supported on an aluminum structure was developed.
- The engine turbopump was powered by the hydrogen that cooled the tie-tubes.

Additional features of the flight engine are described in Table 14.1.

Figure 14.3 Flight engine.

Table 14.1
NERVA Flight Engine Features

- Engine thrust of 334 kN (75,000 pounds) with a specific impulse of 825 seconds.
- Chamber temperature and pressure of 2360 K (4250°R) and 3103 kPa (450 psia).
- Source power to bootstrap time of less than 60 seconds.
- Transient startup and shutdown rates from bootstrap to full power of 85 K/sec (150°R/sec) and 345 kPa/s (50 psia/sec).
- Endurance of 600 minutes with up to 60 cycles at 2360 K (4250°R) nozzle temperature.
- Sixty minutes continuous operation.
- Throttle operation to 65% thrust at 2360 K (4250°R).
- Endurance of 120 minutes with up to 12 cycles at 2500 K (4700°R).
- Reactor decay heat removal during cooldown to accommodate thrust generation with greater than 50% specific impulse or thrust nulling.
- Internal radiation shield to protect components and limit heating in propellant tank as required.
- Nuclear subsystem reliability of 0.997.
- Single-failure modes eliminated or designed to acceptable failure incredibility.
- Capable of emergency-mode operation in the event of primary failures to permit engine thrust of 133 kN (30,000 pounds), 500 seconds specific impulse and 445 MN-sec (10^8 pound-second) total impulse.
- Able to be transported by land, air, or sea.
- Capable of being stored for five years in container, six months on launch pad and three years in space.

NERVA and ROVER Test Program

An understanding of descriptions of the NERVA program at Astronuclear and the ROVER program at Los Alamos Scientific National (LASL, now LANL) can be aided by a brief chronology of tests and the nomenclature used. Design and development progress of the NERVA development programs were closely tied to the reactor and engine test series. Therefore, descriptions of design features or test results refer to the reactors NRX-A1 to NRX-A6. New reactor concepts and design features were introduced in each of the tests.

NERVA development, design, and reactor assembly were performed at the Astronuclear Laboratory at Large, Pennsylvania. Fuel was fabricated at the Astrofuel facility at Cheswick, Pennsylvania. The reactor was assembled at the Astronuclear Laboratory at the Large facility under conditions of continuous neutron monitoring, but with boron-containing wires inserted in many of the coolant channels to keep it heavily subcritical under all conceivable conditions. The reactor assembly was mounted in its aluminum pressure vessel, with a closure plate where the nozzle would be mounted at the test site. It was then shipped across the United States on an open flatbed railroad car, in a steel overpack container. The safety analysis took account of possible flooding, reactivity effects that would follow compaction in case of collisions or falling off the car, and the steel shell was specified to withstand pot shots from deer hunters. Although this was certainly a safe operation, today it would be more difficult to get permission to assemble a live reactor in a suburb of Pittsburgh and ship it intact across the country! Even then, we might have had more trouble with regulators had this not been a government project.

Simultaneously with the reactor assembly, the Nevada site team was preparing the test car, which was designed for firing in the up direction as shown in Figure 14.3. For the NRX tests the reactor was assembled with an Aerojet nozzle and a Rocketdyne turbopump. Teams composed of LASL, Westinghouse, and Aerojet General personnel conducted the tests. Westinghouse had responsibility for the NRX tests, LASL the KIWI tests, and Aerojet General the XE. As stated in Chapter 6, Aerojet General had the prime responsibility for the NERVA engine program. Westinghouse Astronuclear Laboratory had responsibility for the nuclear subsystem, which included the reactor, reactor vessel, and reactor controls.

As the project proceeded a very close relationship developed between the Astronuclear and Los Alamos technical staffs. Information and concepts were freely exchanged and discussed in a very constructive manner. At the start of the NERVA project, Westinghouse people were assigned to Los Alamos for a year to become up to date on the progress that LASL had achieved. This experience led to lasting friendships and a spirit of cooperation.

For example, the Nevada test team was a closely knit group of engineers and scientists from both of these organizations. As tests were conducted the test's director would be from LASL, Westinghouse, or Aerojet depending on who had the prime responsibility for the test.

Following each of these tests, reactor and engine components were meticulously examined to uncover features that could be improved. These postmortems usually resulted in design and fuel element processing changes. Fuel element processing for each reactor was usually designated as a campaign. Processing improvements were made in each campaign, and quality standards became more stringent.

The engine design and tests were to be designated XE-1, XE-2, etc. An XE-1 reactor, equivalent to that in NRX-A5, was shipped to Nevada but never used, because the NRX-A6 type fuel was enough better to be worth some delay and expense of substitution. Thus a second core, equivalent to A6, was shipped and put into the test program. This was called XE' and was the only one tested. Many years later, in the 1980s, XE-1 was disassembled and shipped to Idaho for reprocessing along with much other leftover ROVER fuel material. The test series is shown in Table 14.2.

ROVER tests during this same period were designated KIWI-B4A to KIWI-B4E, Phoebus 1A and Phoebus 2A, and PEEWEE. Nuclear Furnace was a reactor designed specifically for fuel element tests. The NERVA tests are listed in Table 14.3.

The Challenge of NERVA

Reflecting on the many technical challenges faced by the NERVA and ROVER designs teams, the following three stand out.

1. Developing a reactor design to maintain structural integrity when subjected to an environment that included (a) startup from vacuum conditions and thermal shocks from initial liquid hydrogen flow and rapid fuel temperature increases to 2700 K; (b) average fuel power densities of 2900 W/cm^3 with heat fluxes of 300 W/cm^2 with the potential for significant temperature differences across the fuel element support blocks; and (c) thermal gradients of 2200 K/4000°R over fractions of an inch at the interface of the core periphery and the reflector.
2. Development of graphite fuel element with carbide coatings with the capability to operate for 60 minutes in the corrosive high-temperature hydrogen environment. Uncoated graphite or areas with cracks or flakes would be severely corroded in minutes. At the start of the project, it was difficult to produce elements that survived a furnace test for 5 minutes. Significant differences in the thermal expansion of graphite and the carbide coating added to the difficulty of this task.

Table 14.2
NERVA Test Series

Test	Description	Time at Full Power
NRX-A1	Cold flow tests	—
NRX-A2	Full-power tests and restart control tests	5 minutes
NRX-A3	Full-power tests and control tests	20 minutes
NRX/EST	Engine startup tests and control tests	—
NRX-A5	Full-power, fixed drum control	30 minutes
NRX-A6	Full power	60 minutes
XE'	Full-power downfiring engine tests	11 minutes

Table 14.3
NERVA Tests

Date	Test	Facility	Maximum Test Power (MW)	Time at Maximum Power	Maximum Fuel Element Exit (T in K/°R)
7/1/59	KIWI-A	A	70	5 minutes	
7/8/60	KIWI-A'	A	85	6 minutes	
10/10/60	KIWI-A3	A	100	5 minutes	
12/7/61	KIWI-B1A	A	300	30 seconds	
9/1/62	KIWI-B1B	A	900	Several seconds	
11/30/62	KIWI-B4A	A	500	Several seconds	
5/13/64	KIWI-B4D	C	1000	40 seconds	2495 K/4491°R
7/28/64	KIWI-B4E	C	900	8 minutes	2662 K/4792°R
9/10/64	KIWI-B4E	C	900	2.5 minutes Restart	
9/24/64	NRX-A2	A	1100	3.4 minutes	2573 K/4631°R
10/15/64	NRX-A2	A		Restart Mapping	
4/23/65	NRX-A3	A	1165	3.5 minutes	
5/20/65	NRX-A3	A	1122	13 minutes	2723 K/4901°R
5/28/65	NRX-A3	A		<1.5 minutes (28.5 minutes)	
6/25/65	PHOEBUS-1A	C	1090	10.5 minutes	2751 K/4952°R
3/2,16,23/66	NRX-EST	A	1100	1.5, 14.5, and 8 minutes	2751 K/4952°R
6/23/66	NRX-A5	A	1140	15.5 and 14.5 minutes	2723 K/4901°R
2/23/67	PHOEBUS-1B	C	1500	30 minutes	2718 K/4892°R
12/13/67	NRX-A6	C	1100	62 minutes	2823 K/5081°R
6/26/68	PHOEBUS-2A	C	4300	12 minutes	2582 K/4648°R
12/3 and 4/68	PEEWEE	C	514	12 minutes	3023 K/5441°R
6/11/69	XE'	ETF-1	1100	11 minutes	2723 K/4901°R
6/29–7/27/72	NUCLEAR FURNACE	C	44	109 minutes, four tests	2723 K/4901°R

3. Achieving startups from source level to full power in 60 seconds. An analytical and experimental understanding of the fluid flow and the thermal and nuclear dynamics of the core and engine was essential to achieving this objective.

The achievement of these objectives required a close-knit multidisciplinary approach. Each design feature required many iterative nuclear, thermal, and stress analyses before suitable solutions were found. Significant amounts of experimental data were needed to use as design and analysis input.

While Westinghouse's background in reactor development was a solid base for the NERVA development, it was immediately evident that many technological problems required solution. High operating temperatures in a potentially corrosive atmosphere, extreme variations in temperatures across the reactor, and rapid temperature transients made this reactor design a formidable task. The interplay of fuel, coatings, and reactor operating temperature taxed the capabilities of a broad range of materials specialists and reactor designers.

Perhaps the greatest challenge was assembling the skilled specialists and broadening their horizons so that they became an effective interdisciplinary team of designers and analysts. Each team member had to develop sufficient understanding of the others' specialties so that mutual agreement could be reached on key design directions. This need for multidisciplinary understanding can be best illustrated by the following examples of areas that required particularly coordinated effort:

- Nuclear design was strongly dependent on the temperature distributions and hydrogen density in the core calculated by the thermal designer.
- Thermal and radiation analysts were concerned with energy deposited and cooling of reflector and peripheral metal parts, some of which had heat fluxes exceeding that of the fuel elements of water-cooled reactors. Of course, hydrogen at core inlet temperatures is a good heat transfer medium.
- Fluid flow analysts were concerned with the flow patterns in areas where large temperature differences existed between cryogenic core inlet and high temperature outlet made the accurate analysis of these effects essential.
- Structural designers had to cope with large temperature gradients in graphite and metallic reactor components. For example, calculated multidimensional temperature gradients were the key determinant of the stress analysis and resultant mechanical design of the beryllium reflector.
- Control design depended on the characteristics of all of the engine components in addition to the complex reactor dynamics.

The general approaches that were taken for some of these approaches are addressed in the following sections.

The Development Program

Mechanical Design and Development Problems

At the time of the NERVA proposal, we were examining three alternative methods of structurally supporting the core:

1. Hot end (core outlet) graphite support, which used a dome-like structure to achieve compressive forces.
2. Hot end (core outlet) cooled metal support, which was protected by coated graphite insulation.
3. Hydrogen-cooled tie rod design that supported clusters of fuel elements and was supported by an aluminum plate. This support was similar to that selected by LASL for the KIWI-B4A.

In a test on KIWI-B4A run by LASL on November 30, 1962, there was a gross failure of the core due to excessive vibration. This was one of the first reactors to be tested using liquid hydrogen as the fluid, and there were many theories as to the cause of the failure. These included local nuclear instabilities from hydrogen feedback effects, thermal stress, flow- or acoustic-induced vibrations and mechanical vibration induced by leakage flow between fuel elements.

Each reactor test was quite expensive, and when LASL wanted to run another duplicate test immediately, with what they called "quick fixes," Harry Finger, head of the Space Nuclear Propulsion Office (SNPO), refused. He did, however, permit LASL to run a cold flow test. This resulted in mechanical vibration of the fuel elements of the core. However, detailed measurements and even video observation of the core during the test revealed the onset of vibrations and helped explain the design problem. Shortly after the first of these tests, President Kennedy and Vice President Johnson visited LASL. However, the extent of the problem was not known until later.

During design of the NRX-A1 reactor, significant attention was given to the core periphery. The major concern was the corrosion that could be caused by the flow of hydrogen in the interstitial regions between fuel elements.

A seal was required around the core periphery to prevent hydrogen flow in the annulus between the core and the reflector. Such flow would bypass the core. The decision was made for NRX-A1 to put the seal at the outlet end so the interstitial flow, which is the flow between fuel elements, would be radially into the core.

Roman intuitively believed that the reactor peripheral pressures should result in radially inward flow in the interstices. These pressures would bundle the core and reduce peripheral flow. Our thoughts were on reducing interstitial flow and peripheral corrosion of the core. These early considerations of bundling helped us in our later understanding of the vibration problem.

Los Alamos had chosen to put the KIWI-B4A seal at the inlet end of the core. As a result, interstitial pressure caused the elements to move radially outward. As elements moved, interstitial pressure approached that of the core outlet, causing the forces on the elements to become radially inward.

When we saw the extent of the damage to KIWI-B4A and began to understand the cause, the need for a more careful understanding of the forces on the elements and the interstitial flows became evident. An intensive component test program was initiated immediately on this aspect of the design. This approach followed SNPO's fundamental philosophy of component testing prior to committing to a full-scale test.

Rather quickly after the KIWI-B4A test, Mike Manjoine rigged up a component test at Large that showed how interstitial flow could cause vibration and fracture a fuel element. A few days later, we had movies showing fuel elements vibrating in a test rig.

These tests quickly focused attention on the cause of the vibration, proving to the satisfaction of a number of SNPO members and others that we understood that what had happened on KIWI-B4A was a substantial task. Some of the SNPO people questioned the explanation for months afterward.

We asked for help from the Westinghouse research laboratory. They assigned two of their best vibration experts to work with Astronuclear for about six months. They viewed our explanation with doubt and looked for resonance phenomena. Others were analyzing the phenomena as a fluttering flag. SNPO also called on support from the NASA centers, Lewis, Langley Research, and the Marshall Space Flight Center.

A comprehensive experimental and analytical effort was undertaken to prove that the vibration problem was understood and corrected. To demonstrate this, experiments were conducted on assemblies of a few elements to relatively large sections of the core. The component test and analytical program to show that NRX-A1 would not vibrate required about a year of tests at Large before we ran the cold flow test in Nevada.

The tests at Large included experiments on assemblies ranging in size from a few elements, to clusters and a pie segment of the core. (The term *pie segment* refers to a test on a pie-shaped section of the core.) LASL was also doing pie segment tests of its design. Fluid dynamic stability and vibration conditions were established and compared with analytical predictions.

An analog computer model was developed that duplicated the experimental results in which elements were allowed to vibrate. Digital computer codes were developed to calculate interstitial flows and pressures. These pressures were compared with those in the seal region to determine when element motion would occur.

The cause was diagnosed as a nonlinear phenomenon in which higher interstitial pressures caused fuel elements to move radially outward. This motion then caused interstitial pressures closer to core outlet pressures and fuel elements to move radially inward. A few cycles of this type caused the fuel elements to fracture. Today we would call this phenomenon a chaotic non nonlinear vibration.

During this time, we decided to minimize interstitial flow by producing a design that had the core peripheral pressure slightly higher than that in the interstitial space between the elements. We accomplished the desired pressure distribution with a multiple-seal system consisting of 18 rows of seal segments. In addition, a band was added to introduce a rapid pressure reduction of the interstitial pressures. The NRX design was modified to include this distributed-seal system.

In May 1963, a test of the adequacy of the distributed 18-seal design was conducted on a complete unfueled core with fuel channels plugged. Extensive measurements were taken of seal pressure drops, interstitial pressure, and the reactions and displacements of fuel elements.

To test the design margin, seal failures were simulated. No fuel element motion was detected until 12 of the 18 seals failed. The design, therefore, offered a significant margin against vibration.

The final test in this series was the successful NRX-A1 cold flow test conducted at the Nevada test site in 1964.

In addition to the effect of interstitial flow on core stability, these leakage paths were of potential concern as a cause of corrosion. To correct this problem, the elements were carefully machined to reduce interstitial flow between fuel elements. Particular attention was placed on fuel elements at the core periphery, where hydrogen leakage occurred from the reflector region.

Carefully machined filler strips were placed around the periphery of the core, and outside surfaces of the elements along the core circumference were coated. The core periphery on the NRX-A1 contained some partial fuel elements (12 coolant channels instead of 19). Filler strips were added to make the core geometrically round. This area was one of high corrosion and cracking. On later designs, outside surfaces of the fuel elements were coated with NbC.

The hydrogen between the core and reflector, which was at cryogenic temperature and higher pressure, caused the fundamental leakage problem. A thermal gradient of about 4000°F existed over a fraction of an inch. Significant design changes on the NRX-A6 included a graphite foil wrapper surrounding the core. This wrapper prevented in-leakage at the core periphery and eliminated fuel element corrosion in that area.

Structural integrity of the reactor required limiting corrosion and maintaining tolerable stresses on the fuel elements and core components. Bundling of the elements to reduce interstitial flow to minimal values had to be accomplished without developing stresses that would rupture an element. Therefore, an early experimental program was established to verify experimentally the strength of all components at all operating temperatures from cryogenic to high temperature conditions.

Mechanical design and stress analysis of the reactor were strongly dependent on understanding the friction factors and coefficients of expansions over this wide range of temperatures. Significant programs were undertaken to determine the friction factors between components and their coefficients of expansions over the range of operating temperatures.

These capabilities included vibration facilities to check each of the reactor and components to NASA requirements for launch and operation.

Analysis Techniques Development[2]

Throughout the program, great emphasis was placed on pretest prediction and post-test problem-solving analysis. For this reason, extensive techniques were developed by all analysis disciplines—controls and performance, thermal, nuclear, and structural. Computer programs were written for coupled solutions for neutronics, thermal-hydraulics, and controls. Frank Retallick and Joe Gallagher deserve most of the credit for the conception and direction of NERVA analytical efforts.

To achieve good fuel element and reactor design performance required creation of highly sophisticated nuclear and thermal design methods. Three-dimensional heat-generation analysis techniques were necessary. Statistical variations in fuel element dimensions, fuel loading, and other fabrication variables were introduced into calculations of the hot-spot temperature. Each of 32,000 individual cooling channels was carefully orificed to compensate for variations in radial power generation and actual coolant-channel impedance.

Each fuel element was given a thorough quality examination, and one element from each coating batch was tested in the high-temperature corrosion furnace. Loading of an element was measured with a neutron-radiograph technique; pressure drop through coolant channels was measured by a gas-flow test. These data were put into a computer data

---

[2]D.L. Black, G.H. Farbman, and J.F. Wett, "NERVA Program History and Technical Summary," Westinghouse Electric Corporation, presented at USSR Conference, Obinsk, May 15–19, 1990.

base. A continuing nuclear and thermal analysis was under way as each element was selected for a position in the core. Two-dimensional heat-transfer calculations were made of temperature across a cross section of an element. Orifices were selected to achieve a temperature across the element that was as uniform as possible. This process continued until orifices were identified for all elements.

Temperatures resulting from power source distributions were calculated using codes that included two-dimensional conduction through the fuel element. These thermal analysis codes for calculating core-operating conditions were developed by Ken Rieke and brothers Bob and Dick Vijuk.

John Guzek made important contributions to the development of the software that positioned fuel elements in the core and selected one of eight sizes of orifices. These analyses examined the physical data on each fuel element, performed three-dimensional nuclear and thermal analyses, and selected orifices for each of the 19 coolant channels in each element.

NERVA reactor tests verified the overall adequacy of analysis techniques and criteria used in mechanical and structural design. These reactors were extensively instrumented so comparisons could be made between measured data and computed predictions.

Important contributions were made to the development of finite element stress-analysis techniques in elastic and nonelastic regimes. Finite element techniques became the standard method of stress analysis of essentially all reactor components including the reflector, support blocks, and fuel elements. Graphite components and fuel elements had to survive temperature transients from cryogenic hydrogen conditions, where the material was brittle, to high temperatures, where the properties were relatively good.

From the start, the design goal was to keep the graphite in compression. This goal was possible on all fuel elements, but not on the supports for the elements. These supports were the subject of extensive analysis and verification by experiments. When data were available on brittle materials, such as the beryllium in the reflector, Weibull's statistical theory of strength was used in the analysis. This was a probabilistic approach to using materials data. Material strength is fitted to a distribution curve and the probability of failure calculated. These techniques were relatively primitive in 1962, progressing through code development to three-dimensional finite element analysis with simultaneous heat-transfer solutions. Don Miller and Bill Brusalis made important contributions to the development of these techniques. John Swanson developed the first version of the code ANSYS, which he later developed into an industry standard for finite element stress analysis.

Most of these techniques were incorporated into a set of stress-analysis criteria published by SNPO. Definitions used for the stress technology were made consistent with Section VIII of the ASME boiler and pressure vessel code.

Graphite has excellent high-temperature properties with relatively high tensile strengths at operating temperatures. Its ability to withstand thermal shock allowed a design that could be subjected to rapid temperature transients. However, significant advances were needed in available structural design techniques to address properly conditions ranging from liquid hydrogen to 4000°F. Nonelastic-stress conditions had to be analyzed. A wide range of experiments was necessary to obtain materials and engineering-mechanics data needed to support the analytical calculations.

Nucleonics

Initial NRX nuclear analyses were performed with an adaptation of the PDQ diffusion code developed at Bettis. They were accomplished by the analysis people mentioned earlier who were aided by Don Drawbaugh and Carl Saalbach. Software was based on rectangular coordinates and hence was not particularly suited for the NRX geometry. Therefore, a diffusion code with an r-theta geometry was developed by Ravitz; it was suitable for the cylindrical geometry of the core, reflector and control drums. This code became the workhorse of the NERVA effort.

NRX was an epithermal reactor with higher neutron fluxes toward the reflector. Radial zoning of the fuel was used to reduce peak-to-average power generation. The fluxes were still higher toward the reflector after zoning. We wanted a flattened power generation rate, the product of flux and concentration of fuel beads.

Precise analytical methods were required for predicting power density throughout the core. The highly position-dependent neutron spectra required use of multienergy diffusion and transport methods. Also, since the reflector was at a low temperature when at power, the effect of chemical bonding on the reaction cross sections had to be considered. Analytical techniques were verified by critical testing.

Chemical bonding, which was not a major effect, modified the ability of a bound-beryllium or carbon nucleus to change the momentum of a slow neutron when it collided with a nucleus of an atom bound in a crystal lattice. This led in turn to a dependence on the temperature of the crystal lattice itself.

Power distribution changes resulted in transient thermal stresses. If these exceeded local tensile strength, cracking would result. Thus, a proper analysis was needed to obtain three-dimensional time-dependent temperature profiles. While such analysis is routine these days, we were pioneering the state of the art, using first- and second-generation supercomputers that were about the equivalent of today's standard PC.

The nuclear and radiation-shielding design groups developed numerous techniques specific to the NERVA reactor for determining power distribution, control drum worth, and shielding doses by means of theoretical computer codes and experimental critical-facility measurements. Radial and azimuthal power distribution changes during reactor lifetime, caused by rotation of control drums and changing concentrations of hydrogen throughout the core, needed to be understood and predicted. Errors in these predictions would result in high temperatures and gradients in the core, which would affect the structural integrity of the core.

One of the major discoveries confirmed in the tests was the ease with which analog computer models gave accurate predictions of the system's dynamic behavior. Our scientists had been led to believe that the complex dynamics of hydrogen boiling in the shield, coupled with its strong reactivity feedback effect, were going to be the key technical problem of NERVA in the analytical area.

With the help of new large-scale digital computers then available, we were successfully making multidimensional nuclear and thermal-hydraulic calculations. It was assumed that adding the time dimension was going to require more of the same—a task beyond the state of the art. How pleasant was our surprise on finding that system dynamic behavior could be predicted accurately with an analog computer model with just a few nodes!

The nucleonics of the core were affected by the large void fraction that was caused by the 19 coolant channels and absorption by the NbC coatings that protect the graphite from the highly reactive hydrogen coolant. This large void fraction resulted in further aggravation of the undermoderation (low graphite-to-uranium atom density ratio) of the moderator. Niobium is a neutron absorber that takes the reactor further off optimum from a neutron standpoint. Thus, the reactor required many times the amount of U-235 than would a neutronically optimum graphite reactor.

Neutron density was greatest at the outer diameter of the core because of reflection from the beryllium reflector. Fuel element loadings were varied to achieve a uniform power distribution. Reactivity effects in the core were greatly affected by hydrogen conditions in the coolant channels.

The prime objective for the design of the engine shield was to reduce neutron flux at the level of the propellant tank to a value that would not cause heating of liquid hydrogen in the tank sufficient to cause cavitation of the pump. Monte Carlo techniques were used to analyze the heating caused by neutron flux from the reactor. A precise knowledge of the gamma flux was essential to determine the heat deposition, cooling requirements, and thermal stresses in all of the engine components. Mary Ann Capo had a key role in the heat deposition analyses.

Materials and Physical Properties Measurements

As was the case with the submarine program, and many other highly developmental programs, once design and test programs are satisfactorily under way, performance of materials becomes the dominant problem. Even though there is much preliminary work on the selection of materials, after the design is more or less finalized and testing has begun, critical materials problems inevitably occur. In NERVA, it was the fuel. In PWRs, the corrosive effect of hot water caused problems with all parts of the system, especially mechanisms, fuel and steam-generator tubes.

Since the NERVA engine had to operate over a wide range of environmental and physical conditions, there was a great need for measurement of physical properties. Materials data had to be measured—from liquid hydrogen to reactor operating temperatures. Strength and creep properties of graphite at operating conditions were of great importance. Graphite is relatively strong and ductile at reactor operating temperatures, but it is brittle during shutdown and startup conditions. The brittle nature of beryllium also represented an area where much information was needed. There was concern that radiation damage sustained at cryogenic temperatures would be more severe than at ambient or elevated temperatures, because thermal annealing of the damage would be expected to be severely restricted. The design of the NRX-A6 and flight engine beryllium reflectors was modified to accommodate the expected radiation effects resulting from the required longer operating times. SNPO had a cryogenic test loop designed and installed in the Plum Brook Reactor in Ohio. Astronuclear was given the task of making this loop operational and conducting the tests. No severe radiation effects were found in reactor materials. However, significant effects were discovered in instrumentation sensors, such as strain gauges, thermocouples, and accelerometers.

Beryllium played another important role in the program, due to its high cost and to the fact that the program was funded by both the Atomic Energy Commission (AEC) and NASA who were on different fiscal years. Paul Koch, who worked for Hal Faught as NERVA contract manager, would regularly get calls from his counterparts in the government to buy beryllium at the end of one agency's fiscal period to use funds that could not be carried over (which was fine because we would need to buy the beryllium in the next period anyway). These calls would continue to the last day of the period, as the agency rounded up spare dollars from all other programs.

Koch was also the famous keeper of the master annotated copy of our subcontract with Aerojet. After one negotiating session at Aerojet's Sacramento headquarters, he purchased a bottle of California brandy and put it in his briefcase with the subcontract document for the trip home to Pittsburgh. The bottle broke, and the resulting odor was impossible to ignore. After a while, it added a pleasing aura to otherwise very dull negotiating sessions.

Fuel

The prime requirement of any reactor design is a high-performance fuel element. Shortly after we received the NERVA contract, it became clear to Thomas that the technical basis for the reference fuel element design was not adequate to support initial performance goals. The reference fuel element consisted of uncoated uranium-dicarbide particles dispersed in a graphite matrix. Because the graphite matrix was not fully dense, highly reactive dicarbide particles could react with ambient air, especially moist air, during storage. This gives rise to reaction products that lead to dimensional instability.

In addition certain reaction products could cause fracturing of the element when rapidly heated to service temperatures. To prevent such reactions, it was imperative that the uranium-dicarbide particles be coated with pyrocarbon before being introduced into the graphite matrix. In addition, the process for NbC coating fuel element channels, although promising, was not sufficiently developed to be confidently scaled up for reactor tests.

Thomas felt strongly that the reference fuel (i.e., uncoated uranium-dicarbide particles in graphite) should not be used in our reactor tests. He persuaded Krasik to insist on using only elements with coated particles and improved channel coating. It was also considered highly desirable from our point of view to be able to do development work on these and related problems in-house. Our perception was that LASL people were reluctant to see us in the role they felt was theirs. This early difference of opinion on technical interfacing was resolved early in the life of the project. So Astronuclear also was engaged in fuel element materials development activities, and we enjoyed a good working relationship with LASL.

The early difficulties with LASL are quite understandable. LASL enjoyed an unparalleled reputation in the scientific community, and probably because of this came across to Astronuclear engineers as somewhat elitist. LASL also was dealing with a relatively unknown organization. The combination of Astronuclear people proving themselves technically and the influence of SNPO helped improve relations.

Until 1967, NERVA fuel development work was under the direction of Dr. Alvin Boltax of the materials department. The principal engineers in his group were M. Blinn, A. Field, G.R. Kilp, M. Tobin and C. Vogel. It is pertinent to point out that, as the program moved on, the engineers at Astronuclear, LASL, and Oak Ridge Y-12 working on fuel element development were in close contact—exchanging concepts and processing details, corrosion test data, and data feedback from reactor tests. The contributions of Y-12 were mainly in the production area. The majority of the fuel for the NRX reactor series was produced by Westinghouse Astrofuel. Y-12 supplied an appreciable number of elements for these tests to our specifications. Thus, as fuel elements were released for manufacture for each reactor test or engine test, the latest information from all sources was utilized in order to provide the best possible elements.

The Astronuclear fuel fabrication facility at Cheswick, Pennsylvania, called Astrofuel, was originally developed by Nick Pollack who then proceeded to head the reactor assembly operation assisted by Ed Cocoros. Fuel element production was carried out under the direction of Meade D'Amore and Chuck Meuche.

Fuel element corrosion resistance was assessed by observing the performance of full-length elements in a test rig. Elements were electrically heated by their own resistance, while at the same time flowing hydrogen at a prototype rate through its coolant channels. The maximum test temperature was set to equal that of the core hot spot. Design of the test apparatus was under the direction of Ed DeZubay. These tests were originally performed at the Large site, but were soon expanded to a new facility at Waltz Mill, which later became the coal gasification test facility.

The most difficult and pertinent problem faced in the test apparatus itself was the hot-end grip electrical connection that was prone to produce excessive local temperatures. The grips had to be designed so nonprototype stresses were not imposed. In the electrical test, the outside surface of the fuel element was protected by a helium atmosphere, while in actual reactor service fuel element surfaces were exposed to a certain amount of hot hydrogen.

As the program evolved and reactor postmortem examination data became available, it was possible to compare results of electrical corrosion tests with in-reactor exposure. Usually, in-reactor exposure exhibited greater corrosion about halfway through the fuel element, including *pinholing*, i.e., localized attack originating on the outside surfaces of fuel elements. Hot-end corrosion results were comparable in the two environments. By the time of NRX A6, electrical and reactor tests were in good agreement.

Initially the electrical test duration was set at 5 minutes; later it was increased to 10 minutes. Elements from each coating batch were tested as a quality control measure. Gross weight loss was determined as a measure of overall corrosion, while detailed laboratory evaluation yielded an incremental weight loss along the length of the element. Details of other features, such as flaking of channel coating and end damage, were also observed.

The fabrication process developed at LASL for the basic graphite element from graphite flour, furfural alcohol, lamp black, and fuel particles was used at the start of the NERVA project. However, development work was needed to improve the hydrogen corrosion resistance of the fuel elements and to control the processing to provide dimensional tolerances on the order of thousandths of an inch as well as to ensure uniformity of fuel loading required to meet stringent reactor temperature and cooling requirements. Fuel loading and dimensional data were used by nuclear and thermal analysis groups in calculating the position of each element in the core.

In simplest terms, the ingredients mentioned above were well mixed and then extruded to a bar shape having a hexagonal cross section with 19 longitudinal holes. At this stage in the process, dimensions are different from those of the completed element, which are 52 in. in length, 3/4 in. across the flats, with a 0.10-in. hole diameter. The extrusion was polymerized, baked at a low temperature, and then graphitized at a higher temperature (2200°C). At the time, the AEC was carrying out intensive development in technology of pyrocarbon-coated UC_2 particles for other reasons. Therefore substitution of coated spherical particles for bare particles was quite straightforward. The spherical particles, nominally of 150 μm in diameter, had a pyrographite-coating thickness of 25 μm.

The semifinished fuel element described above was then subjected to a high-temperature chemical vapor deposition process to coat the surfaces of the longitudinal channels. In simplified terms, the coating process involved passage of a gas mixture containing gaseous niobium pentachloride, hydrogen, and methane through the channels. It reacted with the graphite to form an NbC coating. The purpose of the coating was to inhibit attack by hydrogen in reactor service. However, since the coating was deposited at high temperature, it would crack as it was cooled. Cracks would, of course, allow hydrogen to reach the graphite in service. To some extent, as a coated element was heated to service temperature, there was a tendency for some cracks to close.

The basic problem was to achieve a match between the thermal expansion coefficients of the graphite and the coating material in order to avoid cracking. This applied not only to coolant channel coatings but also to coatings on outer surfaces of fuel elements, hot-end unfueled tips, and support blocks. Thermal-expansivity matching was possible because graphite's expansivity exhibits a large range of values depending on the raw materials used and processing variables. Thus, selection of graphite flour and subsequent processing adjustments resulted in improved coating integrity.

The following types of fuel elements are identified to aid in tracing their development and relationship to the reactor tests:

Type 1: Graphite extrusion with pyrocarbon-coated fuel particles and NbC channel coating. This type of fuel element was used in LASL KIWI tests and in Astronuclear NRX-A2, A3, A4, and A5 reactor tests, altogether spanning about two years during which all aspects of fuel element processing were steadily improving.

While hydrogen corrosion of the fuel channels was a problem that was expected, there were other corrosion modes that turned out to be more important than had originally been thought. These were corrosion that occurs within the pores of the element and that which occurs on the external surfaces. External corrosion was primarily caused by the hydrogen flow at the exit of the elements and leakage into crevices between fuel elements. At the time of the NRX-A2 fuel production, the adherence at the end of the element was so poor that it was not possible to handle the fuel without chipping or flaking the coating. For NRX-A2, elements were accepted that had 75% of the end coated prior to testing. The adherence of the bore of the element was better.

During fuel production it became evident that the vapor deposition process resulted in more adherent coatings on the unfueled than fueled graphite. This phenomenon was gainfully employed by cementing an unfueled tip to the end of the element. Both the bore and the outside of the tip were coated.

By the time of the NRX-A3, the standard for fuel element acceptance increased significantly. By 1965, one of the standard NRX-A4 production elements was tested for 30 minutes and another for 45 minutes. The bore and ends of these elements were in good condition following these tests.

To combat corrosion from leakage between elements at the core periphery, design changes were made on each of the reactors. The aim was to reduce the hydrogen leakage into the bundled fuel elements. While each design change from NRX-A2 to A5 reduced the leakage and allowed longer running times from A2 to A5, a major redesign on NRX-A6 eliminated peripheral leakage and the resultant corrosion problem.

The vapor deposition process of the niobium carbide coatings required precise temperature control. Since these coatings were deposited at high temperature, the coating would crack as it cooled. The aim was to develop a process that resulted in the cracks being reclosed at operating temperatures. As the project proceeded, evolutionary improvements resulted in fuel elements that exceeded the original performance requirements.

To improve the corrosion performance, the thickness of the fuel coating was increased with each reactor. In addition, since the coefficient of expansion of the NbC coating differed from that of the graphite, the coating process was controlled so that the NbC was deposited at the expected hot end operating temperature. In this manner, any coating cracks would close at the reactor operating temperature. The corrosion performance of the elements improved with each NRX reactor. Similar processing changes improved corrosion resistance of the unfueled tips and the support blocks. By the time of the NRX-A5 test, sufficient confidence was developed that a commitment was made for a one-hour test of NRX-A6.

The yield of elements that met quality assurance requirements was low initially. As experience was gained with the process, the yield gradually increased. By the time of NRX-A3, standards for fuel element acceptance were increased significantly. Elements were rejected that had coating flakes larger than 0.01 in. High-temperature quality control corrosion test time was increased from 5 to 10 minutes, and experiments were being conducted for longer times. By the time of NRX-A4 (NRX-EST) fuel element production in 1965, one of the standard elements survived a 30-minute test and another 45 minutes.

During this period, satisfactory bonded and coated hot-end tips, fuel element surface coating, and support-block coating were developed. By the time of the NRX-A5 test, the NbC channel-coating process had improved, but midband corrosion remained a problem.

It is important to note that the purposes of the series of reactor and engine tests did not include a direct determination of endurance time, such as might be found by running the test until fuel element or core failure. Endurance had to be inferred from reactor core postmortem data and from electrical-corrosion tests.

Type 2: Graphite extrusion with pyrocarbon-coated fuel particles, and molybdenum-overcoated NbC channel coating. LASL found that the cracks in the NbC channel coating occurring on cooldown from the coating-process temperature could be filled by applying a thin coat of molybdenum at a low temperature (200°C) by decomposition of molybdenum carbonyl $Mo(CO)_6$. This had the desired effect of reducing midband corrosion in out-of-reactor corrosion tests of one-hour duration and was incorporated in LASL's Phoebus-1B test conducted early in 1967 and in Westinghouse's NRX-A6 test conducted in late 1967.

The excellent performance of the NRX-A6 test and the encouraging postmortem evaluation of corrosion effects provided the incentive to aim for longer running times. Midband corrosion was greatly improved. External fuel corrosion from peripheral leakage had essentially been eliminated. Most of the fuel weight loss was at the hot end. Post-test examination of the elements from the NRX-A6 reactor test indicated significantly longer core and fuel life than were expected at the beginning of the project.

This observation stimulated the development of fuel elements that could provide life times exceeding one hour with recycling capability and higher operating temperatures. These performance improvements became the objective of the Flight Engine design and a continuing fuel element development program. Some of the elements that were examined are listed later.

Type 3: Graphite extrusion with pyrocarbon-coated fuel particles and molybdenum-overcoated zirconium carbide (ZrC) channel coating. Elements of this type were included, among others, in LASL's PEEWEE-1 test late in 1968. They performed reasonably well in that test and were fabricated by LASL. They were somewhat better than Type 2 elements in the same test. In the later NF-1 test, in 1972, elements of this type using Westinghouse special high-expansivity graphite flour behaved better than any other fuel in that test.

Type 4: Graphite extrusion with solid solution mixed carbide (U, Zr)C with ZrC channel coat. The Nuclear Furnace (NF-1) test by LASL (post-NERVA, but Westinghouse provided some sample fuel elements) showed that since fuel particles were not coated, there was severe fission-product recoil damage in the graphite that occurred in the cooler parts of the element near midband and contributed to corrosion in that region.

In out-of-reactor corrosion tests in hydrogen, this type of fuel element (unirradiated) survived a 10-hour test, which included 63 cycles, with only a 10-gram weight loss per element. A potential solution of this fission product recoil problem would obviously be to coat the particles with pyrocarbon to absorb the recoil fission fragments, so this type of fuel element may well have potential. Note that Type 4 fuel was processed for the NERVA flight engine before the recoil fission fragment problem became known, and after the NERVA project termination.

Type 5: Carbide "spaghetti" elements made of (U,Zr)C with or without coating. This type of element consisted of a hexagonal rod, 0.22 in. across flats, with central longitudinal hole 0.12 in. in diameter for coolant flow, made of the mixed carbide (U,Zr)C and having a ratio of carbon-to-metal atoms of less than unity, 0.90 to 0.975, and a melting point of 3500 K/6300°R. It contained no graphite. Longitudinal and circumferential cracking was observed in the cooler portion of the element in the LASL NF-1 test, but the element was judged to have long-range promise in view of good performance in the hotter portions.

In summary, the type of fuel elements with pyrographite-coated beads (i.e., Types 1, 2, and 3) does not suffer recoil fission-fragment damage. They are temperature limited by the melting point of fuel particles, and corrosion limited by midband attack. Should there be a rebirth of the ROVER program, Type 3 could well serve as the initial reference fuel, while at the same time the remaining types would be attractive candidates for development.

Type 4 suffers from recoil fission-product damage at low temperatures, and potentially it has a temperature ceiling of about 200°C above that of the coated type. The carbide, Type 5, is best at highest temperatures. Further development would involve a sorting out of the advantages and limitations of various approaches. The most outstanding out-of-reactor corrosion performance at project end was a Type 4 element that survived a 10-hour test, which included 64 cycles, with only a 10-gram weight loss per element. Thus, the project was terminated before the full potential of these fuel element types could be realized.

Nonfuel Materials

While fuel materials development work involved difficult problems with a rather limited range of constituent materials, the nonfuel materials effort ranged over a much broader field and a wide range of service temperatures. This effort was under the direction of David Goldberg, aided by F.R. Lorenz, E. Layland, A. Hoppe, and others. Nonfuel materials development involved close communication between design and materials personnel. The bulk of the work concerned metals and alloys selected from among established commercial products.

Selected candidate materials were subjected to appropriate testing, including neutron irradiation where appropriate. Process development was continued, if necessary, to meet requirements. Finally, a formal specification was written to satisfy the rigid demands of NASA's quality assurance system. The formal materials specifications for the project comprise three large volumes. Generally, such materials or items fabricated from them were procured from outside vendors, in which case the in-house materials group was utilized for consultation or other assistance as the need arose.

An unusual case was the composite consisting of boron-carbide particles and titanium-hydride particles dispersed in an aluminum matrix, which became known as BATH. Composition was entirely dictated by shielding physics considerations relating to design of the internal shield. Production was assigned to the materials group, which devised fabrication techniques required to produce a workable piece of hardware.

Control Development

One of the key NERVA challenges was development of reactor and engine control systems. Early questions on the reactor dynamics centered on the need for a fast-acting reactor control system. Besides stringent power-level control requirements, the engine had to achieve full thrust from reactor source level in about 60 seconds. Initial NRX and KIWI startups were made by raising the reactor power from source level to a 1-MW hold and then initiating liquid-hydrogen flow and increasing reactor power level by changing the reactivity by relatively slow movement of the control drums.

The NERVA reactor had a small negative coefficient of reactivity with temperature, certainly when compared to the water-cooled reactors to which we were accustomed. The coefficient with hydrogen density was strongly positive, because the reactor was highly undermoderated. Key questions included how much heat could be picked up by the hydrogen as it passed through the nozzle and reflector, and how the reactor would be controlled.

Reactivity effects of cryogenic hydrogen, as it entered the core, were key issues in the initial years of the program. High-density gaseous or liquid hydrogen entering the reflector or core would have a significant positive reactivity effect. For example, the hypothetical case of liquid hydrogen filling core-coolant channels would result in 60 times prompt critical. Of course, such a condition was not possible.

When flow was initiated, the liquid hydrogen that entered the nozzle-cooling channels picked up energy as it cooled the nozzle and the reflector. As a result, the hydrogen, while at a low temperature, was gaseous as it entered the core. Later in the startup, as power increased, the reflector was heated by radiation and the nozzle primarily from heat transfer from core exit gas. Determining the energy that would be picked up by the hydrogen as it passed through the nozzle and reflector was a key analytical issue.

Extensive analyses were made to understand the transient hydrogen flow and nuclear and temperature effects during startup. These analyses included a coordinated digital and analog computer program. For example, steady-state digital gamma heating calculations were used to determine the functional relations between core power and energy deposition in the nozzle and reflector. Fluid flow and heat transfer analyses would establish the energy transferred to the hydrogen. Resultant hydrogen temperature and density in the reflector and at core inlet were then used to calculate the core reactivity effects. The functional relationships for each of these steps were used to design analog computer models capable of simulating dynamic effects.

One of the principal tools for calculating dynamic effects was a NERVA system model called the common analog model. This model, which used about 650 analog amplifiers, was a joint development—with Westinghouse providing the nuclear subsystem portion and Aerojet General, the engine system and nozzle aspects.

Preliminary analysis in 1961 suggested that the reactivity effect of the hydrogen flow rate in the reactor had the potential of being an effective method of control. Bridging the gap between preliminary analog computer-based analyses and actual engine operation required a significant effort.[3] Closely coordinated analytical and experimental programs were initiated that included a series of tests extending from NRX-A2 to NRX-A5.

Prior to each of the Nevada tests predictions were made of key variables—such as reactor power, hydrogen flow, and temperatures. Subsequent to tests, data were carefully reviewed for differences from predictions. Deficiencies in the digital and analog models were then corrected.

For example, the frequency response of the NRX-A2 reactor was measured at power levels up to full power. Control drums were moved to obtain a sinusoidal variation in core reactivity, and measurements were made of key reactor power, temperatures, and hydrogen conditions. To achieve the required accuracy, the control drum positions were controlled to obtain a 10% sinusoidal variation in reactor power. Esselman recalls witnessing sinusoidal reactor power variations of more than 100 MW in reactor power, which was, to say the least, unnerving, but the data obtained compared well with analytical predictions.

This method of obtaining frequency response information consumed many minutes of test time. To overcome this drain on test time, a cross-correlation technique was developed by Art Wassermann by which small perturbations were imposed on the reactor during normal operation. This method provided data on NRX-A3 that equaled the sinusoidal test information. Bob Uhrig, who was active in the noise/cross-correlation field at that time consulted with us and aided in this effort.

This coupling of NRX reactor test data with the analyses increased our confidence in our predictions to the point that a series of experiments was conducted with control drums fixed. An initial experiment was conducted

---

[3]P.J. Blake and W.H. Esselman, "Nuclear Rocket Engine Dynamic Study," presented at International Symposium on Aero-Space Nuclear Propulsion, Las Vegas, Nevada, October 23–27, 1961.

on NRX-A2 to study the effect of hydrogen flow on reactor dynamics. During this experiment, reactor power was controlled between 30 and 60 MW by the reactivity effects of the changing hydrogen flow rate. Control drums were held in a fixed position, and power was proportional to hydrogen flow rate. This experiment indicated excellent reactor stability and inherent self-control. Comparisons with calculations were good.

These test results and agreement with the analytical predictions supplied the impetus for higher power control studies utilizing the inherent properties of the reactor. During NRX-A3 reactor tests, similar experiments were conducted with power levels being varied from 400 to 700 MW as the reactor flow rate was increased.

Another key experiment on NRX-A3 was conducted to test a fixed drum startup. Hydrogen flow was initiated with the reactor at 1 MW and the reactivity effect of the hydrogen caused the power to rise to 30 MW.

During the NRX/EST test this series of experiments was continued. Bootstrap startup was tested in which the breadboard engine power was increased from a 1-MW hold position to 250 MW by increasing hydrogen flow while keeping control drums fixed. Following the initial test, this startup technique was used for all the subsequent startups.

In June 1966, the NRX-A5 reactor startup from source level to a 1-MW hold point was based on a period control. With this type of control, a desired period is selected and the control drums are automatically moved to maintain this exponential rate of power increase. In this case the desired period was set at about two seconds. This method of startup from source level to the 1-MW hold was first used by LASL on the Phoebus test.

After a brief hold at 1 MW, hydrogen flow was initiated and increased so that the reactor power increased to 800 MW in about 50 seconds. Slower control drum and flow adjustments were then used to achieve 1120-MW full-power operating conditions. This combination of period control up to the 1-MW hold, followed by flow control, demonstrated that the reactor could be brought to full power rapidly in a reliable manner. This is similar to the control of an automobile of a few years ago. The speed of the engine was simply increased by opening the throttle to let in more fuel.

This method of startup and control became standard in operating the NRX-A6 reactor and the XE' engine. Typical startup times were 30 seconds from source level to the 1-MW hold and 40 seconds from the hold to full power.

Control system development required establishment of detailed interface conditions between Aerojet General and Westinghouse. These arrangements, which were made in the first year of the NERVA project, assigned responsibility for engine controls to Aerojet and reactor control and reactor temperature control to Westinghouse.

Test Program

The systems design approach, similar to that we learned at Bettis, was initiated in the design of the NERVA reactor. Each subsystem and component was carefully analyzed and subjected to a comprehensive series of tests to verify the design. Each component was tested to as near operating conditions as possible. Graphite fuel elements were tested at full temperature and hydrogen flow rate. Thousands of fuel element tests were conducted to evaluate the quality of the evolving fuel element processing technique. By carefully controlling the fuel element fabrication and coating process, the performance capability was increased from the initial few minutes of operation to over one hour. Similar high-temperature tests were conducted on clusters and the support systems.

A joint test team composed of Aerojet General and Westinghouse personnel was set up at Jackass Flats, Nevada. Westinghouse had lead responsibility for the reactor NRX tests and Aerojet General the Engine XE tests. During this series of tests the Nevada site organization was managed successively by Bill Henderson, Perry Davison, Dave Bolender, Wade Morris, and Jay Powell. They reported first to Max Johnson, then to Howard Arnold and then to Al Sanderson.

The NERVA test program at Jackass Flats started in 1963 with the A1 cold flow tests. There was a joint test organization with Aerojet's Norv Erickson as the titular head, but Westinghouse held the reactor lead and was in direct line management to run the tests. Aerojet was to run the engine program when it started, and Aerojet's Chuck Rice was in charge of test XE-1', with Westinghouse still directly responsible for the reactor portions.

Max Johnson played an important role with LASL senior management in setting up the Westinghouse Nevada organization, because his naval program and WTR reactor operations management experience was reassuring to them. Before Sid Krasik's death, Sid's stature in the technical community and his evident grasp of the technology caused most engineering arguments to end in Westinghouse's favor. After his death, we had sufficient momentum and Esselman, Arnold, and Roman had sufficient stature by then to continue to carry the day technically. Woody Johnson, who came in as general manager and acting NERVA project manager, exhibited the toughness necessary to keep the Aerojet relationship under control, assisted by Hal Faught, who was a super negotiator, and able to get along well with the administrative people in the customer organizations. Faught also understood the technical aspects of the program better than anyone else in an administrative position and thus couldn't be buffaloed. Roman was universally respected as a talented designer. Esselman was just plain more knowledgeable and diligent than anyone Aerojet had in any senior management position, while Arnold had organized the analytical areas around people excellent in each of their fields. This was one of the most capable teams Westinghouse ever fielded. They believed in themselves and the organization and for good reason.

After Woody Johnson had been in charge for a few months, he redistributed the management talent to address the changing nature of the project. Esselman was appointed project manager and the old three-deputy engineering organization was replaced with straight lines of authority. Max Johnson became czar of all fuel, as the problem of production was beginning to need tight coupling with development. Arnold was put in charge of the Nevada testing program and all supporting activities at the home plant, while Roman headed design engineering. This organization lasted from before the time of NRX-A2 until after NRX-A5, when it was in turn changed as discussed later in this chapter.

While these series of tests were being conducted, an interesting design and test sequence was established so two reactors could be tested per year. While one reactor, for example, NRX-A2 was being tested, NRX-A3 was being designed. The lessons learned from NRX-A1 test were being incorporated into the A3 design. Similarly, the lessons learned on NRX-A2 were incorporated in NRX-A4. This sequence allowed rapid introduction of the information learned on reactor and fuel element performance into later designs. Significant improvements in the fuel element fabrication and coating techniques allowed the improvements in core performance that occurred.

NRX Test Series[4]

Large, Pennsylvania

Preparations for each NRX/XE test included the engineering and assembling of the reactor system at Large. This preparation required significant attention to processes by which the components would be made. Suppliers had to be developed for each of the components. In many cases the manufacturing processes were developmental and required significant Astronuclear input. Fred Reed, who came from the Aviation Gas turbine division, was the key Astronuclear resource on developments of this type. Fred did some very innovative work on coupling of computers and machine tools, precursor of much of today's computer-aided manufacturing (CAM).

---

[4]W.H. Arnold and W.H. Esselman, Private Communications.

NASA quality control procedures were established. Each part received careful quality control examinations and was placed into a bonded storeroom. Fred Henning and Mike Regenda are responsible for many of these procedures.

Astronuclear reactor assembly facilities including the fuel examinations were developed by Nick Pollack who had received core assembly experience at Bettis. Nick had responsibility for assembling of all of the reactors in the NRX/XE series. He was aided in this effort by Ed Cocoros.

Acceptance of the fuel elements was made after careful measurements such as fuel loading, dimensions, and pressure drop through each of the 19 channels in a fuel element and, of course, a corrosion test. Elements were individually selected for specific locations in the core and each channel was individually orificed. Selection of the elements was made by the Analysis group under F. Retallick or Joe Gallagher.

The core was assembled and mounted in the aluminum reactor vessel and then shipped to the Nevada test site for test

Nevada Test Site—Jackass Flats

The core was assembled in the aluminum reactor vessel and shipped to Nevada in a steel overpack as described in Chapter 6. This decision was made difficult by our interface with Aerojet. Since they were responsible for the reactor vessel, they were originally reluctant to ship it to Pittsburgh and allow Westinghouse to include it in their reactor assembly. Instead, they proposed shipment of a "bare" reactor followed by another assembly into the vessel at the point of use. As was usual in such cases, SNPO had to conduct a review and listen to both sides before deciding on the preferred technical choice.

While the reactor was being assembled at Large, the Jackass Flats staff were assembling the test car and readying the nozzle and other parts needed for the test. The Jackass Flats team was composed of LASL, Astronuclear, and Aerojet people as described before.

The NRX tests were conducted with the reactor mounted on a test car in an up-firing position. To carry out the ROVER and NERVA programs it was necessary to construct an array of sophisticated facilities. These included two reactor test stands, an engine test stand equipped to permit firing under space vacuum conditions, and two maintenance and disassembly facilities (MAD) capable of handling highly radioactive test items.

The test car was a railroad car modified to provide a shield region for control actuators and electrical equipment. Following a test, the test car was remotely brought into a hot cell and the reactor was dissembled for postmortem examination. This approach had been developed and applied by LASL for its reactor tests.

The following tests were conducted in the NRX series.

NRX-A1. The NRX-A1 test was a non-nuclear one to prove that the bundled core design did not experience a flow-induced vibration. The nonfueled elements were well instrumented to show that no motions were observed as reactor flow was increased.

NRX-A2. The NRX-A2 test was basically a proof of principle. In July 1964, the NRX-A2 reactor was delivered to the site in Nevada. First criticality was achieved on August 12, and on September 24 a hot test of the reactor was conducted. The reactor was brought to 1050 MW for some 110 seconds and then increased to 1110 MW in the last 40 seconds. Total running time at more than 50% power was more than 6 minutes. On October 15, a second run was conducted, when the reactor was operated at low power for more than 20 minutes. Frequency response and control tests were conducted at several power levels up to 100% power.

This highly successful test was a vindication of the Westinghouse role in NERVA, and resulted in our acceptance into the LASL-dominated fraternity. As had been the custom in the previous KIWI series, our people reaped a reward in the form of radiation-darkened customized ashtrays and other glass souvenirs. It had been the custom to load boxes of such material near the reactor. After a successful test, the souvenirs were surveyed for radiation safety and returned to their owners.

NRX-A3. The NRX-3 reactor, in January 1965, was delivered to Nevada and criticality was achieved April 7. Tests were conducted throughout April in preparation for the full-power run. The test was performed on April 23, 1965. The start of this test was a major disappointment, which later proved unwarranted. Full-power operation was achieved after a hold at 50% power and then one at 80%. The operation proceeded smoothly until a sudden scram occurred after 3.5 minutes at full power.

This scram was caused by a loose connection in a turbine overspeed circuit. The result was a complete loss of hydrogen flow to the reactor. Conditions were as severe as one might imagine in a system of this type. The test cell was designed to start gas flow during a loss of liquid hydrogen; however, prior to this gas cooling becoming effective, tie-rod temperatures rose sharply. Concern was expressed over the integrity of the reactor tie-rod supports and the nozzle. Visual examination of the nozzle indicated a possible melted section.

After a crash analysis by Retallick and Gallagher of reactor conditions during the transient, it was concluded that the maximum temperature attained by the reactor tie rod was 2500°R and that its structural integrity was not impaired. This was tested by a crash component test of the tie rod and its associated graphite element, and it was decided that the core could be restarted. At the calculated time at temperature, it was decided that some reaction had occurred between the Inconel tie rod support and the graphite, but not enough to affect structural integrity. In addition, Chuck Rice of Aerojet gave us the good news that the nozzle we thought was severely damaged was suitable for testing.

In May, the NRX-A3 reactor was restarted and operated for approximately 16 minutes at power, of which 13.1 minutes were at full power. During the test, no major difficulties were encountered with the reactor. About a week later, the NRX-A3 reactor was restarted for the third time, and medium-power mapping and control tests were conducted.

Postmortem examination of the NRX-A3 core was watched with great anticipation. We wanted to know whether the core damage would compare with that calculated. The observed condition of the tie rod assembly showed the temperature reached the calculated level and the Inconel and graphite interaction appeared as observed in the out-of-reactor component test. This experience gave us great confidence in our ability to analyze and predict conditions in the reactor. The original test of NRX-A3 did not proceed in the manner planned, but the experience of a shutdown and the ability to calculate precise core conditions were a great boost to our confidence. We felt we really understood the reactor characteristics.

NRX-A4/EST. The NRX-A4 reactor test series included engine systems tests, and the nomenclature NRX/EST was therefore assigned. This series, conducted in December 1965, was highly successful in meeting its objective of testing the system dynamic behavior of what amounted to a breadboard engine system. These tests, conducted in the up-firing position, successfully checked engine system bootstrap start and restart capabilities.

Ten restarts to full power were accomplished. The total operating time at power was about 116 minutes, of which 24 minutes were at full power. In addition, fixed drum startup tests were continued.

NRX-A5. In June 1966, the NRX-A5 control experiments series completed the series started on NRX-A2. The reactor was brought to a low-power hold position in 40 seconds using the LASL-developed period control and then to 800 MW with control drums fixed. The power level was then trimmed to full power. Total time from source level to full power was about 90 seconds. This method of startup became the standard procedure on later NRX-A6 and XE' tests.

The NRX-A5 was operated in two 15-minute periods for a total of 30 minutes of full-power operation. While fuel element performance remained a problem, their corrosion resistance had improved with each NRX test. The postmortem examination of the NRX-A5 reactor and fuel elements supplied the basis for setting an ambitious goal of a 60-minute run for the NRX-A6.

One incident stands out that typifies the ingenuity and adaptability of the test site people. After the first run in NRX-A5, a bird was observed to alight on the outlet edge of the reactor nozzle, and then drop dead into the reactor outlet cavity. It was deduced that it had not been killed by radiation, but rather asphyxiated by the inert gas that was continually run through the system as a purge to prevent oxidation (helium during the initial stages, and then nitrogen after the reactor was sufficiently cool). Nothing could be seen of the bird, but considerable analysis was performed to determine its possible effects during startup and operation, how much flow blockage would result before it burned up or blew away, and whether or not there would be mechanical damage. Finally Walt Esselman presided over a meeting that decided it must be removed, and the question was how. Within days, the site people had rigged up a vacuum hose attached to a remote manipulator arm that was guided by an operator working behind the test shield wall, viewing through a remote TV camera. The removal was successful, and the bird encased in plastic. For many years it resided on the desk of our resident manager in Nevada, after the radiation technicians had surveyed it.

NRX-A6. NRX-A6 was tested for 60 minutes at full power in December 1967, and the duration goal was achieved. Results of the postoperating examination of the reactor showed it could have been operated for a significantly longer time. Extremely good fuel became available on NRX-A6. The NbC channel coating had been overcoated with molybdenum to reduce corrosion due to cracking of the carbide coat.

XE'. The final tests of the project were complete engine tests. NERVA-XE' was conducted by Aerojet General in September 1969. These tests were run in a specially constructed engine test stand (ETS) in which the engine was fired downward into a simulated space environment maintained at 1.5 psia.

Among the tests were bootstrap startups, multiple restarts, and throttling of the engine. The bootstrap involved bleeding hydrogen from the reactor to start the turbopump as the reactor came up to full power. The engine was started 28 times, for a cumulative time of 3 hours and 48 minutes, of which 3.5 hours were at full power. Fuel elements used in this reactor had NbC channel coating with molybdenum overcoat, as did NRX-A6. All technical objectives were met or exceeded.

This test demonstrated engine operation over its complete design map, including full thrust, multiple restarts, and bootstrap capability. The reactor used in this test contained NRX-A6 grade fuel and hence performed well. A reactor assembly that contained earlier versions of the fuel was not used in the XE test.

Summary of the Test and Development Program

To assess progress made on the many difficult NERVA development problems, a base point of early 1964 can be used as a reference. At that time, no reactor had been operated at the high power and temperature conditions essential to achieve desired rocket performance. Even many feasibility questions existed.

Before this test series, questions existed about the predictability, controllability, and reliability of the NRX reactor and XE engine operation. Other questions existed about restart capability. Would the reactor, and particularly the fuel elements, have sufficient integrity to recycle through several restarts without failure? Tests conducted in this test series resolved all these issues.

Structural integrity and performance were proven by successful power operation of 12 reactors. Full power was demonstrated on both KIWI and NRX tests in 1964. In 1965, capability to restart was shown by one restart to full power in NRX-A2 and three in NRX-A3. Confidence grew, so a goal of 60 minutes was set for reactor operation. This goal was met by tests of the NRX-A6 reactor in 1967. That this reactor could have operated for significantly longer times was shown by postexamination of the core.

The best fuel appeared in NRX-A6 and XE', and both were highly successful tests of reactor and engine. The engine in the XE' was much more like the flight configuration than in A4/EST, but it still was transitional. The program to develop a true flight engine was never authorized. Before termination we were planning an A7 to run several hours under multiple restart conditions—and an XE-2 to further explore the engine operating regime.

The XE engine, which was to be tested, was to have been followed by tests of engines with flight shields, control drum actuators, and on-board control systems. This was to be followed by ground tests of complete flight vehicles, i.e., a flight-configured liquid hydrogen tank rather than test-site spherical Dewar flasks.

Following the test series, development continued on a flight engine design and improvements in fuel element processing. Fuel element development included the development of ZrC coatings and the use of (U,Zr)C solid solution particles in a graphite matrix. Out-of-reactor laboratory corrosion tests were made on the fuel for an operating period of 10 hours, and the corrosion resistance of these elements was exceptionally good.

Flight Engine. During the latter stages of the NERVA engine development, a flight-engine design program was initiated. NASA studies of the application of the optimum NERVA engine focused on its use to propel a third stage of a Saturn vehicle on a Mars mission. Conceptual studies of the stage design considered engines of different thrust levels and operating times. Higher thrust levels required shorter engine running times.

For a time it appeared that an engine of 200,000-pounds thrust operating for about 20 minutes would be required for the Mars mission. The Los Alamos Phoebus reactor tests were based on the potential need for engines of this rating. As reactor testing demonstrated the feasibility of longer operating times, the optimization studies showed the merits of a lower thrust engine with longer running times.

An optimum configuration with a 75,000-pound thrust engine with a specific impulse of 825 seconds was chosen for the flight engine design. The overall reliability requirement was confidently set at an extremely high level of 0.997 at 90% confidence level.

NASA's configuration management and systems engineering techniques were introduced to meet preliminary design review (PDR) milestones. The output of that process produced a set of design requirements and performance specifications for the next-generation reactor and engine designs known as R-1 and E-1, respectively. The PDR of the resultant reactor design met the requirements of a flight engine and the R-1 test was scheduled for early 1973. The program was discontinued before the reactor was tested.

The NERVA flight engine reactor design and fuel element geometries were similar to that tested in NRX-A6 and XE'. Since the reactor thermal power was increased to 1500 MW, about 200 fuel elements were added to make a total of 1978.

Two options were available for the choice of the fuel element. The first consisted of pyrocarbon-coated uranium-dicarbide particles dispersed in a graphite matrix, NbC coated with Mo overcoat (Type 2 fuel element) and a maximum loading of 630 mg/cm^3. The second would have consisted of solid solution (U,Zr)C particles dispersed in a graphite matrix that was Zr coated (Type 4 fuel element). (It was after the termination of the NERVA contract that the recoil fission damage in Type 4 fuel elements was discovered.) Fuel element reactivity loss from corrosion was limited to $1 at the end of life. A dollar of reactivity is the amount of reactivity required to achieve prompt critical.

The number of control drums in the reflector was increased from 12 on NRX to 18 to reduce circumferential flux variations. In addition, a greater number of fuel loadings were specified so that the peak to average power in the core could be further reduced. The control drums were driven by an electrical stepping motor with an epicyclic planetary transmission.

Liquid hydrogen is pumped from the propellant tank to engine operating pressures as shown schematically in Figure 14.1. After the hydrogen passed through the tubes of the regeneratively cooled nozzle and reflector, it entered the shield.

The shield at the inlet end of the reactor shadowed the engine components, vehicle, and payload and protected them from radiation and excessive heating. The shield was composed of boron carbide-aluminum-titanium hydride (BATH) composite supported on an aluminum structure. This design was never incorporated in any of the NRX or XE tests.

A fraction of the hydrogen flow is used to cool the tie tubes. Energy picked up in cooling the tie tubes is use to power the main turbopump. The hydrogen passing through the core is heated to approximately 4000°F when it enters the nozzle.

The reactor design approved at the PDR met the requirements shown in Table 14.1. A synopsis of top-level ROVER and NERVA Test requirements follows.

Weight Objectives. The total nuclear subsystem weight objective for the composite fuel core (not a requirement) was less than 6800 kg (15,000 pounds).

Launching and Safety Program. A significant amount of analysis and experimental work was undertaken to evaluate flight safety issues associated with application of a nuclear rocket engine. This effort, coordinated by a flight safety committee headed by Ralph Decker of SNPO, included efforts by LASL, Westinghouse, Aerojet General, and NASA specialists from Cape Canaveral, Huntsville, and Lewis laboratories. Esselman was an initial member of this committee. He was followed by Arnold, who directed the Astronuclear work on this subject throughout most of the project.

The safety program had two aspects, the ground and the flight portions. Analysis included studies of various launch modes and flight paths. The SNPO office pioneered the use of probabilistic risk assessment, although it was not called that in those days. Probabilistic analyses were made of possible aborted missions. These cases included studies of the potential reactor excursion that would result from a reactor entering the ocean. Activity of reactor particles following engine operation and burnup on reentry were included in the analyses.

The objective of the breakup was to assure sufficient dispersal so no appreciable concentration would survive reentry and lead to contaminated spots on the ground.

If it is sufficiently dispersed in high reentry, the material will be spread through a volume large enough to keep concentration minimal. More recent thinking is to restrict operations to orbits high enough that reentry is just not feasible, as we could not assure reentry burnup of ceramic fuel.

On the basis of directions from SNPO, we never contemplated a launch from the ground mode, all operations being intended to commence after a chemical boost phase put the engine/vehicle into orbit. Neither we nor the AEC could see any insurmountable problem with a launch. The AEC routinely launches isotope power supplies. Nuclear safety review was performed by an office in SNPO, which in turn was reviewed by the AEC division of licensing and regulation.

The potential hazard from nuclear reactors is principally measured by the amount of fission products already created. The best way to think of fission product inventory is simply to compare megawatt-seconds. NERVA was to be 1500 MW(th), but it would run at most a few hours. Thus, you can see that the amount of fission products would be exceedingly small compared with those of a commercial electric power nuclear reactor. Such a 1500-MW(e) reactor, after a year of operation, would have almost 25,000 times as many fission products as a 1500-MW(th) nuclear rocket that had run for an hour, longer than for a flight to the moon and return.

Any possible launch accident would involve essentially no inventory of fission products because of the poison-wire safety system. It is only above zero if a critical experiment is run as part of the quality assurance program, supporting reactor assembly and engine checkout. Even then it is just above zero. Even the unthinkable case of a launch pad criticality accident involves only a few thousand megawatt-seconds before criticality is extinguished by dispersal.

Flight safety was again an area of contention with Aerojet. We had an active lab program and learned much about reentry behavior. We learned how to design the system to obtain the desired degree of breakup in case of accidents, how to pre-safe the system with poison wires to make sure it was inert during all conceivable launch conditions, and how to assure that potential astronauts weren't subjected to undue radiation.

NERVA Flight Capabilities. Performance capabilities of the NERVA engine were the subject of extensive vehicle and mission studies.[5,6] The NASA effort concentrated on using the engine to propel a third stage of Saturn V or an upgraded Saturn V. Many of these studies were performed by Paul Johnson of NASA Headquarters and others by Marshall Space Center staff. Much work had to be done to determine an optimum engine size and stage design.

The performance advantage of the NERVA engine results from its specific impulse of 825 seconds—about double that of a chemical engine. The best way of visualizing the effect of this enhanced performance is by examining a few examples.

One of the interesting studies was a manned lunar mission that avoided the requirement for a lunar orbit. With this mission the crew could land anywhere on the Moon and explorations could be performed at all latitudes. The minimum required lunar weight for this mission included the crew, supplies, and propellant to return and was 31,000 pounds. This mission could not be accomplished with an all-chemical-propulsion Saturn V vehicle. While this vehicle could launch an escape payload of 95,000 pounds, only 28,000 pounds could be landed on the Moon.

At a NASA briefing in November 1964, Finger described the advantage of a nuclear stage for this mission as follows: "If we put the nuclear stage, using the NERVA engine at the performance already achieved on the Saturn V vehicle, we could deliver a landed payload of about 47,000 pounds." This would give enough excess payload over the minimum to conduct a 30-day stay-time mission. A second Saturn V launch could supply another 47,000 pounds and allow significantly longer stay-time.

Planetary missions can benefit from the performance of the nuclear rocket stage. For example, Saturn V with a nuclear third stage could put 30,000 to 50,000 pounds in a Mars orbit or 22,000 pounds in a Venus orbit.

Keen interest existed on the requirements for the manned Mars mission requirements. This mission, which is described in Chapter 6, would start with the assembly of a spacecraft in orbit. For the trip to Mars one stage would be fired to leave the earth orbit. Another nuclear stage would be fired to enter Mars orbit and land. Landing and return to Mars orbit would be accomplished with chemical engines. Return to Earth would be powered by third nuclear stage.

---

[5]Harold B. Finger, NASA Briefing, November 12, 1964.
[6]Harold B. Finger, Testimony before Congress on Aeronautical and Spaces Sciences, U.S. Senate, August 24, 1965.

The spacecraft was designed to carry a six-man crew, support equipment and supplies for the seven and one-half month journey to Mars plus a similar time for the return trip. The time on Mars would be a few weeks.

The weight in earth orbit required to accomplish this mission varies with the year of launch. For an all-chemical system the required weight in earth orbit can vary from about 3 million to about 25 million pounds.

With a nuclear system, the orbital weight requirements would vary from 1 million to 5 million pounds. Therefore the nuclear system reduced the required weight in earth orbit between 2 and 20 million pounds depending on the year of launch. An upgraded Saturn V with a nuclear third stage could put about 300,000 pounds in earth orbit. Hence about 3 to 15 Saturn V launches would be required to support such a mission.

Non-NERVA Contracts and Development

The side of the house that dealt with non-NERVA work was called systems and technology. It encompassed a broad spectrum of technologies that provided a stimulating, exciting environment in which to work. Many even felt it was more interesting than the large, but circumscribed, nuclear rocket project. As previously mentioned, the approach to any given project was to first consider the system in its entirety, and then to consider the individual components. In cases where a subassembly or component was our initial entry, it was nevertheless advantageous to understand the total system in which it was to be used.

For example, consider an isotope-powered thermoelectric system for a space application. It generally comprises an encapsulated radioisotope fuel form, a thermoelectric assembly, a thermal path to a radiator and a power conditioner of some description, at least. Computer codes were developed to handle such systems, and indeed subroutines describing the behavior of each component. Such computer models could be used to optimize the system with respect to power-to-weight ratio, or to some other parameter. And so it was with many of the systems with which we were concerned, a few of which are illustrated in Table 14.4.

Rather than giving a detailed description of the many projects that were undertaken, it will suffice to describe in general terms the technologies that were involved, and to describe a few hardware items that were built.

We became interested in thermoelectrics with a small program involving a tubular thermoelectric module that provided space in the innermost tube for UO_2, the thought being to eventually develop a thermoelectric fuel element. We did not follow the fuel element idea, but instead concentrated on developing a larger tubular module where the innermost tube could be filled with an isotope fuel form, or with hot flowing liquid metal, such as

Table 14.4
Non-NERVA Development Programs

Project	Description	Investigator
Thermoelectric conversion	Tubular module with lead telluride as the active material	J. Kenney
Compact converter	Radioisotope Thermoelectric generator with heat pipe and tubular TE modules	D. Roberts T. Varljen, C. Rose
Artificial heart	Isotope-powered artificial heart for total replacement	D. Pouchot
SNAP 23A	Sr-90 isotope thermoelectric power-generation system for 60-W, 24-V, dc 10-yr life	C. Kim
NERVA/gas dynamic laser	A NERVA-type reactor used to heat seeded helium in a high-powered gas dynamic laser	Esselman
Co-60 heat source	A self-regulating heat source using heat pipes for Brayton cycle remote power	J. Sadler
Weapon effects evaluation	Electromagnetic pulse and effects of same	P. Dickson
Multi-hundred watt	Large power supply for space applications using TE	C. Sinclair
NuEra	Liquid-metal reactor system for aircraft propulsion	A. Bethel
Tantalum alloy development	Alloys T-10W, T-111, T-222 ASTAR-811C, ASTAR-1211C, and ASTAR-1511C	R.T. Begley, R.W. Buckman, R.L. Ammon
Niobium alloy development	Alloys B-66, B-77, and B-88	R.T. Begley, R.W. Buckman, R.L. Ammon

sodium. The outermost tube served as a heat sink, to be connected to a radiator, directly or via a second liquid metal circuit. Thermoelectric washers, alternately p- and n-lead telluride, appropriate insulators, and electrical connectors at the hot and cold sides of the thermoelectric washers were placed between the innermost and outermost tubes. Electric leads emerge from the ends of the module. All of the components were assembled and hot isostatically pressed. The resulting module was void-free and hermetically sealed.

This was undoubtedly the most complex hot isostatically pressed article in the world. Measuring about 1 ft long by 2.5 in. in diameter, each module generated about 1 kW. Analysis of its operation and behavior represented a challenge since the electrical, thermal, and mechanical aspects were closely coupled. Obviously such a thermoelectric module represented a component that could be used in many applications.

Other components or subsystems that were the subject of design, analysis, or experimentation are as follows:

- Isotope heat sources, including launch and reentry safety
- Reactor heat sources, including launch and reentry safety
- Space radiators
- Rankine cycle cesium vapor turbines
- Sterling cycle engines (for heart pumps)
- Rotating electric generators
- Heat pipes
- Liquid metal pumps
- High-temperature semiconductor devices (silicon carbide).

The scope of technology represented was wide range. The thermoelectric technology was referred to earlier as was the refractory alloy program. The latter, resulting in distinct alloy compositions, included not only the melting and fabrication of experimental alloys, but also the determination of high-temperature properties requiring sophisticated test equipment and methods.

Another technological area that required considerable study was the various aspects of liquid metals (sodium, potassium, cesium and sodium-potassium alloy). Employed as heat transfer fluids, the selection of containment materials required a knowledge of their corrosion properties. While much data existed in the literature, experimentation in some cases was necessary. For example, we designed and built a high-temperature loop of the tantalum alloy T111, which used cesium vapor as the working fluid with a special venturi to simulate the corrosion/erosion in a cesium turbine.

A technology that emerged in the 1960s was the heat pipe. It can be viewed as a system or as a component, and we not only developed computer models of thermal behavior, but built many such devices.

These and other technologies often led us far afield. Prior work at the central research laboratories had resulted in an electron gun that could be adapted to welding. A joint program led to an electron-beam welder for the Marshall Space Flight Center, which was later flown in 1975 on the Skylab. Another outgrowth of the electron gun was an adaptation so that vacuum welding could be performed where such welding could not otherwise be done.

Our background in radiation and other nuclear matters led to the study of the electromagnetic pulse (EMP) that results from a nuclear bomb detonation. Our in-depth study led to our serving as a corporate source of EMP effects data, particularly for the aerospace division at Baltimore. An unsuccessful bid to design and build a full-scale EMP test facility capable of testing complete aircraft was made.

We also became involved in an effort to develop an implantable artificial heart. This required not only learning the physiology involved, but also required the development of multifoil insulation for the isotope heat source (Pu-238), a Sterling cycle/electric power supply, and a blood pump—all implantable. Full system implantability was shown to be infeasible, and attention was directed to externally powered systems. Some externally powered heart pumps were tested in live calves. These tests were conducted at a university hospital facility.

The thermoelectric expertise led to the design and construction of an environmental suit for the Army. The suit was heated or cooled by thermoelectrics as needed for the ambient temperature. Another application was air conditioning for the submarine *Dolphin*. These units used thermoelectric cooling, which contributed to low noise submarine operation.

Since one of the factors that had caused the demise of the Aircraft Nuclear Propulsion program was the lack of an aircraft large enough to carry the nuclear system, there was a brief period of renewed interest by the Air Force when the Lockheed C5 transport became a reality. This plane with a million-pound payload capacity was capable of carrying a nuclear plant. We were funded to produce a conceptual design study of a nuclear plant in which the reactor and primary shielding would be centrally located, and would heat liquid sodium or potassium that would be circulated to the combustion chamber of each of four jet engines. For takeoff and landing the combustion chambers would be heated solely by chemical fuel. The study was known as NuEra. We concluded that such a system was feasible, but the Air Force declined to carry the program to the hardware stage.

It was evident from the beginning of the Astronuclear Laboratory that the scope of interest and the scope of opportunity went beyond the scope of space applications of nuclear power. What has been described above summarizes, perhaps too briefly, what the systems and technology part of the laboratory had been doing up to the time of the cancellation of the NERVA project in 1971. Beyond that point further diversification of Astro's efforts were previously mentioned in Chapter 6.

CHAPTER 15

The Commercial Nuclear Technical Story

Although commercial pressurized water reactors (PWRs) were able to build on the technology developed at Bettis for the naval program and Shippingport, much was needed to permit them to play a major role in future generation of electricity. In this chapter a number of miscellaneous technical developments are covered.

The Yankee-Rowe plant had three loops and was rated at 100 MW(e) with the reactor designed by Monsanto and Fluor and was to produce both plutonium and power. However, another loop was added by Westinghouse—with a proportionate increase in power output to 134 MW(e)—when power was the design objective.

During the R&D program, there were many meetings between Westinghouse and Yankee, usually in Pittsburgh. At one of them, Westinghouse advised Yankee it probably could achieve, but could not guarantee, the 134-MW(e) rating and decrease by a foot the height of the core, which was an 8.5-foot-high, right circular cylinder. This would have permitted reducing the reactor vessel height by 3 feet, while still accommodating complete withdrawal and insertion of control rods and followers. Followers, which were attached to the control rods, were made of zirconium and were drawn into the core as the control rods were withdrawn to keep the metal-to-water ratio constant and to prevent coolant water bypassing the fuel. Because of the associated decrease in inventory of reactor high-pressure water, the reactor containment building was also decreased in size.

There was general consensus that the reactor was amply designed to allow for guaranteed output performance. Yankee (Roger Coe) decided that the contractual performance would be reduced from 134 to 110 MW(e). He made this decision based on Westinghouse's stated belief that it would be possible to reach the 134-MW(e) level by means of a two-region core and the extra margin of safety that could be reduced after getting test results from core instrumentation. The core design was decreased by one foot.

Power output increases up to 175 MW(e) were implemented because of conservatively designed components, which in some cases had to be modified. For instance, the turbine generator was upgraded. Licensing increases were possible because in-core instrumentation readings could document improved hot channel factors, and dry and saturated steam was produced at the increased ratings.

Yankee was originally designed and operated at 134 MW(e), and eventually was licensed for and reached 185 MW(e), which confirmed earlier opinions of the margin for more power output.

Power increases were possible at Yankee for several reasons. Boric acid was used in the reactor coolant at power. Multiregion core-fuel loading was used, and new physics and thermal-hydraulic calculations led to new control rod programs. And, of course, experience and confidence from successful operations played a part. Half of the second core was reused with new fuel to form the third core. Enrichment of the added fuel was increased over that of earlier fuel to compensate for reactivity lost by reusing some of core two fuel.

San Onofre-1 and Conn Yankee utilized slightly enriched uranium dioxide in Zircaloy tubes. For these two plants, we were increasing our ability to calculate the flux distribution, which determines the hot spots, and therefore could take advantage of the extra margin of safety. These advances could not be tested out in advance, thus there was considerable economic risk involved.

Design of Yankee was nearly completed, and initial core enrichment had been set with the aid of critical experiments and some data on thermal neutron capture by fission products as they built up. We felt pretty comfortable that we could calculate the effects due to uranium-236, plutonium, etc., but we set the enrichment about 0.1% on the high side of nominal to be safe.

The core was already in fabrication when Howard Arnold came on some information showing that absorption of resonance-energy neutrons by fission products was relatively more important in the spectrum of our reactors than in those from which our existing data came. The calculated effect was about a 20% increase in an absorption cross section, which shortened reactor life by as much as 30%. We weren't too sure about our reactivity calculations anyway, but this was surely a step in the wrong direction. This would have been a severe economic penalty, if we had not overcome the problem by design changes. This was one of several "Black Fridays." For Yankee we just got by, but all of our margin was gone and everything else had to go right with the reactivity. Fortunately, it did, as we had several quotations out with guaranteed lifetimes, which presented a potentially major problem.

Walt Dollard, then an engineer in the fuel department, remembers the day well. It was followed soon after by an emergency Sunday meeting Arnold called soon after I came aboard as a group vice president. Howard Arnold came to this meeting in his uniform, directly from an Army Reserve weekend drill. He still had his pistol holster on, which many people in the room thought was quite appropriate for his self-defense.

Later problems of fuel element corrosion and fatigue failure cracks arose. These appeared to be extremely serious at the time, as they meant shorter fuel life time or perhaps even going to stainless steel cladding, with a staggering added cost, several billions for the contracts already on the books.

Fuel Elements

The original Yankee plutonium-power plant considered uranium-aluminum alloy fuel elements, to facilitate chemical processing of $30-per-gram plutonium. However, electric power economics later motivated the move to higher temperatures than aluminum could survive. Proven uranium-niobium alloys did not accommodate enough low-enrichment uranium inventory to permit long fuel life. Slightly enriched uranium-dioxide pellets were selected. Westinghouse, at Bettis, had proven experimentally that natural uranium-dioxide pellets in zirconium tubes were compatible with corrosive hot pressurized reactor water and were resistant to radiation damage.

Because reactor-grade zirconium was not available commercially, stainless steel tubing was specified for Yankee fuel elements. Slightly enriched uranium-dioxide pellets in stainless steel tubes were tested in the materials testing reactor in Idaho as part of the Yankee R&D program. The Atomic Energy Commission (AEC) questioned paying five million dollars for an R&D program "to put a neutron poison such as stainless steel in a reactor." However, the AEC accepted the economic case made for Yankee fuel elements.

There were several significant advantages to stainless steel. The economics of slightly enriched uranium, extracted from existing diffusion plants, were excellent. Hydrogen production from chemical reaction of zirconium and water does not occur with stainless steel fuel cladding. Effectiveness of the stainless steel for absorbing neutrons in the reactor core is lower because of high uranium loading and the many fission products present at the end of long fuel burnup.

The transition from stainless steel to zirconium cladding occurred in 1965. Experimental fuel elements, furnished by Exxon Nuclear, had been run in the core. These elements were held together mechanically, rather than by brazing, like the stainless steel elements. There were problems with pellets ratcheting and swelling. Later, a different design, in which the fuel assembly was held together by four nonfuel rods in the corners, was operated successfully, and the transition to zirconium fuel cladding was made.

Enrichment could be somewhat reduced with the zirconium cladding. Because of concern over a possible zirconium-water chemical reaction, which might release hydrogen, Yankee reviewed the plant design and provided additional emergency-water-injection capability.

A loss-of-coolant accident may partially uncover fuel before the emergency water injection refills the reactor vessel. When zirconium is not covered by the coolant water but is in a steam environment, it heats and starts to react with the steam. The heat energy from this reaction may equal or exceed that of the nuclear heat from fission products. Under these conditions, the surface temperature of the zirconium might exceed the threshold temperature specified by AEC and agreed to by reactor manufacturers. To reduce the probability of zirconium reaching this threshold temperature, emergency-water-injection capability is increased and time delay is minimized.

There is a borated-water tank pressurized by a high-pressure gas bottle. When the reactor pressure decreases below the tank pressure a check valve opens and the borated water is forced into the reactor. This buys time for the emergency diesel engines to start and inject emergency coolant.

Fuel elements contained 21 metric tons of uranium dioxide pellets in 33 miles of Type 304 stainless steel tubes. Steel washers were crimped and brazed into fuel rods to prevent fuel slumping and yet allow gas-pressure equalization. Fuel rods with stainless steel ferrule spacers were brazed together to form subassemblies. Brazing was a quality control assurance problem also, because excess moisture in pellets caused tubes to balloon during the brazing cycle and be rejected. Use of subassemblies and stainless steel reduced cost of fuel fabrication, recycle, and scrap. Since pellets were loose in the fuel rods, separated only by the washers, the entire length of the rod was available for gas buildup. The washers prevented slumping, but allowed many short lengths of tubing pellets to expand.

A disadvantage of this brazed structure was its inability to tolerate differential expansion of individual fuel rods without distortion of the whole assembly. This would have been a severe limitation on fuel burnup, but was resolved by the SPIC design discussed later.

Exhaustive studies of the economics of the gaseous-diffusion process for enriching uranium made the case for slightly enriched uranium; it supplies long-lived reactivity because of uranium-238 fission plus the production and fission of plutonium, in addition to fission of uranium-235. Stainless steel economics came from low-cost, corrosion resistance to the coolant containing boric acid (which was to be used for reactivity control), resistance to radiation damage, more acceptable heat transfer because of allowed nucleate boiling, and elimination of design requirements caused by hydrogen produced from the chemical reaction between zirconium and water.

An advantage of uranium-dioxide pellets in tubes is that their center temperature is limiting at 4000°F, rather than 2100°F for uranium metal. However, the metal fuel's heat conductivity is far greater. Neutron economy was improved because more uranium-235 per dollar could be available with the slightly enriched uranium. Only about 5% of the neutrons were captured by stainless steel because of competing absorptions in uranium-235, uranium-238, plutonium, and fission products.

Economies resulted from the first core being designed so that uranium-238 and plutonium, respectively, produced about 7% and 23% of the energy. Reactivity lifetime was enhanced by fast fission in uranium-238 and plutonium production. Plutonium, as a by-product, could be greater in value than the electricity generated, as long as the price of plutonium was $30 per gram.

Four fuel elements were left in place in the center of the core for several fuel cycles and achieved a burnup of about 35,000 megawatt days per ton. This was a record and was helpful in advancing PWRs. The burnup was increased with each new nuclear plant. Earlier in most cases lifetime had been assured by test before being used in the nuclear plants.

Fuel Assemblies[1]

After many developments, modern fuel assemblies consist of a square array of fuel rods structurally bound together into a fuel assembly. Control rod guide thimbles replace fuel rods at selected spaces in the array and are fastened to the top and bottom nozzles of the assembly. Spring clip grid assemblies along the height of the fuel assembly provide support for the fuel rods. The fuel rods are contained and supported, and the rod-to-rod centerline spacing is maintained within this skeletal framework.

The bottom nozzle of the fuel assembly controls the coolant flow distribution and also serves as the bottom structural element. The top nozzle

---

[1]The sections on fuel assemblies, the pump shaft seal system, rod cluster control assemblies, and core thermal-hydraulic design are from a paper by George Masche, "System Summary of a Westinghouse PWR," Westinghouse Electric Corporation, 1971.

functions as the fuel assembly upper structural element and forms a plenum space where the heated reactor coolant is mixed and directed toward the flow holes in the upper core plate.

The spring clip grids provide support for the fuel rods in two horizontal directions. Each rod is supported at six points in each cell of the grid. Four support points are fixed: two on one side of the grid strap, and two similarly located on the adjacent side. Two more support points are provided by spring straps located opposite the fixed point. Each spring exerts a force on the fuel rod such that lateral fuel rod vibration is restrained. Because fuel rods are not physically bound to the support points, they are free to expand axially, thereby preventing bowing due to thermal expansion.

Two types of grid assemblies are employed. One type features mixing vanes that project from the edges of the straps into the coolant stream to promote mixing of the coolant in the high heat region of the fuel assemblies. The other, a nonmixing type of grid, is located at the bottom and top ends of the assembly. The outside straps on all grids contain vanes that aid in guiding the grids and fuel assemblies past projecting surfaces during fuel handling or while loading and unloading fuel.

All fuel assemblies employ the same basic mechanical design, although not all assemblies have control rod clusters. Selected fuel assemblies have neutron sources or burnable poison rods installed in the control rod guide thimbles. Fuel assemblies not containing either control rod clusters, source assemblies, or burnable poison rods are fitted with plugs in the upper nozzle to restrict the flow through the vacant control rod guide thimbles. This plug includes an end-flow mixing device to assure that these fuel assemblies have approximately the same coolant flow as those containing control rod clusters.

The Westinghouse "canless" fuel assembly provides optimum performance in an open lattice core by increasing flow stability and reducing the temperature gradient between fuel assemblies. The mixing vane grids increase the heat transfer capability of the fuel rods. Still other features contribute to the reduction in nuclear fuel cycle costs. For example, efficient utilization of the fuel is achieved by decreasing the parasitic absorption of neutrons in the core. In the "canless" assembly design, the only structural materials in the fuel region are a few lightweight spring clip grids, and Zircaloy fuel cladding. Zircaloy is used because it absorbs relatively few neutrons and has good heat transfer properties.

Control Rods

For Yankee-Rowe an alloy (80% silver, 15% indium and 5% cadmium—plated with a nominal 0.0005 in. of metallurgically bonded nickel) was used for the 24 cruciform control rod absorber sections. Hafnium, obtained as a by-product of zirconium production, was not available for commercial reactors. Insignificant amounts of silver were found in the primary coolant during core-one operation.

However, during refueling, probably because of the greater amount of oxygen in the water and imperfections in the nickel protective layer, silver plated out on the refueling-pit walls. Decontamination was not accomplished, because it required the use of potassium cyanide, and the Yankee company doctor considered that an unacceptable health hazard. The silver was removed later by an inverse plating process.

Control rod cruciform absorber blades were replaced with silver-indium-cadmium pellets in stainless steel tubes, which were welded in a cruciform pattern. Many cores later, Yankee replaced these absorbers with hafnium when it was available. After BR-3, SELNI and SENA, control rods were in a cluster configuration similar to fuel elements rather than cruciform.

Control Rod Drive Mechanisms

Commercial reactors enjoy an advantage over naval reactors of remaining vertical at all times. Thus the roller nut mechanisms and spring returns could be replaced with magnetic latch and unlatch jacking mechanisms which scrammed from gravity forces in a fail-safe manner when power is interrupted, plus the additional ability to control with boric acid, which does not have a similar failure mechanism. This is not the case in a boiling water reactor (BWR) where mechanisms are inserted from below, as the top is needed for steam separation. In those reactors, the need to provide scram capability dictates springs or hydraulic reservoirs. As noted earlier, BWRs generally require more control because boric acid cannot be used in control at power.

John Stiefel, the PWR general manager, asked Howard Arnold to examine a PWR in which boric acid would not be used for control to allow the use of carbon steel for all primary loop components in place of stainless. The study showed that the extra control and finer subdivision of the fuel more than compensated for the savings in component costs.

The positive-displacement latch, magnetic-jack mechanism was developed specifically for Yankee.[2] Earlier mechanisms were more costly or—like early friction-type, magnetic-jack mechanisms—were unable to lift the large weight of Yankee cruciform control rod absorbers plus zirconium fol-

---

[2]Before Connecticut Yankee, Yankee-Rowe was referred to as just Yankee.

lowers. These mechanisms have operated successfully for over 30 years, except for some early coil-cooling problems, which were corrected by air coolant ducts. The original five-coil mechanism was invented and modified subsequently by Frisch to operate with only three coils. That simplified mechanism is used now on almost all PWRs.

Chemical Control

Another Yankee innovation was injection of boron into the main coolant water to absorb neutrons and thus shut down the reactor. The use of boric acid for Yankee was originally intended as chemical shutdown assurance, because full-size criticals were not planned, and both Yankee and the AEC wished assurance that the reactor could be shut down under any condition, even with one control rod stuck. The decision was made to use boric acid when Creagan and Witzke agreed to install tanks, pipes, pumps, etc., required to insert and withdraw the boric acid. The use of boric acid for shutdown saved money in several ways: (1) full-size criticals were not required so the same equipment at Waltz Mill used for the smaller BR-3 criticals could be used for Yankee; (2) enrichment specification could be delayed, because reactivity shutdown was assured no matter what the enrichment; and (3) uncertainties in control rod worth, reactivity requirements for fission product accumulation and temperature coefficients could be covered by adding more or less boric acid.

When longer fuel element life was desired in later cores for improved economics, additional fuel enrichment could be added, because reactor shutdown was assured by adding more boric acid to the coolant.

Paul Cohen was just leaving Bettis for CAPA (Commercial Atomic Power Activity), but was not yet involved with chemical shim. Paul, as an observer of the situation in Pittsburgh, was against the Yankee experiment and wanted it done at Saxton first. Yankee listened but went ahead. Later chemical shim was demonstrated at Saxton.

After Yankee and the AEC agreed to use chemical shutdown and after reactor operation, the possibility of chemical control (chemical shim) was considered, because the necessary equipment was already installed and operating. We at Westinghouse were well aware of chemical shim advantages, but we were concerned about possible boric acid precipitation on the fuel rods that might break off and release reactivity.

Long discussions were held with Yankee and the AEC. Glenn Reed of Yankee pushed the idea because he had operated BORAX (a BWR at the National Reactor Testing Station with boric acid shim). It was finally agreed [Glenn Reed, Roger Coe, Tommy Thompson (consultant from MIT), Creagan, Witzke, and Shoupp] that Yankee would operate with chemical shim.

The agreement was that the reactor would be shut down if two-thirds of a dollar of reactivity was unexplained and therefore possibly due to boric acid plated out on the fuel. A "dollar" of reactivity is the reactivity required to reach prompt critical.

The first run resulted in a shutdown, when Thompson considered that 67 cents of reactivity was unexplained. A second run was successful, and the reactor continued to run without incident as boric acid was added and removed from the coolant.

Concern about reduced negative temperature coefficient with high boric acid concentration was appreciated and situations where operational transients might be seriously affected were avoided by strict mandates on allowable concentrations.

Later many (including Radkowsky) advocated fixed burnable poisons to reduce boric acid concentrations, but that involved higher available reactivity for longer burnups. With boric acid (a weaker solution than used to wash out your eyes) and control rods, two redundant and independent means for shutdown were provided that were not subject to a common mode of failure.

To implement the chemical shutdown system, a low-pressure coolant-purification system was developed as a Yankee first. This arrangement facilitated low-pressure addition and removal of boric acid from the pressurized water.

The chemical poison control system became refined into a fully operational system allowing a shim control capability to match fuel burnup and allow the reactor to run through anything except large transients with the control rods positioned optimally for core power shape. This remains a major advantage over the BWR, which must continually move control rods and allow for the less uniform power shape that results. The BWR designers have exhibited considerable ingenuity in shaping enrichment profiles in fuel assemblies and alleviating this problem in other ways.

Chemical shim not only controlled reactivity but improved hot-channel factors by influencing control rod positions; they could be chosen to improve neutron flux profiles and thus increase licensed power output, with boric acid used for controlling reactivity.

Alvin Radkowsky of the AEC, in a meeting of the AEC's advisory committee on reactor physics, informed Howard Arnold that Westinghouse was making a "tragic mistake"—his very words—to go this way rather than pursue burnable poisons. We had studied burnable poisons, and they are used today in commercial PWRs to hold down some of the initial reactivity; this avoids the low negative coolant temperature coefficient when there is too much boron in the coolant. We concluded that the amount necessary to do the whole job would have a negative effect on core economics, because quite a bit would remain at the end of cycle life in a multicycle core. This effect is different in reactors refueled as a single batch.

Milt Edlund at Babcock & Wilcox was advocating a spectral shift reactor, in which the initial coolant was a mixture of heavy and light water. Bleeding out the heavy water as core reactivity declined due to burnup would allow coarse control during life, as did the boron. Theoretically, this would be more economical, as the neutrons wasted in boron would instead make plutonium. However, realistic calculations showed that the economic margin would be small, and the expense of heavy water, including leakage, reenrichment, and its handling would overwhelm the advantages. The marketplace confirmed our choice.

Rod Cluster Control Assemblies

After using cruciform control rods in early reactors, rod cluster control is now used for reactor startup, to follow load changes, and to control small transient changes in reactivity. The control elements of a rod cluster control assembly consist of cylindrical neutron absorber rods, having approximately the same dimensions as a fuel rod and connected at the top by a spider-like bracket to form rod clusters.

Two types are employed: full length and part length. The full-length type incorporates rods of silver-indium-cadmium absorber material, extending the full length of the core. Stainless steel tubes encapsulate the absorber material, isolating it from the reactor coolant. Full-length rod cluster controls provide operational reactivity control and can shut the reactor down at all times, even with the most reactive rod stuck out of the core.

The part-length rod cluster controls, although identical in external appearance and design, incorporate rods with absorber material only in the bottom quarter of the tube. The remainder of the tube is filled with aluminum oxide. The absorber region of the part-length rod is positioned at various elevations in the core to shape the axial power distribution and to control axial xenon redistribution accompanying power level changes.

Each rod cluster control is coupled to its drive shaft, which is actuated by a separate drive mechanism mounted on the reactor vessel head. Reactivity of the core is changed by raising or lowering the cluster in the core.

Each absorber rod in a rod cluster control moves in its own tubular guide thimble. Located symmetrically within fuel assemblies, these thimbles replace fuel rods within the fuel assembly lattice. The thimbles (1) act as guides for the control rods; (2) prevent control rod buckling; and (3) serve as dashpots for terminating control rod motion during reactor trip. In their fully withdrawn position, the absorber rods do not leave the upper end of the guide thimbles. This assures that the rods are always properly aligned, and reduces reactor coolant bypass through the thimbles. The thimbles are perforated over a portion of their length to allow passage of water and permit rapid rod insertion. The lower end of each guide tube is closed and acts as a shock absorber to decelerate the control rod at the end of its drop under reactor trip.

Westinghouse rod cluster control design has contributed greatly to core performance improvement through a more uniform power distribution. Dividing the control material into many small absorber rods provides a more homogeneous means of control. When the control rods are withdrawn from the core, the small resulting water gaps cause no significant power peaking. By installing control rods within the assemblies, the assemblies can be spaced closer together, thereby reducing the reactor coolant bypass flow.

Heat Transfer

Nucleate boiling heat transfer was first allowed in Yankee because of the stainless fuel elements and is now a PWR standard. Even with zirconium-clad fuel, nucleate boiling heat transfer allows higher temperature fuel clad surfaces so that water boils on the fuel clad surface and then allows steam to condense in surrounding water. This process improves heat transfer in many ways, but it was not used in earlier reactor cores because of fear that collapsing steam bubbles would impair the corrosion-proof oxide film on zirconium fuel elements.

Nucleate boiling permits elimination of the hot-channel factor (F-theta), which would have reduced the Yankee heat flux at 134 MW(e) by more than 50%. F-theta is an allowance factor that must be used to reduce heat output so that solid water will stay in contact with the fuel element under all conditions, despite tolerances and other uncertainties.

Long Sun Tong commissioned experimental work at Columbia University and he and Jim Cermak came up with some departure from nucleate boiling (DNB) correlations that continued to be in use for PWR design for many years.

Heat transfer can be improved also by flattening reactor core radial-heat production. This can be accomplished by using multiregion enrichment cores. Such cores were studied during the Yankee R&D program but not implemented until later cores. Zirconium control rod followers displaced water in control rod slots when control rod neutron absorber sections were withdrawn.

Even zirconium followers do not completely suppress the peak. Our design team evolved a follower made of fuel rods, which was employed in SELNI. However, this proved to be expensive and trouble prone. The real solution, appearing in the San Onofre and Connecticut Yankee generation of reactors, was the rod cluster control as described in a previous paragraph. This is now universally employed in PWRs. This invention by Art Thorpe both minimizes the peak and allows the reactor vessel to be shorter in height. This improved heat transfer by reducing hot spots otherwise caused by excess neutrons slowed down in the excess water in control rod slots and by inhibiting water from bypassing fuel heat-transfer surfaces. Even with zirconium followers more than twice the heat energy was produced by the Yankee reactor than from the same diameter Shippingport vessel, and thus reactor vessel costs per kilowatt-electric were reduced more than 50%.

Core Thermal-Hydraulic Design

The basic objective of the core thermal-hydraulic design is to transfer the energy generated in the fuel to the coolant while maintaining the integrity of the fuel rod even under the most severe anticipated transient conditions. To maintain fuel rod integrity and prevent release of fission products, the fuel cladding must be prevented from overheating under all operating conditions. This is accomplished by preventing DNB, because DNB causes a large decrease in the rate of heat transfer from the cladding to the coolant, resulting in high fuel rod temperature.

The measure of assurance that DNB will not occur is the DNB ratio, defined as the ratio of the predicted DNB heat flux to the actual local heat flux. In the reactor core design, both the heat flux associated with DNB and the location of DNB are important. To predict the local DNB heat flux the Westinghouse W-3 correlation[3] includes the nonuniform heat flux effect, and the upstream effects of inlet enthalpy and length.

To maintain integrity of the UO_2 fuel, it is necessary to prevent the fuel temperature from exceeding the UO_2 melting temperature. The temperature profile of the fuel pellet is determined from the linear power density expressed in kW/ft. Calculations show that a rating of 22 kW/ft can be experienced on Westinghouse fuel rods without causing central melting of the UO_2 fuel. Since the maximum heat rate in the fuel rod is less than this under maximum overpower conditions, an adequate safety margin exists.

Three-dimensional nuclear, thermal, and hydraulic design analyses are used to determine the power capability such that the overall power distribution peak does not exceed either the linear power density limit or the minimum DNB ratio at any time during a fuel cycle. Westinghouse does not consider that the axial and radial power distribution factors are sepa-

---

[3] L.S. Tong, "Prediction of Departure from Nucleate Boiling for an Axially Non-Uniform Heat Flux Distribution," *Journal of Nuclear Energy*, Vol. 21, pp. 241–248, 1967.

rable over the period of a fuel cycle. The limiting axial power distribution may not occur simultaneously with the maximum radial power peak. Given a radial distribution, the limiting condition on power is determined by the shape of the axial distribution, and not by a fixed peak-to-average ratio of power density. For example, a small peak in the top of the core on a relatively flat power distribution may be DNB-limited. In the bottom of the core, a much greater peak is allowable and may be limited by linear power density.

Negative Temperature Coefficient of Reactivity

By decreasing water volume in the core below the optimum for reactivity, a negative temperature coefficient of reactivity was designed into the reactor. This caused reactor power to follow turbine throttle demand and also provided negative feedback to inhibit reactivity transients. To emphasize the importance of this coefficient, it should be noted that the temperature and void coefficients of reactivity were positive for the Chernobyl reactor— a major cause of the accident.

Howard Arnold[4] found an article in a technical journal stating that the resonance capture of neutrons in uranium-238 would depend on fuel temperature. Although of negligible consequence in naval reactors and in Shippingport (because there the highly enriched seed dominated the physics behavior), in Yankee this would lead to a new temperature coefficient of reactivity. Although he estimated it to be less than one-tenth that due to the expansion of water as the reactor temperature rose, it would have a time constant much shorter, because it would occur as the fuel heated up— well before water temperature rose.

This was calculated on the analog computer by Clovis Obermesser and, to everyone's horror, this coefficient, whose magnitude was within the error band assigned to the temperature coefficient, had an important role because of its short time constant. It was then concluded that the pressurizer, already on order, was too small. At the end of 1959, Arnold's physics group had $10,000 left in their computer budget, so Bob Daniels used time available over the holidays on the East Pittsburgh IBM 650 performing Monte Carlo calculations to improve our estimate of this effect. This gave enough data to permit the redesign of the pressurizer. Bob was a remarkable individual, a paraplegic from an accident at the age of 12, he obtained his doctorate and later managed the analytical methods development group in our nuclear energy systems group.

---

[4]W.H. Arnold, Private Communication dated June 5, 1992.

In thermal reactors, neutrons from the fission process have energies of about 2 MeV. They are slowed down by the moderator to thermal energy levels. In the slowing-down process, they pass through the resonance region (4000 eV down to a few electron-volts) for capture in uranium-238 and the resonance region for capture and fission in uranium-235. In the resonance region, cross sections for capture or fission have high sharp peaks, which only absorb neutrons at almost exactly the resonance peak energy.

If the uranium temperature rises, uranium atoms have energy increases that cause velocity increases in the motion of uranium nuclei. Uranium velocity relative to neutron velocity changes. This broadens the sharp resonance peak, and more interactions occur. Broadening absorption resonances produces negative reactivity, whereas broadening fission resonances increases reactivity. The net result is negative for uranium-235 and uranium-238. The Doppler effect with uranium-238 is greater.

The peak of the first resonance in uranium-238 is about 20,000 barns at 6.67 eV, and its half-width is approximately 0.027 eV. Virtually all neutrons that hit the uranium-238 nucleus with a relative energy in the 6.67-eV range are absorbed. If the target nucleus is at rest (temperature 0 K), only those neutrons that hit the uranium-238 nucleus with an energy of 6.67 ± 0.027 eV will be captured. At any temperature above 0 K, the energy of the target nucleus ($E = kT$) is added or subtracted from 6.67 eV and thus a broader energy band of neutrons is absorbed.

If the temperature goes up, the uranium-238 nucleus begins to vibrate and, since it is the relative energy that counts, some neutrons outside the range of 6.67 ± 0.027 eV will be liable to capture. At the same time, some neutrons with energies very close to 6.67 eV relative to the laboratory system will now escape capture, since their energy relative to the moving uranium-238 nucleus is now outside the range (6.67 ± 0.027 eV).

When the effective cross section is calculated taking nuclear motion into account, the effective peak of the resonance decreases and width increases in such a way that the area under the curve remains constant. For example, the 6.67-eV peak value (with 6.67 eV being the neutron energy relative to the laboratory) may decrease to 10,000 barns and the width increase to 0.54 eV. Still, virtually all neutrons striking a nucleus having a 10,000-barn capture cross section will be captured. Thus for the higher temperature case, the capture rate in this particular resonance would be almost doubled.

In the case of uranium-235, there are both fission and capture resonances. Broadening of the former *increases* reactivity. Thus, for uranium-235, although the capture process usually wins, the overall Doppler effect is much smaller than it is in uranium-238.

The Doppler effect modifies power level in the reactor, and this causes the water to heat up and cool down at a different rate than predicted without the Doppler effect. Since the pressurizer was designed to maintain reactor pressure, it was sized (i.e., heater capacity, spray capacity and volume) to accommodate changing water volume in the primary loop. Pressurizer heaters produce steam to increase pressure, while the spray quenches steam in the pressurizer above the solid water to reduce pressure when required.

The Doppler coefficient gave assurance that the reactor would be safer on a fast time scale than otherwise, alleviating concerns when the slow coefficient might be at low values with soluble poison in the water. These were of concern when we considered scaling up to much larger reactors. Arnold used the Doppler coefficient to prove xenon oscillations would be damped in even the largest reactors using low-enriched fuel, and thus they could be easily detected and controlled with external detectors and part-length rods.

The Southeast Experimental Fast Oxide Reactor (SEFOR) demonstrated the Doppler shutdown effect on a fast reactor. The large-diameter fuel rods used in SEFOR provided a greater Doppler negative coefficient of reactivity than do small-diameter rods. This is because large-diameter fuel rods have a higher average temperature change and the resonance energy neutron captures increase faster than the resonance energy fissions, thus accentuating the Doppler negative reactivity effect.

When neutrons slow down they can bounce past the resonance energy spectrum in uranium-238 down to the thermal energy range, where the fission cross section of uranium-235 predominates. This was one of the first phenomenon Enrico Fermi discovered, and this is why he went to lumped uranium hunks in the first reactor rather than uniformly mixing graphite and uranium. Neutrons slowed down in the graphite and then hit the uranium hunk. Resonance neutrons were captured in the surface film of the uranium and thus allowed neutrons of other energies either to interact with uranium or to be slowed down further in the moderator. The uranium-238 nuclei were thus not exposed to resonance neutrons and therefore did not absorb them.

The Doppler temperature coefficient alleviated concerns when the slow coefficient might be at low values when there was too much soluble poison in the coolant water. It is also important in damping effects of xenon spatial oscillations. These were of concern for scaling up to much larger reactors.

Doppler feedback is immediate with fuel temperature rise, whereas the temperature coefficient associated with water is delayed until heat in the fuel is transmitted to the water, which expands and thus provides less moderation of neutrons, hence less reactivity. This latter coefficient can be used to increase reactivity lifetime by lowering reactor temperatures. First-core lifetime for Yankee was increased 30% by using this effect. However, this results in slightly reduced plant efficiency as the temperature is reduced.

The Xenon Problem

The xenon problem was not as severe for Yankee as it was for the Mark I and the Nautilus. Yankee incorporated uranium-238 that already absorbed some resonance energy neutrons, and the core was more opaque to neutrons because of stainless steel fuel cladding, uranium-238 and fission fragments. Mark I had a transparent core to neutrons, hence incremental poisons—such as control rods, xenon, or fission fragments—were relatively more effective than in Yankee, where xenon competed with stainless steel, uranium-238, and fission fragments for effectiveness. Also, the neutron flux was lower in Yankee, which makes xenon less dominant. Heavy water reactors fueled with natural uranium may be kept shut down for many hours because they do not have enough excess reactivity to override xenon.

Xenon oscillations, following a nonsymmetrical control rod motion or in conjunction with a flow instability, can cause a geometric tilt of the neutron flux, most easily observed in a large reactor such as Savannah River. There the reactivity is very close to one and is more sensitive to the presence or absence of a neutron-absorbing poison like xenon. Yankee, Shippingport and the submarine reactors did not suffer from xenon oscillation, although some oscillation was observed in Shippingport.

What matters in exciting xenon oscillations is power density and reactor dimensions. In Shippingport, most of the heat—at least at first—was produced in the seed and because of the inner and outer blankets, the seed had the "length" of a racetrack. Thus, one expects oscillations to be more likely in Shippingport than in Yankee.

In-core Instrumentation

Neutron flux in the Yankee reactor was measured by cobalt-alloy wires inserted into tubes in the core; they were later removed and scanned by a gamma counter. One-quarter sector of the reactor core included cobalt wires in each fuel assembly; because of symmetry, each of the other three quadrants had only one fuel assembly containing a wire. Both radial and axial neutron fluxes could be obtained, and thus hot-channel factors, which determine local heat flux, could be documented.

In addition, thermocouples, located with similar radial symmetry, measured localized temperatures of outlet coolant from the core. Thus, hot-channel factors, which enabled measurement of enthalpy rise in filaments of coolant water, could be documented. Documentation of these hot-channel factors was very important in obtaining licenses for power increases.

Considerable controversy raged over whether on-line in-core instrumentation would be needed for routine operation. Early reactors had them for experimental purposes and for documentation of hot-channel factors so that power increases could be licensed, but inclusion as an operating requirement would add complexity, cost and reliability problems. Clovis Obermesser maintained that a well-characterized PWR could be controlled safely with only external neutron detectors at two axial and four circumferential positions, in combination with thermocouples at a number of coolant outlet positions at the top of the core support plate. This was resolved favorably with the addition of flux wires that could be inserted and removed through the vessel bottom during operation of the reactor. These latter are read externally to give axial power shapes and use positions in the reactor replacing fuel rods, just as do the control rods in the rod cluster design mentioned above.

Main Coolant Pumps

Another major advance in systems design, traceable also to Frankie Frisch, was the introduction of shaft seal pumps. The canned motor pumps as developed in the naval program, suffered from size limitations and efficiency disadvantages. Frisch came up with a multiple barrier, hydraulic final stage controlled leakage seal that has enjoyed wide use in PWRs and became a major Westinghouse product line.

The shaft seal arrangement consists of three Westinghouse-designed face-type seals operating in a series arrangement.

1. *No.1 Seal:* The primary seal is located above the lower radial bearing and constitutes the most important element in the seal system. It is basically a film-riding face seal. The film is produced by the system pressure drop across the seal, and therefore does not require shaft rotation for its existence. Normal leakage rate for this seal is three gallons per minute at system operating pressure. Leakage flow is radially inward toward the shaft, and the design is such that the axial pressure forces are balanced. Since this seal rides on a thin film of water, the seal ring does not mechanically contact the seal runner.

2. *No. 2 Seal:* This seal normally accepts the three gallons per minute leakage from the No. 1 Seal at a pressure of approximately 50 psi, and seals it against a back pressure of several feet of water. Normal leakage for the No. 2 Seal is three gallons per hour. A rubbing face-type seal is of conventional design and employs a rotating runner and a stationary carbon member. It is pressure-balanced and spring loaded. Although the seal normally handles only 50 psi on the high pressure side, it is so designed that in an emergency it can

operate with full system pressure across its face in either the rotating or stationary state. Although its life under these latter emergency conditions will be reduced, it will permit limited pump operation or an orderly shutdown without gross leakage.

3. *No. 3 Seal:* The third seal is a smaller, low-pressure, rubbing face-type seal designed to limit leakage into the containment vessel to 100 cc per hour. This small leakage lubricates and cools the seal faces. The No. 3 Seal is similar in design and materials to the No. 2 Seal.

Reactor Vessels

Reactor vessels are fabricated of carbon steel and were required by the Unfired Pressure Vessel Code of the American Society of Mechanical Engineers (ASME) to be designed for a stress 50% greater than expected. Charpy V-notch specimens are cut from the cross section of the reactor vessel at the surface and at each quarter thickness for all steel sections used to fabricate the reactor vessel.

These Charpy specimens are clamped and then hit with a hammer to determine the energy in foot-pounds required to break them. These fractures are measured at various temperatures. That temperature at which the specimen breaks at 32 foot-pounds is called the *nil ductility transition temperature* (NDTT) of the metal. Above this temperature the metal is considered ductile, and below it is considered brittle.

Since irradiation in service may cause embrittlement of the reactor vessel, samples are suspended inside the reactor vessel during operation. In this way the effect of neutron bombardment in increasing NDTT, can be measured by breaking specimens, after the reactor has operated for many years. NDTT in reactor vessels should have an adequate margin of safety below the operating temperature. The above test determines if the margin has been lowered too much.

If the reactor vessel is stressed too much, a crack may start. If a crack starts in ductile metal, the sharp vee at the end of the crack is assumed to dull into a U shape that will not allow the crack to progress. In brittle metal, the vee is assumed to propagate through the metal at the velocity of sound.

Concern is felt that the reactor vessel may operate in the brittle mode and that a crack would permit a major loss-of-coolant accident. However, loss-of-coolant safety systems are designed to handle a major loss of coolant caused by the guillotine-like severance of a main coolant pipe, but not a major rupture of the reactor vessel. Reactors are operated so they never are at stress in the brittle mode. To accomplish this, the reactor is heated up by operating the pumps, which heat the water and thus provide stress only from the result of heat. Even when shut down, a reactor can be stressed—due to earthquakes, for example.

Experimentally a reactor vessel was put under stress when it was intentionally in a brittle state. It did not fracture, so a deep notch was milled into the vessel and again it was put under stress. Eventually the vessel fractured under stress, but only locally.

Experimentally, a reactor vessel that had been exposed to a great neutron flux (and thus presumably the NDTT had been shifted upward) was annealed at red heat so the NDTT was restored to its former value. The vessel was again irradiated to demonstrate that the NDTT did not change any more rapidly than during the first neutron exposure. A reactor vessel was also annealed at about 800°F for a long period. This temperature was slightly above the 740°F critical point of water. Maintaining the temperature constant for a long time was an experimental difficulty, but the results were satisfactory.

This information about possible reactor-vessel embrittlement is given to illustrate the possible problems and solutions. The reactor vessel would be by far the most costly component to replace. In any case, economics are such that it makes a lot of sense to extend reactor lifetime beyond the license period, even if large components such as a reactor vessel are repaired or replaced.

Containment

Shippingport containment consisted of three units, one for the reactor vessel and two others, each containing two loops of equipment. This was done partially to permit maintenance during operation, when one loop was closed off. Spherical containment had been used for the *Sea Wolf* prototype by General Electric at West Milton and for Dresden 1, so Yankee decided to use a 125-foot-diameter spherical vessel, which was believed to be less expensive.

The ASME code allowed a maximum steel thickness of one and one eighth inches for welds without annealing. With a 125-foot-diameter vessel, the thickness required for the design pressure was less than the ASME maximum. The sphere was elevated above grade on column supports filled with concrete, so a railroad spur could run under the sphere. Thus the reactor vessel, steam generators, and other heavy equipment were brought under the sphere on flat cars and lifted into the containment through a bottom opening. Sphere supports were fitted with expansion joints at the top and later with large seismic restraints at the bottom. The steel containment shell was erected by the Chicago Bridge & Iron Company under contract to Stone & Webster.

In the late 1960s, Westinghouse invented and developed an ice condenser containment concept, in which several million pounds of tetraborate ice are stored in the containment. The system is designed to absorb the thermal energy released during blowdown to a nuclear plant containment building as the result of a loss-of-coolant accident. It would be effective up to and including a double-ended rupture of the reactor coolant pipe with substantial margins for energy release and release rate. The effect of ice condenser operation is to limit containment building pressure, resulting from a loss-of-coolant accident, to low values. This effect is obtained in containment buildings having interior volumes that approach one-half of the volume of so-called "dry-type" containment systems.

The chemical composition of the ice inventory allows the melted ice to be used for reactor cooling, and has the further capability of removing fission products from the containment atmosphere both during and after blowdown.

The ice condenser system is an inherently passive safeguard system. No external source of power is needed for operation of the ice condenser during the blowdown period.

We sold 10 reactors of this type and licensed the concept to the Finns to put on their two Russian-supplied reactors. However, the cost of meeting upgraded seismic and blowdown force requirements led to considerable extra costs that more than made up for the size advantage. Utilities were also finding that the smaller size was a disadvantage when scheduling refueling activities in the tight quarters. This system was not used on later designs.

Technological Improvements

Technology was improving, and larger and more advanced reactors were required to take advantage of the economy of scale. The flux distribution was greatly improved with each new design, as was the design fuel lifetime.

Table 15.1 shows a comparison of design features at the time of original design.

Safety Injection System[5]

The safety injection system (SIS) was added to the Westinghouse design at the time of Connecticut Yankee and San Onofre, and was backfitted to earlier commercial reactor designs. Its primary function is to supply borated

---

[5]PWRSD 1976, "A Snap Shot in Time," Westinghouse Nuclear Energy Systems, 1976.

Table 15.1
Comparison of Design Features*

	Shippingport	Initial Yankee	Conn Yankee	Sequoyah[†]
Core average (MWd/ton)	4500	7830	17,000	13,780
kW/liter	40	72	82	105
Average heat flux (W/cm^2)	18	34	43	69
Metal clad temperature (°C)	598	591	635	689

*Data from the *International Atomic Authority Directory of Nuclear Reactors*, Vols. 4, 7, 9, and 10.
[†]This value is a core average, which includes partially unburned fuel due to the multizone cycle scheme. The discharge burnup is of the order of 30,000 MWd/ton.

water to the reactor coolant system to limit fuel rod cladding temperature in the unlikely event of a loss-of-coolant accident.

When in operation, the SIS limits melting of fuel cladding, core geometry distortion, and metal-water reaction to negligible amounts for reactor coolant and piping ruptures up to and including the double-ended severance of a reactor coolant loop. The SIS branch lines are sized such that a break of a branch line will not result in a violation of the system design criteria.

The SIS consists of several independent subsystems characterized by equipment and flow path redundancy inside of the missile protection boundary. This redundancy assures reliability of operation and continued core cooling, even in the event of a failure of single components to respond actively in accordance with design.

Borated water is stored in a tank and backed up with gas pressure bottles. A check valve connected to the reactor opens when reactor pressure drops and permits the emergency injection of water without delay and thus provides time for the emergency diesel to start up and inject water with emergency coolant pumps. A patent was awarded to Al Collier and Bob Creagan. This system is used on all PWRs, Westinghouse and others.

Fast Breeder Reactors

From our review of these various reactor types, the concept emerged of using mixed oxide (about 20% PuO_2) as fuel instead of more highly enriched metal fuel. Mixed-oxide type fuel shows excellent behavior under irradiation and mixed-oxide fuel also exhibits the self-limiting feature of the Doppler effect in a fast neutron spectrum in which the fraction of neutrons absorbed in uranium-238 increases as the temperature of the fuel rises, making fewer neutrons available to continue the chain reaction. Also it has a high melting point (4000°F) with poorer thermal conductivity, which allows it to heat up rapidly to a high temperature. The Doppler effect in mixed oxide fuel thus provides a fast acting shutdown mechanism. The Doppler effect was demonstrated in SEFOR, a 20-MW(th) reactor operated between 1968 and 1972 as an international program consisting of the AEC, a group of U.S. utilities, GE, and the then-Federal Republic of Germany.

Reasons for Not Selecting Other Breeder Types

The steam-cooled breeder was not selected for several reasons:

- Steam coolant requires high-pressure boundaries and more difficult emergency coolant injection.
- High-temperature steam hydrides and interacts with metal alloys, making long lifetime compatibility difficult.
- The breeding ratio at best is very low.
- Control problems are associated with startup and transients leading to water slugs entering the reactor.

Westinghouse also evaluated the helium-cooled breeder, but did not select it for reasons similar to those for the steam reactor. Helium is more compatible with materials than steam, but high pressures are involved. The helium breeder can be designed for higher breeding gain than the sodium-cooled reactor because there is less parasitic capture of neutrons in the helium. However, the technology and engineering base available were at a much lower level than was the case with the liquid metal fast reactor. The neutron spectrum is at a higher energy level, where nuclear cross sections are more favorable for breeding.

In addition to the three reactor types covered, the light water breeder was also evaluated. A thermal neutron breeder reactor core was designed by Westinghouse (Bettis) and operated in the Shippingport reactor. Uranium-233 produces more fast neutrons per thermal neutron absorbed than any other nucleus, therefore it is the nucleus of choice to produce breeding in a thermal energy spectrum.

Experimental measurements made at Shippingport demonstrated that breeding actually did occur. The breeding ratio was slightly above unity. Even theoretically, the breeding ratio for thermal neutrons operating on uranium-233 is not as great as for fast neutrons operating on plutonium-239; therefore, breeding objectives have been concentrated on the fast-neutron spectrum.

This core for the Shippingport breeder reactor was expensive to fabricate and the inventory of fissionable material was high. A further objection to breeding, with the thorium-232 to uranium-233 cycle, is that chemical reprocessing involves highly radioactive byproducts that require heavily shielded fuel reprocessing. When thorium-232 captures a neutron, it emits a gamma and forms thorium-233, which decays with a half-life of 23.5 minutes to protactinium. Protactinium-233 is highly radioactive (27.4-day half-life) and decays into uranium-233, so it is desirable to allow spent fuel to decay into uranium-233 to maximize product and decrease radioactivity. Such a delay however hurts economics.

Another complication of fuel reprocessing is presence of gamma-emitting daughters of uranium-232, which also mandates remote reprocessing. Irradiated uranium-233 containing 240 ppm uranium-232 has been reprocessed at ORNL in the Molten Salt Reactor Experiment (MSRE) program.

Chapter 16

Conclusions and Outlook

Naval Nuclear Program

The naval nuclear program is an outstanding success. In just over 14 years, from the time of the first sustained nuclear chain reaction at Stagg Field, the *Nautilus* went to sea as the most powerful fighting vessel in the world. Shortly thereafter the *Polaris* missile nuclear submarines came along. *Nautilus*, later attack submarines, and the *Polaris* nuclear submarines operated under the polar icecap, and with their missile launching capability they were a critical addition to the U.S. strategic deterrent force. Within a few years the Navy was substantially transformed into a nuclear Navy providing the most formidable deterrent to the Soviet Union or any other possible aggressor nation. The United States is building fewer new nuclear ships now, but the capacity exists and is available if needed. Although a nuclear Navy is no longer needed to protect against a major adversary, it was there when needed and just might have saved the world from destruction.

Space Propulsion

The NERVA nuclear rocket engine development program was successfully completed through the ground testing of engines. This dramatic technological breakthrough is well documented, and is available as needed for a manned mission to Mars or other advanced space missions involving large payloads.

Commercial Nuclear Power

Commercial nuclear power is also an outstanding success although it has failed to reach its full potential. The 419 nuclear power plants—representing

332 GW(e)—that have been placed into operation have had a profound effect on the world's production of electricity. Westinghouse supplied many of these, thereby providing many jobs and a very favorable balance of payments.

The safety record has been unprecedented for any industrial activity of this size. Many nuclear plants have proven to be highly reliable and economic. In some regions, such as the Southeast and in the Chicago area, nuclear plants generate more than 50% of the electricity. This strong reliance on nuclear power exists in many other countries, for example, France, where more than 75% of the electricity comes from nuclear plants. The average performance record continues to improve year by year, even though a few plants have not met reliability and economic expectations.

As a result of the unfortunate accident at Three Mile Island many improvements have been made in the design and operation of nuclear plants, changes that have made them even safer today. The Institute of Nuclear Power Operations (INPO) was formed to provide exchange of information on safety and other operating issues. After the very serious accident at Chernobyl, which involved a plant that had serious design defects, no containment, and poorly trained operators, the World Association on Nuclear Operators (WANO) was formed.

The nuclear industry in the United States is in a hiatus today and will be there for some time. No baseload capacity of any kind is being ordered in the United States today. Power growth capacity requirements are being met by traditional peak load units: gas turbines and combined cycle systems, because at present gas is cheap and in oversupply. History and economics suggest that the situation will change and true base load units will be required.

Electricity consumption in the United States grew about 7% per year from early in this century until the mid-1970s. Total energy usage since that time has increased only slightly, while electric energy has increased at about the same rate as the gross domestic product. In 1970 electricity represented only 25% of total energy; the percentage is nearly 45% today. Similar trends exist throughout the industrial world.

Energy Planning Strategy

It is important for the country to follow a strategy that will provide an adequate energy supply for economic development without endangering the environment or health of the nation. Because no one can predict the future accurately, we must be able to take advantage of possible favorable scenarios and defend against unfavorable ones. The problem is to determine what will affect the future supply and use of energy.

Even with worldwide concerted effort to limit population growth, the world's population might reach 15 billion people by the year 2200 before leveling off—about three times today's population. If by that time developing nations reach a level of energy use that exists in the United States today, world average energy use would be some five times what it is today.[1]

Some possible future situations include:

1. Acid rain becomes as serious an environmental problem as the environmentalists suggest, but I believe this is unlikely and in any case the use of nuclear power would lessen any effect of acid rain.

2. The "greenhouse effect" predictions turn out to be valid, for example, the South Polar ice caps melt and inundate most of the East Coast of the United States. This an extremist position, but its perception as a major problem is a possibility. In any case U.S. electric generators contribute only 7.5% of all the global warming gases generated in the United States.

3. Oil supply becomes a problem within a thousand years, probably within a hundred and possibly within a few decades or less if the Third World standard of living can be raised significantly. U.S. oil reserves are depleted and foreign oil is unavailable to the United States. It will be difficult for the industrial nations to pay for oil they need as it becomes scarcer. This problem will be even worse for the Third World nations.

4. People will realize that the health effects from mining, transporting, and burning coal are more deleterious than those from nuclear plants.

5. The radiation danger from any conceivable nuclear accident turns out to be no more serious than that from Three Mile Island. Bear in mind that there was an almost complete fuel meltdown with no serious consequences to the public and only two individual exposures that were comparable to routine medical x-ray exams.

Except for Chernobyl, the effect of radiation from nuclear plants on cancer incidence turns out to be trivial. Except for thyroid cancer, the incremental cancers due to Chernobyl are so small in number that it will not be measurable within the overall incidence of cancer in the population.

Despite these facts about exposure, the public will still fear the possibility of cancer caused by nuclear plants and this perception

---

[1] Chauncey, Starr, "The Second Nuclear Era," *Economic Perspectives*, January/February 1993, p. 22.

makes it a problem. However, great strides are being made in genetic engineering and biochemistry that lead many well-informed experts to believe that cancer can be both cured and prevented within 10 years and almost certainly within a few decades.

6. AP-600, an improved, NRC-approved nuclear power plant (or similar plants by other manufacturers) is built on schedule and within the cost estimate.

7. The nuclear waste issue is resolved to the satisfaction of the public. Nuclear waste could be buried a thousand feet down in many places with fewer public exposures than from natural ore. A development program is currently under way to prove the suitability of such a site, and studies to date have turned up nothing that would preclude safe storage of nuclear waste. One of the problems is getting 100% proof that there can be no leakage of radioactivity for thousands of years. Certainly there are more important things to spend money on than overensuring against the highly unlikely leakage of radioactivity a few thousand years from now.

8. A battery is developed that provides economic automobile propulsion. Such a possibility becomes more probable with increases in oil cost.

Summary for Commercial Nuclear Power

Nuclear power will become a more viable option if the following occurs:

1. The public is convinced a nuclear accident would be as harmless as the Three Mile Island accident.

2. The next generation of nuclear plants can be built on schedule and for the cost predicted.

3. The public believes a suitable solution to nuclear waste will be implemented.

Nuclear power becomes an even more viable option if acid rain and the greenhouse effect are considered major problems, even though there might still be significant concern about nuclear waste management. Without the nuclear option, clearly there will be an increase in the undesirable consequences of acid rain and the greenhouse effect.

Oil reserves are finite, hence scarcity will be a major problem. Air transportation has no substitute for oil, but ground transportation can substitute electricity from nuclear plants for hydrocarbon fuel without adding to the acid rain or greenhouse effect potential problem.

No single energy source will solve all energy problems. Ultimately nuclear power will be necessary for the economic well-being of the United States and, without it, the Third World can never hope to reach an acceptable economic level, because of the inevitable population growth over the next century. The question is not whether the nuclear option will be utilized, but rather when.

INDEX

A1W, 52, 73, 75, 77–83, 95, 215, 385
Abbott, W., 168, 180
Abelson, Philip, 12
ACEC, 162
Ackerman, Bud, 215
Advisory Committee on Reactor Safety, 170, 171, 358, 387
Aerojet General Co., 133–137, 138, 145–148, 160, 394, 406, 414–416, 418, 420, 421, 424, 425
Allen, Don, 166
Allis Chalmers, 90, 187, 324
Ammon, R.L., 129, 152
Anderson, Clinton, 91, 121
Anderson, H.L., 2
Anderson, W.R., 92
Angra Nuclear Plant, 203
Argonne National Laboratory, 7, 11–18, 21, 27, 29, 31, 35, 38, 39, 71, 98, 118, 132, 158, 161, 165, 168, 224, 227, 287, 288, 296, 303, 315, 329, 339, 348, 349, 352
Arnold, W.H., 128, 129, 131, 134, 137, 138, 141, 148, 154, 155, 163, 168, 169, 170, 177, 199, 205, 206, 240, 243, 244, 248, 416, 417, 424, 432, 433, 437, 439, 443, 445, 459
Ashcraft, Bruce, 8
ASME, 161, 361, 362, 368, 403, 448, 449
AT&T, 90, 107
Atlantic 1 and 2, 218, 221
Atomic Energy Commission, 12, 13, 17, 18, 20, 21, 31, 33, 36–38, 41, 50, 53, 57, 66, 67, 72, 75, 87, 95, 97–101, 109, 112, 115, 116, 120–122, 126–136, 146–148, 157, 159, 160, 164–166, 169–182, 191, 192, 198, 209, 224–227, 230–233, 237–239, 260, 279, 281, 292, 299, 335, 336, 338–347, 349, 362–364, 405, 408, 425, 433, 434, 438, 439, 452
Atomics International Co., 40, 118, 160, 225, 238
Atoms for Peace, 91, 92, 97, 98, 103, 188, 365
Avondale Industries, 272
Axene, Dean, 60

Babcock & Wilcox, 31, 102, 139, 163
Baker, W.R., 88
Bakos, Frank, 199
Baldridge Quality Award, 257
Baldwin Lima Hamilton Co., 77
Barker, J.A., 59, 61
Barker, J.H., 108
Battelle Memorial Institute, 238, 239, 339, 342, 349, 352, 374
Battelle Memorial Laboratory, 352
Battelle Northwest Laboratories, 238
Bechtel Corp., 25, 108, 157, 178, 179, 190, 248, 250, 251, 272, 280
Bechtold, J.H., 129
Beck, Cliff, 170
Begley, R.T, 129, 152, 250
Benedict, Frank, 25
Benedict, Manson, 169
Benz, Carl, 287
Bethel, Albert, 101, 134, 201
Bettis Atomic Power Laboratory, 7, 9, 12, 13, 17–27, 30–39, 41, 42, 46–48, 51, 52, 54, 57, 59, 61, 62, 65–70, 72, 75–91, 94–97, 100, 101, 104, 109, 110, 112, 115–122, 126, 128, 129, 132, 135, 136, 139, 142, 152, 158, 165, 170, 182, 184, 201, 205, 220, 238, 240, 241, 246, 252, 255, 287–296, 299, 306, 316, 320, 323, 329, 331, 333–340, 344–361, 363, 368, 373, 380–383, 403, 416, 418, 431, 433, 438, 452
Betz, J., 218
Birkhoff, G., 293
Bish, Ron, 255
Blake, P., 134, 414
Blankenship, Bill, 129
Blaw Knox Co., 107, 109
Bledsoe, Felix, 81
Blizzard, Everett, 329, 330
Blount Island, 216–221
Bohl, H., 293
Bohr, N., 1
Boiling water reactors, 70, 159–162, 200, 206,

459

207, 258–260, 357, 437, 438, 439
Bolender, D., 416
Boltax, Al, 130, 407
BORAX, 168, 261, 438
Borden, W.L., 85
Bordieu, Andres Martinez, 211
Boswell, Bruce, 241, 246
Bournia, A., 134
Bowman, Mel, 144
Bowman, Robert A., 25–27
Boyer, Keith, 144, 146
BR-3, 162, 163
Branch, Irving, 130
Breden, C.R., 339
Brennan, J.J., 369
Bridges, Joanne, 139
Brill, Tom, 288
Britton, W., 246
Brockett, Del, 202
Brookhaven National Laboratory, 72, 287, 289, 293, 349
Brown University, 367
Brusalis, W., 148, 402
Bucher, George, 3
Budge, William L., 43, 47, 50, 52, 153, 158, 178, 182
Bureau of Mines, 334, 339, 344–347
Bureau of Ships, 4, 6, 7, 9, 17, 50, 66, 361
Burnham, D.C., 183, 186, 191, 202, 214, 278–280
Burris, Harry, 6
Busby, Jack, 159
Byron Jackson Co., 31, 324

CAD, 218, 219
Calder Hall, 98, 103, 109
Caldwell, W., 293
Calihan, Dixon, 288
CAM, 417
Cametti, Benjamin, 46, 323
CAPA, 159, 163, 170, 189, 438
Carborundum Corp., 348
Carnegie Institution, 2
Carolina Power & Light Co., 192, 202
Carolina-Virginia Tube Reactor, 176, 188
Carter, Jimmy, 232, 233, 234, 242, 274
CBI Services, 272

CEA, 205, 206, 279
Cermak, Jim, 441
Cerni, Samuel, 129, 134, 369
Chalk River, 76, 297, 341, 344, 356
Chapman, W., 205, 211
Chave, C., 171
Chernobyl, 263, 264, 311, 386, 387, 443, 455, 456
Cheswick, PA, 31, 85–88, 137, 197, 255, 256, 324, 355, 393, 407
Chirigos, John, 361
Chooz, 178
Cisler, Walker, 224
Clinch River Breeder Reactor, 154, 201, 230, 232, 233, 239, 240, 252
Cloud, R., 272, 361
Cobean, Buss, 60
Coe, Roger, 165, 166, 168, 171, 182, 431, 438
Cohen, Karl, 310
Cohen, Morris, 80
Cohen, Paul, 48, 224, 356, 357, 438
Collier, A.L., 213, 215, 451
Columbia National Co., 348
Combustion Engineering Co., 31, 70, 77, 87, 180, 186, 203, 211, 367
Commonwealth Edison Co., 157, 159, 161, 164, 194, 195, 197, 202, 230, 232, 233
Compagnie General d'Electricite, 206
Compton, Arthur H., 3
Comtois, William, 70
Conley, George, 60
Conn Yankee, 112, 179, 185, 432, 450
Consolidated Edison Co., 159, 177, 191, 192, 194
Conway, John T., 79, 181
Conwell, William, 100
Cooper, W.E., 361, 362
Cotman, D.A., 88
Cotter, F.P., 112, 116–119, 121, 126, 127, 179, 181, 230, 243, 277
Cox, Ellis T., 79
Creagan, Robert J., 17, 63, 70–72, 158, 160, 161, 163, 165–171, 209, 224, 261, 284, 286–288, 306, 328, 380, 438, 445, 451, 459
Cresap, Mark, 41, 92, 117–121, 133, 134, 183, 186
Creusot Loire, 205, 278
Cronkite, Walter, 110
CRUD, 342, 356

INDEX 461

Cunningham, Richard C., 45, 46, 48, 52, 55, 64, 101, 107, 118, 130, 134, 306, 315
Curtis Wright Corp., 139

Daly, Paul, 361
Danforth, D.D., 248
Daniels Pile, 8, 10–13, 17, 334
Daniels, Farrington, 7–13, 17, 296, 334, 443
Daniels, R., 443
Danko, J., 129, 152
Danzker, Milton, 313
Daugherty, H., 241
Davenport, J., 178, 179
Davis, Kenneth, 127, 166
Davison, Perry, 138, 146, 163, 177, 416
de Toledo, Carlos Alvarez, 211
Debartoli, R.A., 299
DeHuff, P., 177
Department of Defense, 97, 153
Department of Energy, 153, 233–237, 239, 240, 241–252, 267, 274–276
Detroit Edison Co., 157, 164, 171, 224, 378
DeZubay, E., 139, 407
Dick, Ray, 7, 34, 134
DiPetro, W., 83
Dollard, Walter, 191, 433
Doncals, R., 134, 149
Dow Chemical Co., 157, 164, 168
Dravo Co., 107–109
Drawbaugh, Don, 134, 403
Dreadnought, 87, 88, 129, 205, 210
Dresden, 161, 162, 164, 195, 449
Duffy, Leo, 244
Duke Power Co., 202
Dunford, James, 7
Dunn, Thomas W., 44
DuPont Co., 35, 249–251
Duquesne Light Co., 100, 107, 109, 110, 202, 379
Durham, Frank, 144

East Pittsburgh, PA, 5, 13, 22, 130, 157, 170, 235, 335, 350, 443
Eck, John, 337
Eckert, R., 214, 218
Edgerton, Harold, 4
Edison Volta, 162, 178
Edlund, Milton, 163, 440
Ehricke, Kraft, 115

Einstein, Albert, 1, 2
Eisenhower, Dwight, 15, 97, 98, 165, 171
Eisenhower, Mamie, 15, 106
Electric Boat Co., 41, 44, 46, 64, 67, 88, 120
Electric Power Research Institute, 189, 265, 269, 270, 272, 273, 456
Electricite de France, 205, 206
Ellis, W.R., 70, 79, 88, 101
Elmendorf, W., 46
Engineering Test Reactor, 176, 297
Enterprise CVN-65, 73–79, 81, 82, 83, 84, 185, 215, 359
ESADA, 240, 241
Esselman, Walter H., 46, 47, 53–62, 112, 115–118, 128, 131, 133, 137, 138, 141, 145, 147, 239, 265, 266, 306, 308, 315–319, 414, 416, 417, 421, 424
Etherington, Harold, 11, 17, 27
Exxon Nuclear Co., 242

Fast Flux Test Facility, 201, 224, 226–228, 236–252
Faught, Hal, 129, 147, 148, 405, 416
Fawn, J., 88
Fermi, Enrico, 1, 2, 24, 224, 261, 445
Fillnow, R.H., 331, 343, 344, 358, 359, 361
Finger, H.B., 131, 132, 134–137, 147, 148, 153, 398, 426
Fisch, J., 134
Fleger, Philip, 100, 108
Florida Power & Light Co., 189
Fluor Corp., 164, 165, 168, 247, 431
Foote Mineral Co., 335
Fort, Tomlinson, 189, 190
Foster Wheeler Co., 189, 379
Framatome, 205, 206, 210, 211, 256, 278, 279
France, L., 129
Frisch, Ehrling (Frankie), 48, 133, 158, 168, 169, 177, 180, 315, 438, 447
Furnas Centrais Electricas, 202, 203

Gabriel, Dave, 153
Gallagher, J., 138, 144, 148–250, 401, 418, 420
Gardner, Richard, 81
Geiger, Lawton, 37, 107, 109, 120
Gelbard, Ely, 293, 294
General Advisory Committee, 13
General Atomics Co., 139, 202

INDEX

General Dynamics Co., 41, 120, 121
General Electric, 9, 11, 27, 31, 32, 76, 99, 100, 107, 108, 117, 120, 122, 133, 160–162, 180, 184, 187, 188, 192, 193, 195, 200, 206–210, 223, 224, 262, 378, 379, 386, 449, 452
Gilligan, W., 88
Gipson, Joe, 358, 360
Giraud, Andre, 279
Glennan, Keith, 116, 132
Goldberg, David, 129, 130, 153, 412
Goldman, K.M., 339, 340
Goldsmith, Mark, 293
Goodman, Leo, 171
Gordon, Robert, 76, 101, 118, 221, 340, 345, 347, 349, 350, 353
Graves, Harvey, 163, 168, 285
Gray, John E., 100, 109, 111, 128, 281
Green, S.J., 299
Greenewalt, Crawford, 249
Grigg, Jack, 320
Grove, Don, 235
Groves, Leslie, 2
Gue, Edward, 100
Gulf Oil Co., 202

Hafstad, Lawrence, 18
Hageman, L., 293
Hahn, Otto, 1
Hallam, 160
Hamilton, William H., 46, 58, 77
Hanford, WA, 4, 13, 31, 113, 154, 161, 180, 226, 227, 237, 238, 239, 245–252, 291, 297, 343, 351, 386
Hanstein, H.B., 2
Hardigg, George W., 70
Hayden, Vernon, 46
Hayes, E.T., 339, 347
Hellens, Bob, 292, 293
Henderson, Bill, 416
Henning, F., 129, 134, 418
Henry, A.F., 4, 35, 63, 70, 72, 121, 170, 283, 291, 295, 309, 380, 445
Hesse, Martha, 243
Hiroshima, 1, 5, 103
Hitachi, 207, 208
Hodnette, J.K., 183
Holifield, Chet, 116, 121, 179, 181, 225
Hollingsworth, R., 239
Holmgren, J., 139

Hooten, William, 81
Hope Creek, 221
Hoppe, Al, 116, 130, 412
Horan, F.T., 44, 64
Horton, Jack, 178, 179
Hosmer, Craig, 121, 225
Hulm, John, 235
Hunter, Don, 46, 182
Hurlbert, G.C., 221
Hurwitz, Henry, 35
Hutcheson, John A., 26

Indian Point, 191–195
INFCE, 232
International Nickel Co., 360
Ireland, Mark, 60, 326
Iselin, D.G., 100, 108
ITER, 235

Jackass Flats, 137, 138, 144, 146, 147, 240, 416, 418
Jacksonville Electric Authority, 221
Jacobi, W.M., 137, 152, 170, 237, 246, 248
Jansen, Carl, 109
Japan Atomic Power Co., 207, 209
Jensch, Sam, 171
Johnson, Earl, 120
Johnson, Lyndon, 116, 121
Johnson, M.T., 57, 81, 138, 146, 153, 154, 176, 240, 416, 417
Johnson, W.A., 65, 101, 337, 340, 348, 384
Johnson, W.E., 44, 47, 125, 143, 153, 158, 182, 189, 211, 235, 416, 417
Joint Committee on Atomic Energy, 18, 20, 38, 67, 68, 79, 85, 90, 91, 96–98, 112, 116, 126, 127, 132, 159, 166, 179, 198, 225
Jones, Charles, 100, 158, 230
Jones, Reg, 230

Kaiser Wilhelm Institute, 1, 2
Kansai Electric Co., 180, 207
Kaplan, Stanley, 294, 296
Kaufman, Albert, 12, 334
Keilly, J., 178
Kelly, Leslie, 60
Kelly, W., 205
Kennedy, Bobby, 121
Kennedy, John F., 121

INDEX 463

Kerze Jr., F., 339
Keyes, Roger, 50, 279
Kilp, G., 129, 407
Kim, C., 152
Kintner, Edwin E., 48, 52–54, 67
KIVA, 149
Klein, Milton, 132, 145, 153
Knabenschuh, J., 241
Knapp, Sherman, 181, 182
Kneppel, D.S., 339
Knolls Atomic Power Laboratory (KAPL), 27, 31, 35, 120, 122, 361
Korea Electric Power Co., 211
Kouts, Herbert J.C., 72
Kraig, Howard, 139
Kramer, Lloyd, 17, 118, 130, 152, 350
Krasik, Sidney, 48, 52, 55, 56, 65, 101, 115–118, 121, 125, 126, 133, 134, 137–143, 147, 170, 183, 189, 291, 316, 348, 406, 416
Krause, 166
Krauss, Jim, 241
Kreh, Edward J., 46
Kroll, W.J., 334, 335, 344
Kroon, Ryan, 128
Krsko, 211
Krupp, 219
Kyger, Jack, 54

Langer, Bernard, 48, 52, 55, 304, 361, 362, 368
Langley Research, 399
Large Ship Reactor, LSR, 70, 79
Large, PA, 77, 85, 136, 137, 147, 158, 175, 179, 226, 260, 332, 336, 393, 399, 407, 417, 418
Lawrence Livermore National Laboratory, 130, 131
Layman, William, 60, 62, 177
Lee, W.B., 215
Lego, Paul, 248
Lessman, G., 129
LeTourneau, B.W., 299
Lewis Research Center, 98, 109, 132, 147, 153, 399, 424
Libbey, Miles, 7
Lide, Basil, 46, 316
Lockhead, 131, 430
Long Beach CGN-9, 73, 78
Lorenz, F.R., 152, 412
Los Alamos Scientific Labortory, 116, 122, 123, 131, 132, 137, 143–149, 393, 394, 398, 399,
406–420, 424
Ludwig, Leon, 24
Lusk, Robert, 26
Lustman. B., 65, 101, 297, 331, 343, 346, 355, 385
Lynn, Clarence, 26, 39, 40, 41

MacArthur, Douglas, 208
MacNeil, J.B., 22
Magel, T.T., 339
Mairson, Ray, 241, 319
Makita, 208
Mallinkcrodt Chemical Co., 255
Mallory Sharon Corp., 348
Mandil, I.H., 35, 44, 76
Manhattan District, 2, 3, 5, 7, 8, 12
Manjoine, M., 136, 137, 360, 361, 399
Mare Island Naval Shipyard, 67
Mark I, 20, 21, 26–28, 33, 41–47, 51–57, 60, 62, 63, 66, 68, 82, 86, 99, 110, 157, 158, 302, 303, 327, 329, 330, 335, 337, 338, 341, 344, 348, 350, 353, 356, 382, 446
Mark II, 20, 21, 26, 43, 45, 46, 62, 66, 68
Marlowe, Orv, 293
Mars, 115, 122, 123, 150–155, 382, 391, 423, 426, 427, 454
Marshall Space Flight Center, 132, 399, 426, 429
Massaro, Eileen, 243
Maytham, Walter, 178
McCone, John, 121, 132
McCormack, John, 116
McCreary, H., 88, 129, 134
McCullough, C. Rogers, 8
McCutchan, D., 134
McDonnell Douglas Co., 235
McKim, William, 83
McNamara, Robert, 78, 79
Mechlin, George, 69, 152, 153
Mei, A., 152
Meircord, M., 211
Mershon, J.L., 362
Metals & Controls Co., 87, 355
Meyer, John, 294
Meyer, Orv, 308, 319
Military Liaison Committee, 13
Miller, Don, 148
Miller, Ernie, 8, 13, 14

Miller, William, 8, 13, 14, 86
Milligan, Joe, 44
Mills, Earle, 7, 13, 17
Mims, Stuart, 70
Minton, George, 168
Moberly, Larry, 129
Monsanto Chemical Co., 8, 11, 157, 164, 165, 167, 431
Monteith, A.C., 26, 40, 159, 169, 182, 183, 191
Moore, James, 250
Morgan Jr., D.W.R., 157
Morris, E.T., 80, 175, 416
Morrison, B.W., 182
MUFT, 292
Murray, Peter, 227, 236, 238
Murray, Thomas E., 53, 99, 227, 236, 238
Murrin, T.J., 243, 248

Nader, Ralph, 277, 278
Nagasaki, 5
National Aeronautics and Space Administration (NASA), 116, 122, 126, 131–136, 147, 148, 150, 153, 176, 399, 401, 405, 412, 418, 423, 424, 426
National Bureau of Standards (NBS), 309
National Waste Policy Act, 274, 275
Nautilus, 12–15, 26, 31, 33, 41, 43–46, 51, 52, 57–68, 75–79, 86, 92, 95, 106, 114, 117, 157, 164, 168, 177, 185, 283, 293, 302, 306, 309, 310, 312, 316, 317, 323–327, 330, 334, 339, 341, 354, 355, 362, 378, 387, 446, 454
Naval Proving Grounds, 4
Naval Reactors Branch (NRB), 17, 23, 31, 34–37, 41, 42, 46–48, 60, 75, 77, 83, 95, 99, 100, 108, 306, 317, 320, 339, 342, 344, 348, 349, 357, 380, 384, 385
Naval Reactors Testing Station (NRTS), 33, 37, 43, 44, 47, 63, 209, 224, 261, 359
Naval Research Laboratory, 12, 320, 365
Navy Department, 12, 20, 21, 22
NEPA, 115, 274
New England Electric System, 164, 165, 166, 168
Newport News Shipbuilding & Dry Dock Co., 60, 81, 83, 84, 215, 220, 276
Ney, Edward, 83
Nichols, Fred, 361
Nicholson, Jack, 60
Noble, L.D., 310
NOK Beznau, 269

Nolan, John, 248, 319
North American Aviation Co., 127, 133, 134, 157, 225, 265, 272
North Pole, 92
NPOC, 272, 273
Nuclear Fuel Services, 172, 240
Nuclear Materials Co., 31, 87, 355
Nuclear Regulatory Commission, 173, 174, 181, 222, 267, 270–275, 457

Oak Ridge, TN, 4, 6–12, 17, 70, 71, 235, 237, 242–250, 288, 291, 296, 333, 346, 347, 407
Oak Ridge National Laboratory, 11, 17, 31, 118, 159, 163, 165, 242, 243, 288, 322, 329, 330, 453
Obermesser, Clovis, 443, 447
Oldham, Melvin, 100, 109
OMRE, 160
Oppenheimer, J.R., 171
Osborne, Latham, 22, 41

Pacific Gas & Electric Co., 157, 281
Palen, Charles, 81
Palladino, N.J., 8, 13, 14, 17, 101, 296, 369
Panoff, Robert, 34, 44, 46, 67, 306, 320
PDQ, 293, 294, 381, 403
Pearson, Denning, 88
Penfield, M., 88
Pennsylvania Advanced Reactor (PAR), 159
Pennsylvania Electric Co., 177
Pennsylvania Power & Light Co., 93
Peoples, R.S., 339
Piqua, OH, 160
Plant Apparatus Department, 85, 86, 88
Plumbrook reactor, 176
PLUTO, 130
Point Beach, 168, 192
Polaris submarines, 35, 69, 70, 454
Pollack, Nicholas, 129, 134, 337, 407, 418
Pomerantz, Herbert, 12, 334
Porse, S., 369
Portland General Electric Co., 202
Portsmouth Naval Shipyard, 67, 358
Pottmeyer, E., 243
Povejsil, D.J., 189
Powell, E.U., 158, 416
Power Research Development Corp., 224
Pratt & Whitney Co., 133
Pray, H.A., 339

Preliminary design review, 423, 424
Prentice, Bruce, 161
Pressurized water reactors, 27, 32, 68, 70, 72, 75, 80, 95, 98, 99, 115, 159, 160, 161, 178, 193, 197, 199, 200, 201, 205–207, 211, 224, 250, 257–261, 294, 298, 357, 363, 365, 368, 369, 373–377, 381, 437, 441, 447
Price, Gwilym, 7, 11, 15–19, 21, 23, 41, 42, 65, 89, 100, 107, 119, 120, 121, 156–159, 170, 185, 225
Price, Melvin, 121, 225
Princess Sophie of Greece, 104
Project Management Corp., 231, 232, 233
Project Wizard, 13, 17, 18
Public Service Electric & Gas Co., 157, 194, 214–218, 220, 221
Public Service of Northern Illinois, 157

Queen Frederica, 104
Quinton, Harold, 178

Radkowsky, Alvin, 35, 61, 75, 313, 348, 381, 439
Raisbeck, Frank, 81
Ramey, James T., 20
Ravetz, J., 138, 149
Redding, Arnold, 129, 134
Redmond, W.C., 329
Reed, Fred, 168, 417, 438
Reed, Glenn, 168, 417, 438
Rengel, J.C., 68, 101, 108, 109, 158, 189, 214
Rentschler, Harvey C., 3
Retallick, F., 129, 137, 138, 144, 148, 199, 418, 420
Reynolds Aluminum Co., 92
Rice, Chuck, 145, 416, 420
Rickover, Hyman G., 4–61, 65–70, 71, 74–92, 96, 100, 103–105, 108, 109, 111–113, 117–122, 132, 135, 156, 158, 164, 171, 172, 177, 205, 224, 288, 306, 321, 330, 334, 335, 337, 340, 348, 378
RIFT, 131
Roberts, D., 152
Robertson, Andrew W., 3
Robinson, R., 192, 315
Rochester Gas & Electric, 192
Rockwell, Al, 225
Rockwell International Co., 248
Rockwell, Theodore R., 32, 35, 225, 243, 247, 248
Roddis, Cdr. Louis H., 7, 70, 71, 111, 166, 177, 249, 283
Roderick, C., 179
Roissman, A., 281
Rolls Royce Associates, 87, 88, 280
Roman, Walter, 70, 115, 118, 128, 137, 138, 315, 398, 416, 417
Roosevelt, Franklin D., 2
Rose, C., 152
Roseberry, C., 159, 161, 168, 190
Ross, Philip N., 8, 13, 19, 26, 34, 35, 70, 78, 79, 101
Roth, E.B., 34
ROVER, 122, 131, 132, 143, 144, 393, 395, 412, 418, 424
Ruppel, Henry, 81, 83
Ryan, Pat, 83

S3W, 67
S4W, 67
S5W, 68, 69, 95, 96, 357
Saalbach, Cael, 403
Sadler, J., 152
Salem, 194, 195, 197
San Onofre, 112, 172, 178–183, 185, 187–189, 192, 193, 260, 262, 432, 441, 451
Sanderson, A.C., 54, 83, 319, 416
Saturn, 131, 136, 150, 152, 391, 423, 426, 427
Saxton, 177, 188, 261, 438
Schasseur, Robert, 206
Schlesinger, James R., 210, 230, 231
Schrag, Fred, 337
Schrieber, Raemer, 144
Schroeder, 147, 153
Schultz, M.A., 101, 175, 308, 316
Schwanekamp, Karl, 46
SCOTTR reactor, 177
Seaborg, Glenn, 115
SEAC, 309
Seawolf, 32, 99
Secret Six, 35
SEFOR, 310, 445, 452
Seitz, Frederick, 11, 80
SELNI, 185, 187, 188, 437, 441
SENA, 185, 187, 188, 205, 437
Seran, Paul, 358, 359, 360
Shapiro, Z., 335, 336, 337

Sherman, J., 168, 181, 369
Shippingport, 57, 63, 68, 73–80, 92, 97, 99–114, 119, 135, 157, 159, 160, 163–166, 169–171, 180, 185, 188, 255, 290, 293, 300, 346, 363, 366–369, 373, 375–379, 381, 384–387, 431, 442, 443, 446, 449, 450, 452
Shiring, P.B., 182
Shor, S.W.W., 100
Shoupp, W.E., 7, 19, 65, 101, 152, 158–161, 165, 168, 183, 438
Siegel, Sidney, 8, 17, 288
Siemens, 93, 183, 200, 205
Simon, Sidney, 9, 23
Simpson, John W., 4–14, 22, 23, 26, 33, 34–44, 47, 48, 51–56, 59, 60, 62, 65, 68, 69, 70, 76, 80, 82, 83, 89, 90–92, 94, 101, 104, 107, 109, 111, 116–121, 125–128, 133, 134, 135, 137, 138, 164, 169, 171, 179, 181–189, 191, 196, 201, 202, 205–211, 214, 215, 218, 224, 225, 227, 230–235, 236, 239, 254, 259, 261, 262, 274, 276–283, 317, 334, 341, 343, 348, 368, 374, 375, 394, 433, 456
Sinclair, C., 153
Sizewell, 197, 200, 269
Skate, 66, 68, 92
Slack, Charles, 24, 39, 40, 68
SLAM Missile System, 130
Smith, Cyril S., 31, 80
Smith, McGregor, 31, 80, 189–191, 218, 221, 272
Smith, R., 218, 221
Smith, Sherman H. Jr., 272
SNAP, 128
SNUPPS, 197
Sodium Reactor Experiment (SRE), 160
Sorg, Lou, 358
South Carolina Electric & Gas Co., 202
South Texas Partners, 202
Southern California Edison Co., 178, 179, 181, 202, 262
Southern Co., 221, 272
Space Nuclear Propulsion Office, 123, 132, 134, 135, 139, 143, 147, 148, 153, 391, 398, 399, 402, 405, 406, 418, 424, 425
Spanier, Jerry, 294
Spence, Rod, 144
Spencer, Douglas C., 46, 68
SPERT, 198
Spurrier, F., 134
Sputnik, 110, 114, 116

Squire, A., 68, 85, 335, 336
Stackpole Carbon Co., 129
Stadelman, Joe, 218
Stagg Field, 3, 15, 24, 454
Stanojev, Albert, 100
Stein, Stanley, 293
Stern, Theodore, 189, 214, 246
Stevens, Harry, 9, 35
Stiefel, John T., 26, 34, 68, 78, 79, 81, 189, 437
Stone & Webster, 107, 108, 166–168, 171, 172, 182, 191, 228, 449
Strassmann, Fritz, 1
Strauss, Lewis, 57, 98, 109, 112
Submarine Fleet Reactor (SFR), 50, 57, 66, 67, 68, 69
Submarine intermediate reactor (SIR), 32, 71, 161, 288
Submarine thermal reactor (STR), 17, 21, 41, 43, 47, 51, 52, 57–60, 62, 68, 71, 82, 307, 319, 321, 349, 361, 362
Sunnyvale, CA, 69
Surry, 194, 195
Swanson, John, 148, 402
Szilard, Leo, 2

Taub, John, 144
Taylor, Hobart, 230
Taylor, John J., 35, 48, 189, 201, 230, 283, 284, 286, 313, 322, 328–330, 333, 382
Teller, Edward, 63, 64, 387
Tenneco Inc., 215, 217, 218
Thee, W.B., 182
Theilacker, John, 255, 331, 350
Thiokol Co., 133
Thomas, Charles, 8
Thomas, D.E., 8, 70, 79, 116, 118, 129, 131, 133, 138, 153, 331, 337, 339, 340, 347
Thomas, Dick, 144
Thomas, R., 139, 144
Thompson, T.J., 169, 438, 439
Thorpe, Al, 177, 442
Toepfer, Adolph, 8, 13, 14, 17, 287
Tokomak, 235
Tokyo Electric Power Co., 208
Tolman Committee, 5
Tong, Long Sun, 177, 205, 299, 441, 442
Toshiba, 200, 207, 208
Tribble, J., 172
Troy, M., 299

Truman, Harry, 41, 43, 105, 249
Turkey Point, 190, 191
Turnbaugh, Marshall (Bill), 34, 37, 67
Turner, James, 217

U.S. Congress, 38, 96, 98, 104, 112, 126, 127, 159, 171, 179, 232–234, 241, 274, 275, 310, 426
Union Carbide, 12, 237, 242, 347
Union Electric Co., 157, 197
United Engineers & Constructors, 270

Van de Graaff generator, 7, 19
Varga, R., 293
Varljen, T., 152
Vickers-Armstrong, 87, 88
Vogel, C., 407
Voysey, Albert E., 46, 47, 53, 55, 80, 158, 168

Wade, Elmer, 11, 416
Wah Chang Co., 348
Wallin, H.N., 50
Waltz Mill, 137, 139, 163, 173–177, 197, 226, 252, 253, 407, 438
Watkins, James D., 277
Watkins, F.T., 65, 228, 277
Weaver, Charles H., 19, 22–26, 34, 39–41, 47, 48, 62, 65, 68, 70, 85–88, 91, 101, 118, 121, 134, 143, 156–159, 164–167, 178, 181, 183, 189, 335
Weaver, J., 134
Weber, Les, 248, 250, 335
Webster, W.E., 107, 108, 165–169, 171–173, 182, 191, 228, 449
Weiler, Robert, 129
Weinberg, Alvin, 11, 12, 17, 31, 165
Weiss, A., 299
Wells, R.L., 125, 168
Welner, Eric, 101, 108
Westinghouse Electric Corp., 3–27, 31–41, 44, 47–52, 64–67, 69, 81–89, 91, 92, 94, 98–100,
107, 109, 110, 112, 114–122, 125, 128, 129, 133–139, 143, 144, 147, 148, 152–169, 172, 173, 175–189, 191, 193, 194, 196, 199, 201–215, 218, 220–227, 230, 231, 235–257, 260, 267–269, 271–273, 276, 278–281, 287, 296, 316, 320–324, 335, 336, 340, 347, 350, 351, 356, 360, 362, 363, 379, 394, 397, 399, 401, 407, 410, 411, 414–419, 424, 431–441, 442, 447, 450–452, 455, 459
Westinghouse, George, 156, 185
Westinghouse Reactor Evaluation Center, 163
Westinghouse Testing Reactor, 138, 175, 176, 416
Wigner, Eugene, 2, 11
Wilcox, G.L., 225
Wilkinson, E.P., 60, 310
Wilson,, 248, 264
Witzke, R.L., 157–167, 168, 191, 438
Wolf Creek, 197
Wood, Roger, 83
Woodruff, Owen J., 46, 70
Woods, P., 241, 246
Wright, F.L., 243
Wright, J.H., 129
Wroughton, D.M., 101

Y-12, Oak Ridge, 7
Yadon, Mel, 62, 182
Yankee, 93, 112, 119, 161–189, 192, 193, 254, 255, 260, 261, 431–439, 441–446, 449, 450
Yasinsky, J.B., 155, 242–246, 294
Yates, Perry, 178
Youngholm, Dave, 3
Yucca Mountain, 247, 249, 274–276

Zechella, Alexander P., 70, 80, 81, 82, 83, 215, 221
Zerbe, J.E., 299
Zinn, Walter H., 2, 21, 71, 161, 165, 288
Zion, 194, 195

AMERICAN NUCLEAR SOCIETY
BOOK PUBLISHING COMMITTEE

Roger W. Tilbrook, *Continuing Technical Contact*
Argonne National Laboratory

Tunc Aldemir
The Ohio State University

Ahmed Badruzzaman
Chevron Petroleum Technology

John A. Bernard, Jr.
Massachusetts Institute of Technology

Winfred A. Bezella
Argonne National Laboratory

Robert A. Krakowski
Los Alamos National Laboratory

Alan E. Levin
U.S. Nuclear Regulatory Commission

Robert L. Seale
University of Arizona

Raymond Waldo
Southern California Edison Company

Lewis A. Walton
B&W Nuclear Environmental Services